CONCEPTUAL FOUNDATIONS OF QUANTUM MECHANICS

Second Edition

ADVANCED BOOK CLASSICS
David Pines, Series Editor

CONCEPTUAL FOUNDATIONS OF QUANTUM MECHANICS

Second Edition

BERNARD D'ESPAGNAT
Université Paris-Sud
Orsay

CRC Press
Taylor & Francis Group
Boca Raton London New York

CRC Press is an imprint of the
Taylor & Francis Group, an **informa** business

ADVANCED BOOK PROGRAM

First published 1976 by Westview Press

Published 2018 by CRC Press
Taylor & Francis Group
6000 Broken Sound Parkway NW, Suite 300
Boca Raton, FL 33487-2742

CRC Press is an imprint of the Taylor & Francis Group, an informa business

Visit the Taylor & Francis Web site at
http://www.taylorandfrancis.com

and the CRC Press Web site at
http://www.crcpress.com

Library of Congress Catalog Card Number: 99-60031

ISBN 13: 978-0-7382-0104-7 (pbk)

Cover design by Suzanne Heiser

Editor's Foreword

Perseus Books's *Frontiers in Physics* series has, since 1961, made it possible for leading physicists to communicate in coherent fashion their views of recent developments in the most exciting and active fields of physics—without having to devote the time and energy required to prepare a formal review or monograph. Indeed, throughout its nearly forty-year existence, the series has emphasized informality in both style and content, as well as pedagogical clarity. Over time, it was expected that these informal accounts would be replaced by more formal counterparts—textbooks or monographs—as the cutting-edge topics they treated gradually became integrated into the body of physics knowledge and reader interest dwindled. However, this has not proven to be the case for a number of the volumes in the series: Many works have remained in-print on an on-demand basis, while others have such intrinsic value that the physics community has urged us to extend their life span.

The *Advanced Book Classics* series has been designed to meet this demand. It will keep in-print those volumes in *Frontiers in Physics* or its sister series, *Lecture Notes and Supplements in Physics*, that continue to provide a unique account of a topic of lasting interest. And through a sizable printing, these classics will be made available at a comparatively modest cost to the reader.

Bernard d'Espagnat has been a leading figure in the study of the foundations of quantum mechanics for over three decades, and the publication of his lucid lectures on this topic have made it possible for his ideas to reach their deserved large audience. Published in a second edition in 1976, *Conceptual Foundations of Quantum Mechanics* has continued to attract the attention of all those interested in the important subfield of theoretical physics. Its appearance in the *Advanced Book Classics* series means that d'Espagnat's lectures will continue to be readily available to the ever-growing community of theorists and experimentalists who wish to learn more about the conceptual foundations of quantum mechanics, and I am pleased to welcome d'Espagnat to the ranks of ABC authors.

David Pines
Tesuque, NM
January 1999

1999 Special Preface

Basically, the present edition of this book is a reprint of the 1988 Second Edition, which was itself just a re-issuing of the original edition, now dating back twenty-two years.[1] In spite of this fairly long space of time it seems fair to state that, still today, the book covers the hard core of the conceptual advances produced by quantum physics and the problems raised by them. In particular, the recent developments do not overthrow the most general conclusions that emerge from the book, namely that multitudinism (materialistic atomism) is dead, that separability cannot be kept as a feature pertaining to mind-independent reality and that indeed, science is presumably not in itself a sufficient tool for gaining full access to such a reality, although it brilliantly accounts for the relationships between observed phenomena. The first two points follow—convincingly, I believe—from the investigations here reported on in Parts Three (Quantum Nonseparability) and Four (Measurement Theories). Concerning the third one, I consider that, combined with other issues (especially, questions involving relativity theory) the just mentioned first two make its plausibility very great (see Part Five).

But this, of course, does not mean the developments that took place after the writing of the book are irrelevant. Quite the contrary, some—such as the so-called Quantum Zeno Effect—are most surprising and therefore quite interesting for their own sake. Others throw valuable additional light on the questions alluded to above. Such, in particular, is the case concerning the renowned Decoherence Theory, which constitutes quite a significant indication that the quantum rules are universal. Still others are worth a thorough study for quite a different reason, namely, because they were viewed by some of their advocates as a possible means of escaping one at least of the above stat-

---

1. The only significant changes appearing here have to do with the last paragraphs of Section 13.1.

ed conclusions: to wit, as restoring separability within quantum physics. This, for example, was, at one time, the case of the so-called Consistent Histories Theories (Griffiths, Gell-Mann and Hartle, Omnès...). Eventually it was shown, however, that in fact no such restoration takes place. To summarize these debates—and, more broadly, to study the implications of all these recent developments—I wrote, in 1995, a new book, *Veiled Reality*, the content of which is explained by its subtitle: *Analysis of Present-Day Quantum Mechanical Concepts*. Among other items, this book contains detailed accounts of Zeno's paradox; of outcome dependence versus parameter independence; of the various forms of the Bell theorem; of decoherence theory; of the "and-or" difficulty that, even when decoherence is taken into account, still besets quantum measurement theory; of the problems met with when trying to reconcile quantum collapse with relativity theory; of the various ontologically interpretable theories, their nice features and their drawbacks; and, last but not least, of the philosophical problems raised by this all.

The book *Veiled Reality* should be considered a complement to *Conceptual Foundations of Quantum Mechanics*. It is written in the same spirit (indeed, the introductory chapters are practically the same in the two books), and it is of use to those who need be "in the know" concerning the more advanced aspects of the present-day developments. On the other hand, it contains somewhat elaborate analyses that may make it a little hard reading for nonspecialists. *Conceptual Foundations of Quantum Mechanics* brings its readers more directly into contact with the basic features of the problems at hand and also analyzes a number of questions—such as, for example, the role of non-Boolean "quantum logics"—that are not touched upon in the more recent book. In these respects it should remain just as useful as it may have been twenty years ago.

A few words are here in order concerning the guiding idea that inspired this book. It is that quantum mechanics can be formulated axiomatically, that, for clarity sake, it is of course quite appropriate to do so, but that the axioms in question then have to take the form of (precise and general) "rules of the game," serving to predict what will be observed. This is a difference with classical mechanics, the axioms of which (Newton's laws and the rest) are most simply expressed as statements bearing on the structure of some mind-independent reality. It is a fact that attempts at doing the same in quantum physics quickly lead to conceptual muddles ("Are wave functions real?," "Is collapse real?," etc.), while, in contrast, viewed as a set of observational predictive rules, quantum mechanics is crystal clear. The rules in question must therefore be considered as being—by far—what is most solid in quantum physics. And it is for this matter-of-fact reason—and not because of any a prior allegiance to positivism, empiricism or what not!—that it was here found advisable to begin by just stating these predictive rules and investigating their consequences. Since no allegiance to phenomenalism is made, the question of the possible interpretation of the said rules in terms of some underlying reality of course remains significant. In fact such a study constitutes, in a sense, the very purpose of the present book. But the corresponding analyses must—and do, here—come in only in a second stage, after the rules have been duly stated and examined.

Note that the just explained standpoint is precisely the one that gives us maximal freedom concerning interpretation problems, since it bars out any a priori prejudice relative to what constitutes reality. Within it, we are not, right at the start, forced to conceive of reality in terms, either of waves, or of particles, or of "wavicles," or etc. Any way of thinking of it is a priori admissible, provided only that, in the end, it turns out to be compatible with the observational predictions yielded by the basic quantum rules. But, as will be seen, this condition proves to be a demanding one. It does not leave many vistas open. Indeed the book shows that such an approach gently leads to quite definite ideas concerning the conceptual foundations of the incredibly powerful science that is called quantum mechanics.

Bernard d'Espagnat
February 1999

Vita

Bernard d'Espagnat

Director of the Laboratoire de Physique Théorique et Particules Elémentaires, Orsay, was born in Fourmagnac, France on August 22, 1921. d'Espagnat received his Ph.D. in Theoretical Physics from the Sorbonne in 1950. Professor d'Espagnat was a Research Physicist at the French National Center for Scientific Research, at the Fermi National Accelerator Laboratory in Chicago (1951-52), at the Niels Bohr Institute in Copenhagen (1953-54), and at CERN in Geneva (1954-59). In 1959, d'Espagnat joined the University of Paris, where he was a Professor at both the Paris and Orsay campuses. Professor d'Espagnat became Director of the Laboratoire de Physique Théorique et Particules Elémentaires, Orsay, in 1970.

Contents

PEDRO'S DEBATE

(Preface to Second Edition)

The ecological movement was still in its prime when I first met Pedro. I think I remember we were marching together for the defense of some Mediterranean beaches. If we were not, this conjecture nevertheless sets the tone of our exchange of ideas. We soon found that both of us were physicists, and it turned out that he had read the first (and preliminary) edition of the present book.

Pedro is interested in general ideas. On the following days, he told me about his own questions, judgments, and guesses. Such "half-baked" conceptions are not easily written down. Nevertheless, let me try to do so. Perhaps it is not entirely accidental that these questions and opinions reflect rather well both the motivations of the present work and some of its most ambitious (and—alas!—also most uncertain) conclusions. Apparently, there are several Pedros nowadays in the world of physics, all differing from one another on important or minor views. This essay will exemplify their problems.

Mathematics and Physics

Pedro began by recalling that a fairly complete consensus exists among physicists as regards the "practical" use of the theoretical algorithm, including the methods for predicting new experimental results. On the other hand, he said he also shared a view that is the backbone of the present book: namely, the opinion that such a general working agreement really conceals considerable differences. He pointed out, for instance, that even among theorists those who call themselves phenomenologists and those we call axiomatists have widely separated points of view. In particular, he stressed the existence of a large body of physicists who are "die-hard pragmatists" in that the role of the theorist is—in their opinion—*merely* to try to relate the new experimental data to the older ones. Pedro claimed, however, that this was definitely *not* his personal view about science. For him the scientific truth is something more than just a collection of technical recipes of such a kind.

"Now," he said, "does this mean that I should be, at heart, an axiomatist?"
He claimed that it does not. The axiomatist holds—or, at least, is supposed to
hold—the following opinion: "A deep understanding of what lies at the roots
of modern physics (quantum mechanics and its offsprings) requires *only* a
quest for more general mathematical algorithms. The *mere* use of the latter
gives *all* the substance of such understanding of these subjects of which we are
capable. Hence any statement expressed in plain language and not ancillary
to the mathematical formulation is merely superfluous talk." Pedro denied
this. He said that of course mathematics of an ever more elaborate variety is
necessary to physics, but it should be ancillary to the search for increasingly
refined *concepts*, instead of the reverse being true. To convince me further, he
pointed out that this had been Einstein's own conception of science, as is
shown by his statement about E. Mach: "His [Mach's] action was particularly
healthy in that he made it clear that the most important physical problems are
not of the mathematico-deductive kind but are those which bear on the fundamental concepts" (Einstein-Besso correspondence).

"But," I asked, "if mathematics has to be supplemented by something,
should it not be by formal logic?" Pedro granted that some extraordinarily
interesting attempts had been made in this direction. Indeed, he even prompted me to include in this new edition a discussion of the bearings of these contributions. Unfortunately I have been able to do so only in a very partial and
concise way (see Chapter 12); but I have tried to emphasize at least one point:
that an approach along these lines can be used to generalize what is called here
"nonseparability" (other authors call it "nonlocality"). On the other hand,
Pedro pointed out to me that the very first generalization ever made of the
"obvious" nonseparability pertaining to "orthodox" quantum mechanics concerned the hidden variables theories, and that this important result had *not*
been obtained by the use of formal logics. Instead, it emerged as a consequence of an elementary analysis made with the usual tools of mathematical
physics supplemented by clear *concepts* (see Chapter 11).

Physics and Existing World Views

On one evening Pedro was confidential and described to me the intense
surprise he experienced when he discovered the existence of the modern
epistemological problems. At the university his professors had—as is usual—
carefully concealed such facts from him. But, he said, they had some excuse.
Such matters can be introduced fruitfully only to advanced students; and
the latter are, as a rule, more inclined toward laboratory and/or mathematical research than toward an effort at reflection bearing on foundations and
generalities. Consequently, Pedro already had some competence in manipulating group theory and S matrices when he found out (with real dismay) in the
existing literature that some ideas he had always considered as true are not

compatible with the modern physics in which he was trained. Until then, he said, I was convinced that, when the old metaphysics of the Gods, fairies, and demons was overthrown, it was replaced by no metaphysics at all, but simply by objective science. But then I suddenly discovered that this was not true after all. I realized that I—and most other people as well—had unconsciously believed in *another* metaphysics, namely, that of Democritian atomism, in the philosophical sense of the word; and that—what is more—the set of the general ideas underlying that metaphysics (strong objectivity plus separability and so on: see Chapter 11, 12, and 20) could nowadays be shown to be erroneous in a way that leaves hardly any room for doubt. It was a shock to realize that this popular view of "up-to-date" man is not up to date after all, and as a consequence I had to change my whole outlook.

Previously, Pedro continued, I went on with my work without giving any thought to epistemology. When appropriate (i.e., quite often), I would use the *methods* of pragmatism and logical positivism. In particular, I would frequently refer to the notion of measurement, to the criterion of usefulness and so on, and I would carefully avoid such unfashionable words as "existence" and "reality." But that was merely imitating the elders and bringing into play sound technical recipes. I would never ponder about all this, however, safe as I felt in the comforting idea that in the last resort all I did necessarily conformed to the general conception of a vague but "obvious" multitudinism. By that word I mean, as you do (Chapter 19), any world view in which the universe is considered as being ultimately composed of a great number of very simple parts, all of which have properties attached locally to them. My unspoken belief was that everything we could find out could not but fit into this picture, and that the picture itself was lying *outside* my field of research. Such a belief had given me a superiority complex toward anything that might look philosophical. But now I have no substitute for such security. In particular, as I said, I also can not accept—at least not uncritically—the view that a mere mathematical formalism could constitute an acceptable substitute for my former metaphysics. Nowadays, such a formalism always refers (implicitly or otherwise) to man-made measurements (Chapter 14 to 18), so that this attitude would amount to nothing other than an unspoken and vague idealism. This means that I would be making a philosophical choice, and—what is worse—I would be making it implicitly, without even noticing it! My former experience (the one I just told you about) prompts me to avoid this.

This, then, is how I became convinced that a serious study of the philosophy of science was necessary. Consequently, I read quite a number of books, which apprised me of the considerable advances made in that field. Of course, what I learned there I cannot summarize in just a few sentences. The highlights were the elaborate theories of *scientific reality* and *scientific truth* which have been built up by the positivists and their followers. I mean the theories in which scientific reality is considered as a *construct*, and in which scientific truth is no longer viewed as an "adequation of our intellect to some pre-existing primitive reality," as in the former theories, but rather is regarded as a kind of *creation*, satisfying the criteria of verifiability, consistency, and so on. By emphasizing the role of man's observation and action in the very *definitions* of reality and of truth, these conceptions do in fact reflect quite faithfully the evolution of modern science—in particular, they go a long way—it seemed to me—toward the solution of some of the conceptual difficulties of quantum mechanics, which appear as paradoxes when the transcendent reality of the things is kept, as in the older theories.

On the other hand, I was always worried by what seemed to me to constitute real difficulties and ambiguities in the conceptions of positivism and of the related theories (instrumentalism, operationalism, and so on). I need not describe the nature of these difficulties and ambiguities, since they are precisely those that you yourself recalled in some detail (Chapter 20). Admittedly, many of the difficulties in question are not new, and they have been discussed—at least in general terms—by the positivists themselves. However, the advent of quantum mechanics has undoubtedly brought them into sharper focus; and, consequently, none of the books I read ever gave solutions that really satisfied me. Moreover, I agreed with the observation that the opponents of positivism have sometimes made, according to which the philosophy in question does not, on the whole, favor the elaboration of new questions. This is due to the fact that, as a rule, questions that are basically new (in that they open avenues for scientific reflection) seem at first sight not to be meaningful according to the criteria of positivism, for their novelty prevents us from imagining the corresponding modes of verification and of insertion in "the known." Hence positivism, as presently used in our daily practice, keeps physics pure but, at the same time, tends to keep its advance horizontal. Indeed, one of the main problems with which this science has to cope is how to go beyond strict positivism in a way that will not lead to science fiction.

For these reasons and some others, my studies of the existing epistemological literature (which is centered mainly on positivism) definitely did not solve my problems. I had to go on, and I was happy to find that several physicists had published—even in highly respectable scientific journals—investigations that went quite distinctly *beyond* the frontiers of verifiability, and so on, imposed by positivism.

What name should then be given to the investigations in question? Are they science or philosophy? This is a matter of semantics and not, therefore, a primordial one. If a scientist insists on calling them "philosophy," let him have his way. What he calls "science" is then but a set of rules connecting past observations to predicted future ones, and he is quite free to restrict in that way the meaning of the word. On the other hand, it is just as reasonable to restrict also the meaning of the word "philosophy," so that it covers only the views which—rightly or wrongly—many thinkers have believed they could elaborate *directly*, without the help of outside experience. Then the investigations alluded to are not philosophy either. Still they can make sense. To take an elementary but clear-cut example, let it be assumed that at some stage the only possible way to unify physics would be by means of a theory that leads to *no* new prediction. Would that theory be philosophy, or would it be science? Wavering is legitimate. Since questions of methods are decisively important, perhaps the more reasonable choice is to call such theories "*scientific*," as long as they make use of scientific facts and methods. Then, since the theories in question can conceivably exist, it is legitimate to assert that the scope of science is larger than what most pragmatically minded scientists take it to be. Indeed, it extends over much of the territory formerly ascribed to general philosophy.

Incidentally, this is, for us, a good reason to become acquainted with the works of the general philosophers (those who are not epistemologists). This is even *necessary* in order to widen our views. Nevertheless, it should not induce us to adopt their language, particularly the language of those who are more or less our contemporaries. Implicitly or explicitly, most of the latter accept as a fact that any philosophical truth incorporates an essential ambiguity, which prevents it from being expressible except through sentences that can have several meanings. As has been said (somewhat nastily perhaps) by their critics, these general philosophers "misuse a language that was created precisely for that purpose." This practice has been imitated in recent times by many linguists, psychoanalysts, and ideologists of various specialities, writing on generalities. Unquestionably, it endows their texts with a difficulty of access that is often identified—correctly or incorrectly—with genuine richness of thought. For *ethical* reasons (we might almost venture to say!) such language should not be used by a scientist. As already stressed, the latter must use his own methods. For this reason, his argumentation may well look clear—hence infantile!—to the average professional philosopher and to his disciples; but it must be noted that, conversely, the deductive methods of the last-named group often look infantile to scientists. For the present, these difficulties of communication must unfortunately be accepted.

On the following night, Pedro once more took up the general problem of how to go farther than positivism. Thinkers for whom the domain of science lies entirely inside the realm of positivism would call this the problem of the metaphysics lying beyond physics. Pedro said that the best testimony to its importance is the great amount of literature on the subject which has appeared in the past and is still appearing. But he also expressed the view that, considered from the viewpoint of present-day physicists, this literature is on the whole rather misleading. He elaborated his opinions as follows.

First, he pointed out, we scientists must acknowledge the fact that some philosophers of the past have had an intuition of our present results. Admittedly, men like Descartes, Kant, and even Hume said many things that we know now to be dubious or even false. But some of their negative statements have turned out to be entirely right. I refer particularly to the doubts they expressed that any *naive* metaphysics of reality could in the long run be found to be the correct one. The multitudinism, particularly the multitudinistic atomism, we mentioned before was a recent example of such a naive metaphysics. The modern history of its rise and fall thus constitutes a confirmation of the appropriateness of the doubts in question. Similarly, when some Buddhists claimed that space and time—as we apprehend them—do not "really exist," or when Kant asserted that the latter entities are but modes of our *own* sensibility, these thinkers were making guesses that, in the light of quantum physics and of nonseparability, appear to us retrospectively as having been oriented in the right direction. But the question is then as follows. On the one hand, we are prepared to go beyond the narrow positivist orthodoxy. The most straightforward way of doing this is to consider that human observations will serve as *criteria* but not as *definitions* of reality and truth: as discussed in your Chapter 20, this seems to resolve a major ambiguity in these doctrines, and at the same time it discards any "hidden idealism." It also implies that "something" (let us refer to it as "reality") exists that does not *reduce* to man's observation or will. On the other hand, we have just acknowledged, like many philosophers of the past, that such a reality is not describable in terms of a naive metaphysics, built up with familiar concepts. Can we therefore hope to know at least some general features of reality, and, if so, what are they?

It is at this point, Pedro continued, that, as scientists, we must assume an attitude of interest but also of deep reserve with respect to philosophical research. I am now thinking particularly, he said, of the "pure" (i.e., nonepistemological) philosophical developments which have occurred in our times. These developments have taken many forms, which are related but different. It is therefore almost impossible to describe and to critisize them in a manner that would be both precise and fair without being unduly lengthy. A kind of first approximation of one of their most common general tendencies can, nevertheless, be obtained. For this purpose we may, for example, consider an opinion which appeared rather early during the nineteenth century: that, somehow, order and harmony are mere appearances (which belong to the realm of phenomena or, in other words, are projections from our minds), whereas, on the contrary, primitive "reality," the "real thing"—whatever that may be—is governed by contradictions.

Retrospectively, it seems possible to understand why such a view became the cornerstone, so to speak, of the conception of the World held by a considerable number of persons, mainly in intellectual and literary circles. It seems to be based on two attractive ideas, which (unfortunately) are also misunderstandings. The first one has to do with the meaning of the word "contradiction," and this is so vast a subject that I suggest we set it aside for the time being. The second one is the "intuition" (it was never stated quite explicitly, but it is transparent in *The Birth of Tragedy* of Nietzsche and in some works of Engels and others) according to which *primitive reality* should be somehow identified with *primitive human life* or with prerational consciousness. The primacy of the notion of contradictions then follows, since prerational consciousness is indeed dominated by contradictions (which, incidentally, may have a high emotional value, as in tragedy). But the said "intuition" itself is clearly quite controversial. To support it, Schopenhauer is sometimes quoted. And indeed, if we adhere literally to the ultimate view expressed by the very title of Schopenhauer's famous book, *The World as Representation and Will-to-Live,* we may consider the said intuition as a great truth. The difficulty is that the thinkers who were most influential in spreading abroad the idea under discussion (namely, the opinion that basic reality is governed by contradictions) in general did *not* agree with the philosophical idealism which so obviously underlies Schopenhauer's conception. It is therefore quite difficult to understand how they could consistently substantiate their claim. Indeed, we cannot discard altogether the conjecture that such an idea remained for them all a mere *unanalyzed* (or not entirely analyzed) intuition. Presumably its incorporation into the system of dogmas of most of the intelligentsia of our own times results, at least in part, from a phenomenon of similar type. The idea is emotionally quite attractive. It can be given a kind of justification based on the premises of idealism, as we saw; and this justification was indeed supported by great authors of the past who somehow accepted these premises. For some of us who reject idealism the temptation can be great to remember *only* that the idea in question (*a*) is attractive and (*b*) received some justification, and to forget on what premises that justification is based. But, of course, as soon as we become aware of the inconsistency that is inherent in such an attitude, we can no longer retain it.

In particular, Pedro continued, this makes me suspicious concerning a great deal of what the upholders of dialectical materialism have written about the problem of reality. There is no question of the two of us indulging here and now in a systematic, critical survey of that doctrine! Nevertheless, in view of its appreciable impact on the conceptions of many human beings, including some scientists, I would like to tell you why I, personally, could not find in it a solution for my own problems. The main reason is that I find it to be inconsistent in the way I have just described. For example, I have read many weighty books (both old and modern) which complacently describe the contradictions of capitalism or of societies in general and which proceed from these—in a sweeping generalization! —to the law of contradiction (or of negation, as they sometimes put it), considered as a basic law of the Universe. In fact, more abstract and apparently more serious argumentations often follow the same line, as I eventually found out. The Being, it is argued, is also

Non-Being, because otherwise it would be static. This contradiction is re-solved (synthesis) by going over to the Becoming; and immediately examples are given, borrowed from phenomena and, more precisely, from systems in nonstationary states, which are nearly always living systems. All living beings must die, that is, go back to Non-Being, and so on. In other words, the neces-sary distinction is never made between, on the one hand, the *changes* of a nonstationary system (changes that, as we all know, do occur, even accord-ing to physical theories which are free of internal logical contradictions) and, on the other hand, alleged logical contradictions of the type "A is A but A is also Non-A." In most cases the examples chosen have an emotional impact (life, progress, germination, death, and so on) so as to make the "argumenta-tion" more impressive. And when I tried to disentangle the general ideas, about reality, of the author I was reading from all these faulty illustrations, nothing was left!

Having said this, Pedro remained silent for some time. Then he added: No, this is not *entirely* true. To be fair, I must acknowledge the fact that the "dialecticians" did a great deal to popularize a view which the development of science has confirmed on a grand scale. This is the idea according to which the concepts and other intellectual tools that mankind has at its disposal at any given time never exactly fit its needs (be they psychological, social, or even scientific), so that *apparent* contradictions, due to deficient conceptions and formulations, always turn up. These apparent contradictions disappear, however, as soon as new intellectual tools are found, which are better fitted to describe what has been observed; but then, of course, new facts are dis-covered which are not necessarily all describable, even with the help of the intellectual aids thus elaborated, and therefore the process goes on for ever. The impossibility of describing the observed subatomic particles in terms of the old classical concepts of "particles" or of "waves" is perhaps the most spectacular of such apparent contradictions, due, as I said, to the deficiencies of the old concepts. If such an evolution is "dialectical thinking," then of course "dialectical thinking" has some correct applications.

Similarly—but much more generally—some physicists claim that Bohr's thought was of a dialectical nature, when he stressed, for example, that we are not only spectators but also actors on the theater of the world. Indeed, it seems possible that to some extent our wishes for a better understanding of reality are baffled (for the time being) by the following fact: we have not yet found the intellectual tools that can allow us to overcome some apparent contradictions related to the pair of concepts "description"–"action." This indeed seems correct if the concept of reality is identified with that of empiri-cal reality, but let us not discuss such difficult matters for the moment. The fact remains that, if Bohr's thought is to be considered as dialectical (although he himself did not apparently consider it as such), then, again, dialectics has some interesting aspects. However, I cannot really understand what we gain by such assimilations. On the contrary, I can understand what we may be losing, for they could induce us to believe that dialectics is a general and reliable method in science, and we could decide to apply it to our problems. No scientific discovery was ever made in this way, and the very ideas underlying dialectics are so hazy and partly childish that we must doubt whether one ever will be made.

At that stage I recalled to Pedro that dialectical materialism has another feature which he might consider interesting, especially since he found "multitudinism" to be deficient. The feature in question is the emphasis laid on the concept of a Unity underlying diversity. He agreed, but at the same time he pointed out that the Unity of which the dialecticians speak—the "dialectical Unity of the opposed"—seems again to occur essentially between concepts, aspirations, theories, wills, and so on: that is, between entities the very definitions of which must necessarily refer to man. The relationship of this to the nonseparability introduced by modern atomic physics certainly is not obvious and presumably is nonexistent.

Pedro then enumerated a few ideas and customs that the dialecticians seem to cherish, the validity of which he regarded with considerable doubt. They are (a) the use of the word "scientific" to qualify mere philosophical intuitions; (b) the idea (previously mentioned) that philosophy, particularly dialectics, can serve as a guide to science; (c) the undifferentiated use of the word "matter" to describe *either* "Being in its totality" *or* some parts of it *or* even mere phenomena (and the faulty deductions which result from such confusions); (d) the reduction (related to the foregoing point) of consciousness to "matter"; and (e) the reduction—inconsistent with (d)!—of primitive reality (or Being) to human *praxis* that some dialecticians perform from time to time when convenient. He concluded that dialectics is, on the whole, a bad method when applied to science and to its extensions toward philosophy. It incorporates some views that are correct but are rather trivial in that field; moreover, it expresses them very poorly, so that it mixes them up with other ideas which are clearly wrong. For example, Pedro said, Hegel had the intuition that logic cannot be entirely independent of its field of application and that perhaps the old logic of Aristotles could and/or should be modified. This was, in a way, good foresight, as shown by the advent of quantum logic (although that logic is of limited usefulness: see, e.g., Chapter 12). But unfortunately in his time Hegel could have no idea of how this might be done soundly. The result was a frightful mess, which the episode of Hegel deducing from his pure logic the nonexistence of a large class of asteroids illustrates quite vividly. It is somewhat distressing to think of all the brilliant minds that have been led to similar (if less blatant) absurdities by the prestige and apparent deepness of the "method." In regard to science, it is perhaps unfair to attribute to the deficiencies of this method the fact that a fraction of the dialecticians glorify scientific research unreservedly, whereas another fraction condemn it bluntly. But it may be a consequence of the use of dialectics that neither group seems able to state whether such a difference bears on the practice or the applications of science or whether it bears on the value of science itself.

At this point I told Pedro that he was too critically minded, since the dialecticians did, after all, discuss fundamental questions of considerable interest, instead of just ignoring them, as is the general rule. This he granted, saying that for this reason he was indeed quite hopeful. By comparison with the unnecessary obscurities of dialectics, however, he stressed the much clearer way in which some contemporary thinkers who are *not* using that method succeed in expressing their views on many basic problems, including even fundamental and difficult questions about reality, its meaning, and man's relationship to it. In particular, he mentioned Piaget.

This was not a surprise to me. Indeed, every scientist must be interested in, at least, the *purpose* of the founder of genetic epistemology, even if the non-specialist can of course make no judgment on its results in any particular case. For some 200 years the empiricists and their followers claimed that all our knowledge comes from the senses, and yet they never thought of studying experimentally and in detail the process of the *formation* of this knowledge. Genetic epistemology now bridges this gap, at least to some extent. In particular, it has quite important things to tell us about such subjects as the formation of our logical structures and of our notion of "number." But I asked Pedro in what respect he thought it was really relevant to the very fundamental problems that we had been discussing up to this point: I recalled to him that, just like the positivists, Piaget does not attach any considerable importance to the concept of a fundamental reality or of general truths of any kind, and that he seems to share the common view of most scientists that only particular truths exist.

Pedro's answer was that two aspects of Piaget's views seemed to him to be most relevant to our problems. The first one is rather well known: it is Piaget's observational discovery that the concept of separate, localized objects is a construct. The very young child builds it up as an element of a theory that is useful to him in accounting for the regularities he observes and in coordinating his actions. But the observations and actions of the child are of course confined to macroscopic effects of the most common types. We have no guarantee, therefore, that the construct in question is fitted for describing any other type of effects. This shows that the principles on which philosophical atomism is built are mere *assumptions,* which have a priori no greater likelihood than any other ones. It shows also that the idea according to which any macroscopic object necessarily occupies some definite region of space—to the exclusion of all other regions—is not an obvious (and hence unquestionable) truth, but rather an element of the definition, useful in given circumstances, of the word "object." Correlatively, it also gives us hope that, even if our atomistic descriptions—or our descriptions in terms of "objects"—of basic reality ultimately turn out to be inadequate, this is no reason for us to abandon the attempt to describe said reality, and to become "diehard positivists." Perhaps other concepts will emerge which will be better fitted for this purpose. But, above all, the fact that the notion of localized objects is but a construct liberates us from the so-called commonsense view (which was also that of Aristotle) according to which individual macroscopic objects obviously—and therefore *necessarily*—exist as individuals independently of ourselves; hence it also liberates us from the apparent necessity of considering them as more basic than numbers, logical structures, and so on.

On the other hand, genetic epistemology seems to indicate that what is true in regards to the concept of object is in fact quite general: most of the "properties" that we attribute to the outside world, or simply to "necessity," are really mere "projections" of our *own* possibilities. According to Piaget, the Platonic theory of Ideas is just another example of this general tendency. Indeed, even the theory that the essence of fundamental reality is pure mathematics is for him just another example of the same (illegitimate) mode of thinking: the mathematics is projected from us.

This is the second point in Piaget's general conception that Pedro finds especially stimulating, for the reason that it is not arbitrary and that, on the contrary, while not *proved* by science it has at any rate *some* experimental indications in its favor.

It should be stressed that the "projection" theory does not make Piaget a positivist. On the contrary, this author expresses the view adhered to above that the requirement of strict verifiability is too stringent and sets somewhat arbitrary limits to the domain of possible scientific investigations. This is another aspect of Piaget's standpoint that Pedro appreciates quite highly. Indeed, after listening to all his favorable comments about the work of that philosopher-scientist, I thought he had at least found there firm grounds on which he could stand. But he answered negatively my question to this effect.

Own Guesses

Quite often, Pedro said, even those writers who try to show the importance of the role that human *action* plays in our knowledge really minimize that role. This is true of all those who *merely stress* the fact that in order to know something we must not only observe but also make experiments: for experimentation could well be necessary even in order to know a world which would then turn out to be describable by means of logical structures and concepts already existing in our minds (the "clear and distinct ideas" of Descartes). There is a strong indication, given both by quantum mechanics and by genetic epistemology, that the role of human action is much more important, in that it is also responsible for the *formations* of the logical structures and of the concepts themselves. Again, we cannot overstress the importance of Piaget's contribution in this area.

On the other hand, Pedro continued, it seems to me that the general conception of this author still resembles positivism in that it cannot be taken completely seriously without leading to some idealism (human action is *the* primitive reality): an idealism that could even lead to solipsism (the speaker is the only reality) when the difficulties raised by the problem of intercommunications are not skimmed over. I know, of course, the conventional philosophical manner of bypassing objections of this kind. It consists in making use of the difference between the two possible conceptions of reality that the existence of the two German words *Realität* and *Wirklichkeit* expresses so well in that language. More precisely, it consists in asserting in succession (*a*) that the concept of a *Realität* (i.e., of a "primitive reality") is uninteresting and of doubtful validity, (*b*) that in fact the possible human actions lie at the basis of any conceivable *Wirklichkeit* (i.e., of any empirically reachable reality), and (*c*) that this has nothing to do with idealism since it does not assert anything about the existence or nonexistence of the *Realität* (because it considers the question as basically uninteresting). But it seems to me that this procedure for not taking sides in an excesively difficult debate does not pass candid examination. *Either* we consider that no meaning whatsoever can be given to the concept of a (known or unknown) *Realität* having existed before mankind (and probably due to exist after it), *or* we consider that such a concept may be meaningful. In the first case nothing at all can be even conceived as being more basic than man. That view is distinctly of the idealistic variety: no verbal trick, no hectic denial, can prevail over such a simple logical truth. In the second case, either the said concept corresponds to something or it does not. If it does not, then again nothing would exist that would be more basic than man and we would have idealism. Hence, if we want to avoid idealism, we must make the assumption that the concept of *some* (hitherto undefined) primitive reality (*a*) makes sense and (*b*) reflects the existing situation. But then what the procedure for not taking sides summarized above amounts to is really only an assertion that primitive reality exists but cannot be approached by any experimental-deductive method and should not therefore interest scientists. I believe that this pronouncement—even if it approximates the truth, since primitive reality is so remote from our senses—is too radical and too pessimistic.

But, Pedro went on, I would be very sorry if you were to consider that these remarks I make constitute a criticism. Actually, it seems to me that we are here at the very "heart of the matter." If it is true that Piaget and others *do* try to avoid or to circumvent a question, they have—I believe—quite a good reason for doing so. The reason is that the question constitutes one of the oldest and hardest of all the philosophical puzzles into which men have ever had to cope. Most certainly, there is no easy answer to such a question, and presumably we do not even know as yet how to formulate it quite correctly. Hence, when we observe such authors as Bohr, Heisenberg, and Piaget approaching it and then suddenly retreating, or when we notice how they cautiously try to circumvent the matter, or how they keep silent at the most interesting point, we must certainly not blame them. The fact is just that they do not consider themselves as being in the councils of the gods and they do not want to induce in us the illusion that they are. On the whole, this is a good example that they give us, and we would be wise to follow it.

I was unhappy at this last sentence of Pedro's because I thought that we were at last just reaching the interesting points. I told him so, but he had very good arguments in favor of *not* saying anything more. He stressed again that, although public opinion is very much in favor of philosphers expressing mere *opinions*, it always blames scientists who venture to do so (the idea presumably being something like this: these people are specialists; let them keep then to their specialities). He said that even scientists react in this way. He said also that, if I were to act as a wicked newspaperman and let his problems and guesses be known, his identity would be discovered and he would no longer be taken seriously by any colleague. He underlined jokingly the tacit agreement existing among scientists that only a Nobel prize gives the right to formulate general views. More seriously he insisted that he had only mere opinions to set forth, that these opinions could change, and that they were scarce and scanty (since for him science sheds only a little light on fundamental problems, although that very little is essential).

I answered by pointing out that up to this point he had merely expressed criticisms, that criticism after all is easy, that nowadays too many young people like him limit their intellectual activities to such purely negative—if safe—aspects, and that he, at least, should have the courage to advance his own opinions as to how the problems he raised could conceivably be solved, even if these opinions were most uncertain and unorthodox.

For response, he said that he would comply, although I would be deeply disappointed by the *reduction* he would suggest of an immensely overworked field to very sketchy, simple views that—to him and for the time being—contained everything essential.

First of all, he said, I am not an idealist. I agree with thinkers as different from one another (in other respects) as Planck and Lenin in believing in the existence of "something"—for the time being, you may call it "Matter," "God," "Reality," "Tao," "Aeon," or what else you like—that is, in a way, prior to Man's mind (i.e., existed before it and so on; I explained this point more fully a moment ago). Please notice that I start by *not* specifying anything about this primitive reality except that, contrary to Heidegger's *Dasein,* it cannot be reduced to man. Hence it could a priori be a very "trivial" one (atomism) or a very "remote" one (Plotinism). It could be changing or eternal. This is so because my main motivation in believing in its existence is that I need a basis to account for the regularities we observe (as discussed in Chapter 20 of your book); and for that I do not need to ascribe to it any attribute in particular, except just the one that (precisely) it does have at least some relationships—even if quite indirect ones—with our observation (this differentiates it, perhaps, from the conceptions some philosophers have had about the Thing-in-Itself or about similar entities).

As a preliminary remark, we should note at this stage that even such a general concept of a "something"—of a "Being," unique or plural, immutable or changing, but one that would *be* even if we did not exist, nothing else about it being specified—even such a *minimal* assumption has been questioned (implicitly at least) and made disreputable by many philosophers. The process began during antiquity, when thinkers like Plato introduced the concept of various degrees between Being and Non-Being. But it is now clear that to a great extent these distinctions were based on semantic difficulties (which are not fundamental) and on overemphasis of the role of the logical a priori. More radically, Heraclitus is reputed to have discarded such a concept, and after him many other thinkers thought they could do the same. But I consider that neither Heraclitus nor his followers really succeeded in that endeavour. When they thought they discarded the "Being" as a concept, as a matter of fact they discarded only some of its hitherto assumed attributes. Nietzsche, who in our times started the fashion (although he seems to have wavered between sharp rejection and more mitigated views), did in fact nothing more, perhaps without fully realizing it. For example, when (using Heraclitus as a symbolic spokeman and herald) he asserted that this philosopher "beheld Law within *Becoming*" and perceived "the *plays* of Zeus in the *Changes* of the Universe," he *seemed* to emphasize Becoming and Changes at the expense of Being. But to formulate his statements he could not avoid referring to the *Law* or to *Zeus,* that is, to entities which are admittedly more remote from our familiar viewpoints than are *atoms* or *matter,* but which are taken as entities—beings—all the same.

Many modern philosophers have tried to follow the same path, with the result that they have only made the issue unduly complicated. For example, they take due notice of the fact that we, as scientists, are not satisfied with a mere *description* of phenomena, and they object to such a standpoint. Following Husserl or others, they ask us, "Why do you introduce—along with phenomena—a "reality" that you assume is underlying them? Are you not making the picture unnecessarily complex? Since you grant that this reality is extremely remote from the phenomena we see, why don't you use Occam's razor and suppress it? Why don't you go back to the "plain things" (as we see them)? You say that a reality is necessary in order to explain the regularities of the phenomena, but you grant that an evolution does take place: this is an inconsistency. Indeed, these alleged regularities are illusory. As Hume has pointed out, we do not really know whether the sun will rise tomorrow."

Such are, in their essence, the objections most often formulated by general philosophers (I do not speak now of epistemologists, whose case was considered before and is dealt with also in your Chapter 20). I believe that these objections arise most often from unawareness of very simple scientific facts. Many of these thinkers simply were never told that compact formulas such as the Maxwell equations underlie a tremendous diversity of "happenings." More generally, they do not know that a simple equation that does not change, such as the Schrödinger equation, may quite well—without any inconsistency!—account for complex processes of *evolution*. To me, the kind of economy they suggest when they scoff at the concept of Being (or of reality) greatly resembles the one that would consist—for a scientist—in trying to give up Maxwell's equations in the hope of "simplifying the whole picture." This is nothing but an analogy, of course. Still, it is a significant one in this context. Finally, the regularities are not illusory, and the vindication of the inductive inference method—though never totally convincing—is appreciably more so (see Chapter 20) within a realistic world view than without it. Hence the whole situation is such that we should not be too much impressed by the apparent profundity of many of these general philosophers. A moment ago we discovered, as you may remember, that the very general problems bearing on reality and so on do *not* (to the extent that they have a meaning) fall outside the realm of our competence as scientists. Even if their solutions are not testable experimentally, we are qualified to make assertions about them as long as out *methods* can be used for this purpose. In other words, in doing so we are not amateurs and should not be overawed by alleged "professionals." Similarly, here we realize that in some crucial aspects the level of the argumentation of many general philosophers in that domain may lie, not "above," but "below" our own, in spite of the apparent depth that brillant or abstruse language often bestows on their utterances.

In particular, we must therefore not be worried by the objections that most modern philosophers raise against *any* concept of Being. Some minutes ago (when discussing a few aspects of Piaget's philosophy), I gave you some reasons for considering that, as soon as we discard idealism, we are under the logical necessity of accepting the concept of a reality anterior to ourselves. In other words, we must believe in the "existence of some Being" (without of course making any a priori assertion about the attributes of this entity). Although these reasons could be refined, I think that basically they are sound; and I believe therefore that the objections to the concept just alluded to are either misunderstandings, or else mere virtuosities, based on ambiguities and on no substance.

Let us then bring this discussion to an end, and let us turn to the question that is now the essential one: Can we assert something about reality, or Being (I take these two words as synonymous)? Most philosophers of the past—let us think of Plato or of Spinoza, for instance—thought that they could do so quite independently of experience (observation and experiment were considered by them as deceptive). We have good reasons to be quite skeptical about such claims. On the other hand, observation and experiment inform us regarding *phenomena,* lead us to the invention of *models,* and in short (as we now know better since the discovery of quantum mechanics) do not give us any *positive* information on facts that we could surely ascribe to reality itself. Nevertheless, we are not completely deadlocked because experiment—and, more generally, science—do provide at least some *negative* information. They give us, for instance, some very strong indications that Einstein's principle of separability (as you call it) is a mere projection in Piaget's sense, or—more precisely—that Being itself is nonseparable (in space and also, presumably, in time). This, as we noted in our preceding discussions, is a very important piece of information. Up to now, many philosophers had spoken favorably of "holism" but (*a*) they could not prove their claim and (*b*) nothing could be done with it. Now we have (*a*) scientific indications in favor of holism and (*b*) a formalism, namely quantum mechanics, which is the first one to incorporate "wholeness" into a workable scheme, leading to verifiable predictions.

Let me mention one thing that I find most interesting and amusing in this connection. It is that, although Plato's and Spinoza's methods were far from those of science (as we just said), the only conception of basic reality that science does not finally reject (into the category of mere "models") is one that is "transcendental" in the way those of Plato and Spinoza were—that is, in the sense that no attribute of *empirical* reality really fits it: that most of the concepts we have built up as children (in Piaget's view), or inherited from our ancestors, can simply *not* be accommodated to it. Indeed, this negative relationship of primitive reality to our "trivial" set of action-motivated concepts is precisely one of the most distinctive features of Spinoza's "substance," which he also calls "God" or, in some other places, *Natura Naturans* (as opposed to phenomena: *Natura Naturata*). Personally, I rather like these two denominations used by Spinoza, and I hesitate to make a choice between them. They are undoubtedly much better than "Nature" or "Matter," since both of the latter names tend to confuse reality with "observed things." Moreover—and consequently!—the use of either of these latter names to designate what hitherto we called "reality" or "Being" has a major drawback. Implicitly, it leads to faulty attributions—to that reality—of qualities which can only be *elements of empirical models,* and which therefore cannot be attributed without qualification to primitive reality. Admittedly, the second objection also holds good in regard to the use of the word "God" insofar as that term conveys to most minds the idea of exact anthropomorphic attributes such as omnipotence and activity. In that respect the denomination *Natura Naturans* is better. On the other hand, its disadvantages are that it leaves less place for attributes considered as elements of naive but fruitful (and hence significant) models (God's "love" should be conceived in that way), that it sounds more pedantic than the name "God," and that it conveys less immediately the idea of transcendence. As it is easily explainable, there is—for the latter purpose— no adequate substitute for the *emotional understanding* (Whitehead) brought about by very old words.

Anyhow, even in that field, too great importance should not be attributed to matters of semantics. What is fundamental is that—contrary to the opinion of many upholders of the concept of Being!—the Being that science allows us to consider is remote and is yet to be searched for. Hence several approaches to it are equally legitimate: its existence does not bar such diversity but, on the contrary, justifies it. Similarly, no priest and no head of state or of party can assert that he knows *the* Being! These elementary remarks should, in my opinion, allay the prejudices of many with respect to the use I am making of this concept.

Eddington's famous parable (about the footprints that we discovered in a cave, that we analyzed meticulously, and that we finally discovered to be ours) remains, of course, quite valid. But it should be interpreted in a somewhat restricted sense: not that primitive reality does not exist, but that (for the reasons, such as nonseparability and so on, which we have already discussed) it cannot be identified with the phenomena (the footprints) which are mere projections *we* make of it. On the other hand, these projections do exist as such, with all their tremendous diversities. How can we account for them? The general principles of quantum mechanics *plus* the knowledge of the initial state vector do not give the necessary determinations. Hence we need supplementary ones. For the time being, I, personally, can imagine only two solutions.

One of them is to believe in a hidden variables theory. The one of Louis de Broglie and of Bohm constitutes a good—since explicit—example. Such a theory does provide the supplementary determinations we need and at the same time, since we know that it must be "contextualistic" and nonseparable (see Chapter 11), it incorporates holism. Hence it shows how diversity and holism can be reconciled, but of course it has considerable shortcomings as well (relativistic aspects and so on).

The other "solution"—if it is one—is more philosophical and therefore more vague, but it corresponds somewhat better to actual physical practice. It consists in supplementing the set of "general principles of quantum mechanics *plus* the initial data" by a new version of an old and profound philosophical concept, known as the "separation between subject and object." This is more or less the "orthodox" point of view. The question of whether it really leads to a solution free of inconsistencies (or, more precisely, the question of under what conditions it does this, in regard to our requirements concerning the very concepts of understanding and knowledge) is a complex one, which your book and others study and to which I do not want to give here any oversimple answer. I might just point out parenthetically that, as soon as the concept of Being is taken seriously, such a world view can again be compared to Spinoza's philosophy, with its Substance or God (or primitive reality) possessing two attributes that are (as Bohr would have said) its "complementary" aspects: *res extensa* (the phenomenological space-time and matter) and *res cogitans* (the mind which sees a signal or the instrument, *defined with reference to the properties of our minds* which does the same). Under this conception "thought" does not appear as subsidiary to "matter." These two aspects of reality are, in truth, complementary, and neither of them makes sense in the absence of the other. It may well be that what the reduction of the wave packet is trying to tell us in modern language is just that this very ancient view is, in fact, true.

Would that be dualism? Certainly not. Dualism is the postulate of the *separate existence,* "in their own rights," of entities belonging to different realms (the body and the soul, the earth and the sky, etc.). It is often accompanied by the postulate of a *conflict* taking place between the members of these pairs. Hence it should be clear that both of the conceptions sketched above are quite opposed to dualism. Moreover, it should also be clear that they are not as opposed to one another as might be believed at first sight. Presumably, the existence of the supplementary entities of the hidden variables theory can not be checked by experiment; these entities are not localized in the usual sense—indeed, they have, in a way, a metaphysical existence. Also here, the empirical complementarity between the observer and the observed remains appreciable (if only because of the difficulty of giving a general meaning to the concept of a localized chain of events taking place within primitive reality). On the other hand, whichever solution is found preferable, the superiorily of the quantum-mechanical complementarity over the vague old philosophical "object-subject separation" should also be noted. It comes from the fact that the "cut" between the observer and the observed can be arbitralily moved within limits that are precisely determined by the conditions of observation. It comes also from the fact that intersubjective agreement on the results of observation can be accounted for—in quantum mechanics—without postulating the localized absolute existence of the observed quantities (in this connection see Chapter 23). Hence in the new physics the old saying of Protagoras that "man is the measure of things" both acquires a precise, restricted meaning and at the same time loses its most paradoxical aspect. The precise meaning is that, in it, the word "things" should be identified with "phenomena" or "happening events" *not* with "Being." The paradoxical aspect *was* that, if the particular objects, the particular events, and so on had no absolute existence (independent of the subjects), one could not understand the intersubjective agreement. The quantum-mechanical formalism gives us an example of how this agreement is possible even within the realm of a theory which does *not* postulate such an absolute existence and indeed denies it (see again Chapter 23).

At this stage I told Pedro that I considered his discussion of the matter incomplete. In fact, he had practically ignored the most fundamental problem in the field: Is (primitive) reality eternal, or is it changing? If it is changing, as scientific discoveries seem to indicate, then all the references we make to given "general laws of Nature" or to "Being" seem to lose much of their meaning. Indeed, when we speak of laws of Nature we necessarily always refer to some kind of perpetuity, and the same holds true with respect to the concept of Being (since we so closely associate it with Parmenid's vision). I asked Pedro what he had to say about that.

It is true, he answered, that this is a significant problem. It is hard to find convincing arguments in favor of either of the two possible answers: mutability and unchanging eternity. This is unfortunate since admittedly, the choice determines practically everything in regard to our feelings of relationship with the world. In this respect, I can only advance what might be called a "motivated preference." However, before fulfilling my promise to tell you about it, let me point out that the scientific indications you mentioned in favor of mutability are not entirely convincing. First of all, we may wonder whether some of them are not based on an elementary mistake which we discussed just a moment ago. It consists in ignoring the fact that there can be evolutionary processes which are governed by laws that do *not* change. If we forget this obvious fact, we will have to consider that the time evolution of the stars (or of the galaxies) *implies* a corresponding time evolution of the Maxwell equation, for example. However, such a deduction is obviously incorrect. More generally, we know that the "big bang" cosmological theory is probably right, and we can even compute approximately the time when that event took place. But on what basis do we make such an estimate? Not by assuming that the great laws of Nature changed, but, on the contrary, by assuming that they did not. If these laws do not ultimately refer to us, then they refer, as we have already discussed, to a primitive reality of some kind. Hence, if they do *not* change, it seems that this immutability gives the said reality at least some features of eternity.

On the other hand, the assumption that the laws—and the universal constants—suffer no change at all is of course extremely difficult to verify; indeed, it has been questioned by leading scientists. It seems to me, therefore, that here we have reached a point at which, for the time being, we should just say, "I do not know." In Heraclitus' saying that "everything is in an eternal metamorphosis," the word "eternal" may be just as essential as the word "metamorphosis." Or it may not. We never bathe in the same river, but it may just be that the laws of hydrodynamics remain unaltered nevertheless. Or, alternatively, they may be (slowly) varying.

If the principle of separability were valid, there would probably be no real difficulties in thinking of a changing Being. On the other hand, if nonseparability holds good for time as well as for space, then time as well as space appears to some extent as being a mere projection, a mode of our sensibility as Kant would have said, or else a mode of our action. Under these conditions the concept of a Being that would be both independent from us (as follows from the definition) *and* embedded in time seems to present some inconsistency. Admittedly, this argument is weak because it is too general. Still, it is the only one that I can develop in order to decide between mutability and eternity of Being, and it favors the second possibility. Within that conception the universe of phenomena with all its stars and galaxies is but a transient appearance: the *Natura Naturata* of a *Natura Naturans* that is eternal. Such a view thus coincides (more or less, of course) with some very old intuitions of mankind. Admittedly, this is not an indication that it should be right. Neither is it—except to very shallow minds—an indication that it should be wrong.

Life now gets inserted into this; and, with it, sensitivity, love, struggles, contradictions, transgressions, and also pictures, action-motivated concepts, conflicting aims, and conflicting models. The error of the dialecticians was an inability to avoid the pitfall into which they said their main opponents were falling, namely, a naive anthropomorphism that made them consider as the basic element of reality itself some of the most conspicuous characters of life in general and of human life in particular. But in regard to *these* characteristics what they say is partly correct. What is more, since empirical reality is, to a great extent, built up by our possibilities of action, several features of what they say about the world do apply to empirical reality. And, if we look further back in history, we discover many thinkers and poets who advanced the same general themes, thereby inspiring great emotions and (sometimes) useful actions.

Primitive reality being so remote, it makes sense to overlook it completely when we consider practical purposes. This, as we saw, is true in regard to science. A fortiori, it is true also in regard to action. Moreover, to some extent it is true in regard to philosophical thinking itself. At any rate, the view that evolution dominates everything, life and universe, is a deep and beautiful one; and it is also quite undoubtedly a *true* one if by "everything" we mean every phenomenological object or, in other words, all that comprises *empirical* reality (reality referred to man). On the other hand, if a man has become convinced—for example, by arguments similar to those above, or (as is more usual) by intuition—of the existence of an eternal Being, he may reasonably direct some of his thoughts to it.

As a last question, I asked Pedro what he had to say about the possible means of apprehension of his "primitive reality." His answer was that, strictly speaking, none exists. This, he pointed out, is what differentiates most markedly our modern science from *both* the science of the so-called classical period *and* the former theology. Admittedly, classical physics and classical theology had little in common. Nevertheless, they shared *one* view, namely, that Being or what "is" could be adequately known, at least in part, although they agreed neither on the methods for reaching this goal nor, of course, on the conclusions. Classical physics, in particular, identified what "is" with matter, atoms, fields, and so on. In other words, it identified primitive reality with empirical reality. It agreed that a full knowledge of that reality could presumably never be acquired but thought that it could be, at least, asymptotically approached; and it believed that some partial features of reality, at any rate, could be known "as they really are," even within finite times. Nowadays, a physicist who believes in the existence of primitive reality must grant that this reality is extremely "remote": so remote, indeed, that quite possibly it is not rationally apprehensible except as a limit of a sequence of theories none of which has any feature bound to last forever. On the other hand, Pedro said, if I accept these views, I can hardly oppose those who say that the sequence of theories just mentioned could very well not be unique, and moreover that each sequence could lead asymptotically to one aspect only of an otherwise unknowable reality. This is the way in which I would reformulate the views of Niels Bohr and of the so-called Copenhagen School. This is also the way in which I understand the theory of *epistemological levels,* according to which each level (atomic, thermodynamical, biological, and perhaps also psychological) has its own mode of description adapted to it, and which asserts that *none of these modes is more basic than any of the others,* although none is complete and self-sustaining.

It is only in a very weak sense such as the one just expounded, or in a rather negative way (nonseparability), that I should like to speak of an intellectual apprehension of primitive reality. Still, we can reasonably believe that advances along these lines will always remain possible. We can also think that, although the distinction between empirical and primitive reality is essential, it should not be made sharp. For example, we may conjecture that it would be erroneous to set excessively strict verifiability criteria on the first one and to relegate indiscriminately all the questions rejected by these criteria into the domain of those that are either meaningless or relevant to primitive reality and therefore "scientifically uninteresting." It is much more proper to think that one proceeds smoothly from the domain of questions bearing on *empirical* reality to the one of those bearing on *primitive* reality: this is a much more promising attitude, and this is why it is reasonable to consider that the interest shown by many physicists in, for example, the relative state theory (see Chapter 23) points in the right direction.

Now your original question could also mean, in regard to primitive reality, "Do I believe in the possibility of modes of apprehension other than scientific ones?" Here the answer is: of course I do! Why not? It is so pleasant a belief, and at the same time it is one that can so obviously *not* be refuted that I do not see why I should not indulge in it. Yes, when I look at some works of art (not all of them, by a long way!) or at some beautiful manifestation of Nature, when I listen to music that appeals to my taste, I do believe that this gives me some glimpse—although quite personal and noncommunicable—into primitive reality. Why should my impressions be necessarily mere projections of an ego whose *separate* existence is, after all, *also,* to some extent a myth? What I do *not* believe in is the thesis of the old rationalism, according to which primitive reality is within reach of our intellect, without any reference to outside experience. For me, as for Allan Watts and his followers, the role of the senses is, on the contrary, quite essential. Hence Plato's myth of the cave takes on a somewhat unorthodox meaning: the prisoners should not *disregard* the shadows they see, but rather try to interpret them as testimonies bearing on the real. Friends of mine who make love, naked, on barren dunes say that, when they look at the sea afterwards, they sometimes "see" a reality that is not otherwise attainable; I feel incline to believe that their opinion is correct!

On the other hand, Pedro went on, we must absolutely eliminate two possible misunderstandings about these matters. The first one bears on the question of whether the personal, extrarational, extrascientific glimpses into primitive reality just postulated can be utilized for practical purposes. There the answer must be a final "no." These glimpses do not bear on phenomena. Action is strictly the domain of empirical reality. No witchcraft can be justified by the belief I have expressed. The second conceivable misunderstanding consists in thinking that these glimpses offer an easy way to profound truth. In recent years, many people have believed this. However, it is here that I part from Allan Watts and his pupils (here and also on the identification they seem to make of primitive and empirical reality). I do not agree that these personal illuminations are in any way what is most essential. On the contrary, I feel quite convinced that no *real* enduring progress in such a matter as gaining knowledge about any kind of reality can be achieved without some effort—in particular, without the collective effort of science. Science is insufficient, and, in respect to the knowledge of primitive reality, it ultimately fails. But it is only by looking at the barren territory lying *beyond* its frontier that we, as a society, can find some significance in the object of all these quests.

This is a faithful but succinct summary of what Pedro told me. On every subject dealt with above, he said more than I could properly report. Of course, by trying to make his sayings concise and readable I have oversimplified them; this is unavoidable. Moreover, as repeatedly stressed above, all these sayings are mere opinions. They are only slightly better founded than the corresponding assertions of philosophy. It would be both a misinterpretation and a great pity if some readers should consider these statements as expressing alleged 'established truths'. On the other hand, they may conceivably—even in this simplified version—contribute just a little to the process of unifying our problems as scientists and our problems as plain thinking men. In other words, they may help slightly to bridge the gap between science and general culture, perhaps by stimulating reactions and contrary views. Such are the hopes that prompted their publication.

Readers who, although primarily interested in *general* views, accept none but *exact* ones should not ponder too long on Pedro's guesses and intuitions. Instead, they may turn directly to Chapter 12. Most parts of that chapter can be understood even without detailed knowledge of the content of the earlier parts of the book.

To conclude this preface, let us finally note that the present edition incorporates several other new chapters and sections. The overall effect of these additions is that now the main general conclusions of the book should not be considered as being subject to any significant limitation associated with the finiteness or infiniteness of the number of degrees of freedom of the systems.

Paris and Orsay, 1975

PREFACE

(To First Edition)

If the reader provisionally tolerates some oversimplification, he will presumably accept the assertion that science has always had two purposes. One is to organize our perceptions and thereby to enhance our power. The other one is to help us to understand the world at large and our relation to that world. These two purposes are clearly complementary but distinct. However, during the main part of our century, a growing emphasis was laid on the first one. The reason for this is obvious. The advent of relativity and of quantum physics necessitated such a dramatic revision of notions formerly accepted as basic that the concepts of perception and of action were left as almost the sole ones whole relevance and validity could quite confidently be believed.

Partly as a consequence of that development, scientists now have to cope with a situation which, both internally and externally, seems somewhat worrying. Internally to science, we must observe that most of the very recent theoretical advances are only formulated in terms of "rules of the game" that are but partial recipes and consequently shed very little light on the fundamentals. Externally, that is in our relation with laymen, we have to confess that—lost as we are in our techniques—we are, as a rule, neither willing nor even able to produce consistent sets of concepts and relations that would compose legitimate descriptions of that very world which people believe we study. Indeed, the very subordinate role imparted to basic science in the evolution of ideas and impulses that we now witness, and even the fact that this evolution sometimes has the aspect of a reaction *against* science, probably have no other origin. We feel that trend should be reversed, and we are ready to face the consequences of the corresponding change of outlook. In particular, to anybody who claims that science is but a set of rules which connect past and future observations, we willingly grant that books like the present one are not purely scientific under such a definition. Rather, they should then be described as attempts at exploring the borderlines between science and that which, for lack of a better name, we may call natural philosophy, meaning by this nothing else than the investigation and classification of the possible sets of views, concepts, and relationships that compose acceptable descriptions of the world revealed by the facts, and of our relation to it.

Being immensely too vast, the question is drastically specialized right from the start: we deal only with quantum mechanics. It is probably unnecessary to dwell on our reason for doing so, for quantum physics is obviously a central part of man's knowledge. Moreover, for the purpose here considered, a correct understanding of the quantum measurement process is undoubtedly at the heart of the matter. This is why a critical investigation of the various quantum measurement theories currently in existence is carried out. Its essential purpose is to determine under what conceptual scheme each of them is valid. It leads therefore to a critical examination of several epistemologies.

Somewhat drastic limitations must be mentioned: we study only elementary quantum mechanics. The specific features of relativistic quantum mechanics are not considered in detail, second quantization is hardly touched upon, and the modern developments dealing with systems having an infinite number of degrees of freedom are not discussed. All these limitations are serious ones for none can be really justified on theoretical grounds. In particular, it is conceivable that the latter theories will contribute something essential to our understanding of the measuring instruments, which have necessarily a macroscopic size. However, we believe that there is some good practical motivation in attempting an investigation of the elementary aspects of the theory first. This is especially true since the somewhat unsatisfactory aspects of several of the conclusions reached might stimulate further research bearing on more elaborate theories.

The scientists and the science students who are interested in such problems are split up among many diverse branches of physics, chemistry, and other disciplines. For that reason (and also because the field of investigation makes it possible), we tried to write a book whose access would be quite easy to any reader who has only a very elementary background in modern physics.

Hence, the first two Parts are dedicated to a description of the elementary tools of the theory. Hence, also, the principles of quantum mechanics are not stated in their most compact form. Instead, we thought it appropriate to formulate them in a way already familiar to a great variety of readers, although with more care than is exerted in most textbooks. This introduces the main contents of the book. Part Three deals with nonseparability, hidden variable theories and several related problems. Part Four deals with measurement theories. On all these problems we feel that there exists a considerable lack of agreement even among experts. This lack of agreement is of course not due to differences bearing on the mathematical aspects of the theory. We therefore try to analyze the implicit philosophical postulates that differ from one author to the next and that lead to such divergences. This is the subject of Part Five. It deals essentially with objectivism, with positivism, with Bohr's and Heisenberg's ideas, with Wigner's approach and with the relativity of state theory. Then, a conclusion is given, which incorporates a description of a set of relationships and concepts that could compose a legitimate view of the world. The only prerequisites of the book are an elementary knowledge of the fundamentals of quantum mechanics and of the Dirac notation (bras and kets).

On several of the topics studied in this book the author has greatly benefited from discussions with Prof. J. S. Bell, with Prof. A. Shimony and with other participants to the 1970 Varenna course on the foundations of quantum mechanics. He also wants to thank Dr. M. K. Gaillard, both for her constant technical help and for her very useful theoretical suggestions.

Bernard d'Espagnat

Conceptual Foundations of Quantum Mechanics
Second Edition

PART ONE

ELEMENTS OF QUANTUM MECHANICS

The formalism of quantum mechanics was introduced some 50 years ago. Since then it has been successfully used to account for an extraordinarily large number of newly discovered phenomena. As a matter of fact, it has become a universal tool in physics. Even the theoretical descriptions that deal with the elementary particles are based on its fundamental principles. This is a remarkable fact indeed, since the energies involved are some billion times larger than those that characterize the domain of atomic physics, in which quantum mechanics was born.

As a counterpart to this universality, some of the basic ideas of quantum mechanics bear no relation whatsoever to the concepts used in our daily life. As such they are worth investigating, and our program is indeed to do so. However, before we engage ourselves in this enterprise, we must remember that simple basic experiments are at the origin of all these surprising notions. Therefore Chapter 1 and 2, which are introductory, briefly recall how the quantum formalism could be abstracted from some crucial experimental facts. The principles of quantum mechanics are then formulated (Chapters 3, 4, and 5). Since quantum mechanics is a coherent, self-contained formalism from which a vast number of consequences and predictions of different kinds can be deduced, it appears pertinent to formulate these principles in an axiomatic way, that is, as a set of rules whose ultimate justification lies in the agreement of its predictions with the results of physical observations.

The problem of trying to understand what physical interpretation can be given to these rules is the main subject of this book. It is a highly nontrivial one. It should not be confused with the simpler and more immediate problem of stating precisely what these rules *are*, independently of any interpretation that should, or could, be given them. In Chapter 3, 4 and 5 the principles are simply formulated as a set of rules that make it possible to predict—in a statistical way at least—the results of future observations from a knowledge of the results of past observations on physical systems. It should be stressed that this method of exposition does not imply a preference for a particular inter-

1

pretation or a particular epistemology. It is, rather, a provisional step: we must *first* state precisely (independently, if possible, of any assumption and with economy of words and symbols) a set of rules that lead to experimentally verified predictions for observed phenomena. Later (but only later) we will assume that it is interesting to try to construct from these rules a real description of the physical world; for example, we can try to reformulate the rules in a language not involving the concept of an observer. The decision as to whether such a reformulation is desirable (assuming that it is possible) is essentially, in the view of many, a philosophical option; for this reason it is preferable to defer it to later chapters.

Finally, it is important to note the following fact: some of the most significant principles to which a detailed analysis of the foundations of quantum mechanics leads are in fact more general than is suggested by the manner in which they were obtained. Indeed, they also follow from just a few very qualitative ideas *plus* the results of some recent experiments. We refer here in particular to the principle of nonseparability. Although the latter can be deduced (see Chapter 8) from the general axioms of quantum mechanics, *plus* the assumption of completeness, it can also be shown (see Chapter 12) to be simply a consequence of extremely general conceptions applied directly to properly selected facts.

CHAPTER 1

MATTER WAVES

Classical physics identified matter with particles and radiation with "waves." Although this description accounted for a large number of very important facts, it became apparent around the turn of the century that other phenomena, such as blackbody radiation and photoelectric effect, are satisfactorily explained only if radiation is described in terms of particles. Some time later the hypothesis that, conversely, matter has wave properties was proposed by Louis de Broglie and was confirmed by the Davisson-Germer experiment of electron diffraction. In this experiment and similar ones, an electron beam was passed through a regular lattice (crystal), and subsequently showed the well-known diffraction patterns that are characteristic of waves. These results confirmed the theoretical predictions that a monoenergetic electron beam should be described by a plane wave and that the wave vector $\mathbf{k}$ of the wave should be related to the momentum $\mathbf{p}$ of an individual electron by the formula:

$$\mathbf{p} = \hbar\mathbf{k} \tag{1-1}$$

where $|\mathbf{k}| = (2\pi)\lambda^{-1}$, λ being the wavelength ($\hbar$ is Planck's constant, divided by 2π).

The simplest possible form for the $\mathbf{x}$ dependence of a plane wave with wave vector $\mathbf{k}$ is

$$\psi(\mathbf{x}) = e^{i k x} \tag{1-2}$$

However, as the particle propagates, it is natural to associate it with a traveling wave, whose direction of propagation is that of the particle. This is achieved by writing for the wave function of the free particle

$$\psi(\mathbf{x}, t) = e^{i(k x - \omega t)} \tag{1-3}$$

where, by definition, ω is the circular frequency of the wave. Now, if we assume that the Planck-Einstein relation

$$E = \hbar\omega \tag{1-4}$$

between the energy of a photon and the frequency of the corresponding wave is also valid for "material" particles such as those described by ψ, we can deduce from Eq. (1–3) an equation for $\psi(\mathbf{x}, t)$ simply by noting the identities

$$i\hbar \frac{\partial \psi}{\partial t} = \hbar\omega\psi = E\psi \tag{1-5}$$

$$\hbar^2 \Delta\psi = -\hbar^2 k^2 \psi = -p^2 \psi \tag{1-6}$$

and making use of the relation

$$E = \frac{p^2}{2m} \tag{1-7}$$

between the (nonrelativistic) energy and the momentum of a free particle. The result

$$i\hbar \frac{\partial \psi}{\partial t} = -\frac{\hbar^2}{2m} \Delta\psi \tag{1-8}$$

is the Schrödinger equation for a free particle. It can be deduced formally from the corresponding classical equation (1–7) by the replacement

$$\mathbf{p} \to \frac{\hbar}{i} \nabla, \qquad E \to i\hbar \frac{\partial}{\partial t} \tag{1-9}$$

and the prescription that both members of the operator relation thus obtained should be applied to ψ.

Remark: Phase and Group Velocities

The velocity of the traveling plane wave (1–3), that is, the velocity of motion of the plane of constant phase

$$\mathbf{k}\mathbf{x} - \omega t = C$$

is (setting $k = |\mathbf{k}|$)

$$v_\phi = \frac{\omega}{k}$$

which is different from the classical velocity v of the particle. The "group velocity" of the wave, however, defined in the usual classical way as the velocity of the center of a "wave packet" made of waves with wavevectors centered around k, is

$$v_g \equiv \frac{d\omega}{dk} = \frac{d\hbar\omega}{d\hbar k} = \frac{dE}{dp} = \frac{p}{m}$$

It is obviously equal to v.

In the general case when the particle is not free, the Schrödinger equation, that is, the equation for the wave function $\psi(\mathbf{x}, t)$, is obtained formally by applying Rule (1–9) to the identity

$$E = H(\mathbf{x}, \mathbf{p}) \qquad (1\text{–}10)$$

where H is the *Hamiltonian function* or total energy of the system, written in terms of $\mathbf{x}$ and of the conjugate momentum $\mathbf{p}$ (the rule being to symmetrize in $\mathbf{x}$ and $\mathbf{p}$ whenever this proves necessary). In other words, the Schrödinger equation is

$$\mathbf{H}\psi(\mathbf{x}, t) = i\hbar \frac{\partial}{\partial t} \psi(\mathbf{x}, t) \qquad (1\text{–}11)$$

where $\mathbf{H}$, known as the Hamiltonian operator, is obtained from the classical expression $H(\mathbf{x}, \mathbf{p})$ of the total energy by making the substitutions

$$\mathbf{x} \to \mathbf{x}, \qquad \mathbf{p} \to \frac{\hbar}{i} \nabla \qquad (1\text{–}12)$$

after the appropriate symmetrization of $H(\mathbf{x}, \mathbf{p})$.

That the Schrödinger equation (1–11) is correct is, of course, not a deduction. It is really a physical assumption. The validity of this assumption, however, is demonstrated beyond any reasonable doubt by the truth of its consequences. If a system has a definite energy E, the substitution

$$\psi(\mathbf{x}, t) = \phi(\mathbf{x}) \, e^{-i(Et/\hbar)} \qquad (1\text{–}13)$$

in Eq. (1–5) leads to the "time-independent Schrödinger equation"

$$\mathbf{H}\psi = E\psi \qquad (1\text{–}14)$$

When several particles, with coordinates $\mathbf{x}_1, \mathbf{x}_2, \ldots, \mathbf{x}_N$ are present instead of just one, the wave function ψ is a function of $\mathbf{x}_1, \mathbf{x}_2, \ldots, \mathbf{x}_N$ and the time t. The Schrödinger equation for ψ is obtained from the expression for the total energy of the system, $H(\mathbf{x}_1, \ldots, \mathbf{x}_N, \mathbf{p}_1, \ldots, \mathbf{p}_N)$, in exactly the same way as above.

CHAPTER 2

SUPERPOSITION AND LINEARITY

Modern quantum mechanics developed from wave mechanics by abstracting from its most fundamental features. These are the superposition principle and the linearity of the equation of motion.

2.1 SUPERPOSITION OF WAVES

Any theory finds its justification from experiment, but all the aspects of a successful theory do not gain similar weights from such corroborations. The dynamical development of a theory, such as the transition from wave mechanics to the more general scheme of quantum mechanics, can thus be fully understood only if a few important experimental facts are kept in mind, namely, those whose description really requires that particular sets of principles be valid. The chance is good then that these principles will remain as guides in the development of the theory.

Although the experiments that corroborate wave-mechanics are very numerous indeed, most of them are quite similar in their essence. It will be sufficient, therefore, for what we have in mind, to recall merely one. The experiment of electron diffraction may be chosen, or, alternatively, the conceptually even simpler experiment of particle diffraction by two holes, the principle of which is briefly sketched in Figure 2-1.

The result of these experiments is that the relative number of particles found in the neighborhood of some point z on screen B (number of particles in that neighborhood divided by total number of particles) shows, as a function of the location z of that point on the screen, a diffraction pattern that is characteristic of waves. More precisely, let us, for instance, assume that in a first stage of the two-hole experiment hole 2 is shut and that a density repartition $\rho_1(z)$ on screen B is then measured; that in a second stage hole 1 is shut and a density repartition $\rho_2(z)$ on B is then measured; and, finally, that in a third stage both holes are opened and a density $\rho_3(z)$ on B is then measured. The experimental

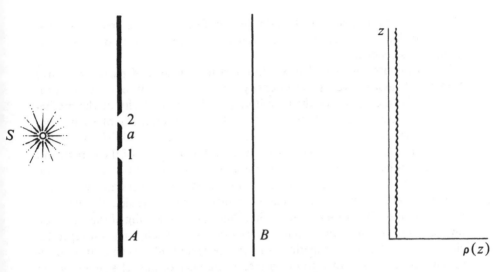

Figure 2-1. The two-hole experiment.

Particles emitted from source S go through screen A, which has two holes, 1 and 2, separated by some distance a, and then fall on screen B, where they leave a permanent mark. Source S is so weak that no more than one particle is on the way between S and B at any given time. Each particle leaves *one and only one* mark on screen B. The density distribution $\rho(z)$ of these marks after some time T nevertheless shows the interference pattern that is characteristic of waves.

densities ρ_1, ρ_2, ρ_3, turn out to be such that

(i)
$$\rho_3(z) \neq \rho_1(z) + \rho_2(z) \tag{2-1}$$

and

(ii) two (complex) functions $\psi_1(z)$, $\psi_2(z)$ exist which are such that

$$\rho_1(z) = |\psi_1(z)|^2 \tag{2-2}$$
$$\rho_2(z) = |\psi_2(z)|^2 \tag{2-3}$$
$$\rho_3(z) = |\psi_1(z) + \psi_2(z)|^2 \tag{2-4}$$

Because of the fact that the setup in the third stage of the experiment is a simple combination of what it is in each of the first stages, one naturally expects that at least some of the functions that describe the experimental results in the third stage should be simple mathematical combinations of those that describe them in the first two stages. The occurrence, in the description of the third stage, of a function that is just a sum of what it is in each of the first two stages is therefore not at all surprising. But, surprisingly enough, what the experiment shows is that, contrary to what would be the case if the particles obeyed classical physics, this function is *not* identical to the density of presence, and that, more precisely (see point ii above), the density of presence is its modulus *squared*.

This, of course, is just what would happen if, instead of particles, source S emitted (classical) waves (the intensity, or energy content, of a wave is proportional to the square of the amplitude), and this similarity of the mathematical structures accounts for the fact that in experiments of such kind, just as in experiments with waves, interferences are observed. The similarity is only partial, however, since, for instance, if the duration of the experiment with both holes open (i.e., the time of exposure of screen B) is kept constant and the intensity of source S is sufficiently lowered, what eventually happens is very different from what a classical wave theory would predict. The law $\rho(z)$ of distribution of the observed intensity, that is, of blackening of the emulsion on the screen, *becomes appreciably altered* in a way which (*a*) is not reproducible (it varies from one experiment to the next) and (*b*) is such that it corresponds to the impact of zero or an integer number of particles in any given small area. This shows that the assumption that S emits just classical waves instead of particles, and that what is observed on screen B is simply the local effect of these waves, is certainly not a correct hypothesis, although, if the time of exposure is then increased so as to allow for large numbers of impacts in most areas of screen B, the cumulative density of impacts $\rho(z)$ again takes, within the usual approximation of statistics, the form $\rho_3(z)$ predicted by the wave picture.

Of course, the discussion above is very sketchy (more elaborate treatments can be found in many textbooks). It is, however, sufficient to show that these diffraction experiments of particles raise problems that will not easily yield to an interpretation in terms of familiar concepts, such as those of waves or particles. Situations of this kind are not really exceptional in physics; and when they occur, it is usually not rewarding to try to force some model description upon them. A much better procedure, at least in the first stages of the inquiry, is to make no assumption whatsoever about the mechanisms at work and to simply describe the observed facts, it being understood that this description should be as accurate and as general as possible. The simple features of the description thus obtained are then assumed to be valid in similar cases involving different experimental settings and so forth.

If this is done in regard to the experimental facts recalled above, three general ideas or principles immediately emerge. They are as follows.

(i) *Elementarity of Microstructures*
Objects are normally endowed with properties that cannot be split indefinitely; consequently entities—called "particles"—exist which, in any experiment, manifest themselves as integer wholes.

(ii) *Probability Principle*
In some situations at least, functions $\psi(x)$ (usually called "wave functions") can be associated with particles in such a way that the probability of finding a particle in a small region of volume dv around point $\mathbf{x}$ is

$$P(\mathbf{x})\,dv = \frac{|\psi(\mathbf{x})|^2\,dv}{\iiint|\psi(\mathbf{x})|^2\,dv} \qquad (2\text{–}5)$$

where the summation extends over all space.

(iii) *Superposition Principle* (static linearity)
If $\psi_1(\mathbf{x})$ and $\psi_2(\mathbf{x})$ are two possible wave functions (i.e., if they are functions associated with particles in possible settings of the apparatus), then any linear combination of $\psi_1(\mathbf{x})$ and $\psi_2(\mathbf{x})$ is in general also a possible wave function.

It should be pointed out at this stage that the principles thus formulated are merely rules that have proved useful. As such they are probably true. They do not, however, commit us to any interpretation. In particular, we are not yet at this stage especially committed to the indeterministic interpretation. Because of the context, the word "probability" which appears in the formulation of principle (ii) is, clearly, to be taken in the elementary operational sense of "relative frequency." If N is the total number of particles emitted by the source, then, for sufficiently large N, $NP(\mathbf{x})\,dv$ is simply the number of particles one expects to find, upon measurement, in the volume element dv. Thus principles (i), (ii), and (iii) do not *imply* that the motion of an individual particle is not determined (by sufficiently complicated unknown forces). We do not pursue this problem further here since it is discussed at some length in Chapters 4, 7, and 11.

As regards the superposition principle, it should be observed, of course, that it is not a particularly quantal idea since it is a general feature of all wavelike phenomena. It could even be argued that in a restricted sense it is also present in classical particle physics: even there a function exists that satisfies a modified form of principle (iii). This function is just the distribution function $\rho(\mathbf{x})$. However, $\rho(\mathbf{x})$ is necessarily nonnegative, so that principle (iii) cannot be satisfied in complete generality. An ensemble of $\psi(\mathbf{x})$ and an ensemble of $\rho(\mathbf{x})$ have for that reason mathematical structures that are extremely different from one another. One of the farthest-reaching discoveries of quantum mechanics is indeed that the linearity requirement, in the form of principle (iii), underlies physics at an extremely deep level. Finally, a difference between $\psi(\mathbf{x})$ and the classical wave amplitudes may also be noticed. The difference, which at this stage may be considered as a rather technical one, is that, whereas the amplitude of a classical wave is given an absolute meaning, $\psi(\mathbf{x})$ in the formalism of wave mechanics is given only a relative one. No physical predictions are modified if ψ is multiplied by a nonzero factor independent of **x**.

2.2 THE EVOLUTION IN TIME

Let us now forget about the particular arrangement of a two-hole or diffraction experiment and imagine, much more generally, arbitrarily complicated physical instruments, incorporating fields varying with time and so on. Let us further assume, tentatively, that the time-dependent Schrödinger equation is valid. Since this partial differential equation is of first degree in the time variable, it determines ψ at any time t if ψ is known at one particular time t_0 and satisfies some general conditions (Cauchy-Kowaleska theorem). Symbolically,

$$\psi(t_0) \rightarrow \psi(t) \qquad (2\text{-}6)$$

Then, if, in view of principle (iii), we consider a linear superposition

$$\psi(t_0) = \alpha\psi_1(t_0) + \beta\psi_2(t_0) \qquad (2\text{-}7)$$

where α and β are (complex) numbers, we may write

$$\psi_1(t_0) \rightarrow \psi_1(t) \qquad (2\text{-}8)$$

$$\psi_2(t_0) \rightarrow \psi_2(t) \qquad (2\text{-}9)$$

and inquire about the corresponding $\psi(t)$. Because of the fact that the Schrödinger equation is linear in ψ, we obtain as a consequence of Eq. (2–7)

$$\psi(t) = \alpha\psi_1(t) + \beta\psi_2(t) \qquad (2\text{-}10)$$

The fact that Eq. (2–7) entails (2–10) is, as later chapters will show, of utmost importance for all the discussions that bear on the possible interpretation of quantum physics. Here it has been deduced from the time-dependent Schrödinger equation. It is apparent, however, that only very general features of the equation are involved in the proof (as a matter of fact, it turns out that only linearity is essential). We may therefore abstract the result and turn it into a principle which now appears as a *corollary to the superposition principle.*

(iv) *Linearity in Time Development*
If the wave function $\psi(t_0)$ at time t_0 is a linear superposition

$$\alpha\psi_1(t_0) + \beta\psi_2(t_0)$$

of two wave function $\psi_1(t_0)$, $\psi_2(t_0)$, then the corresponding wave function $\psi(t)$

at time t is the *same* linear superposition of the functions $\psi_1(t)$, $\psi_2(t)$ corresponding to $\psi_1(t_0)$, $\psi_2(t_0)$, respectively.

In particular, one may apply this corollary to the times $t_0 = -\infty$ and $t = +\infty$. The corollary is then applicable to theories with an even wider scope than quantum mechanics, for instance, to S-matrix theory.

2.3 THE CORRESPONDENCE PRINCIPLE

Classical physics leads to correct results in many cases. It is therefore to be expected that from the quantum theory we should be able to deduce classical mechanics and electrodynamics by taking suitable limits. As to what these limits should be, we may notice that, in general, classical theory correctly accounts for the observed phenomena when the quantum discontinuities can be considered as infinitely small. Hence we are entitled to require that *under this condition the predictions of the new theory bearing on practically observable facts should coincide with those of classical physics.* This requirement is known as the correspondence principle. That it is satisfied by quantum mechanics is shown in most textbooks with the help of general theorems such as the Ehrenfest theorem and by making use of the relationships that exist between the Schrödinger equation and the Hamilton-Jacobi equation.

Formulated as above, the correspondence principle had a very important place in the whole elaboration of atomic physics. For the construction of quantum mechanics, a slightly different idea was also used, namely, that a *formal* detailed correspondence should exist between classical and quantum physics. (See Messiah [1],* Ch. 1, Sect. 12.) This is considered as a second form of the correspondence principle, a form we shall not use explicitly.

Finally, we shall at one moment need a relativistic extension of the correspondence principle. The explanation of what this generalization consists of will be found at the place where it is used (Section 8.3, footnote).

2.4 THE NECESSITY FOR A MORE GENERAL FORMALISM

Starting from the premises described in Chapter 1, wave mechanics can be developed into a coherent and unambiguous formalism. This formalism accounts with remarkable precision for practically all atomic and molecular phenomena, that is, essentially for all phenomena that take place in nature around us, where, because of prevailing conditions of temperature and so forth, most particles are stable. The stability of the number of the particles constituting a system makes it possible to describe the system by a wave function $\psi(x_1, \ldots, x_N, t)$ with a fixed number of variables.

---

*Numbers in brackets refer to items in the reference list at the end of the chapter.

It is clear, however, that if one considers, for example, a phenomenon in which the number of constituent particles varies with the time, a formalism which gives such an importance to the *coordinates* of the constituent particles automatically leads to difficulties. The production or suppression of some coordinates in a function is not (except in the most special case of *projections*) a standard mathematical operation. Now, such phenomena exist. A familiar instance is the production of photons by any source of light or their absorption by matter. Other examples are the production of pi-mesons in proton-proton collisions, and the production of electron-positron or of proton-anti-protons pairs. As a matter of fact, such phenomena can affect particles of any kind. It is therefore clear, even if only for this reason (and there are many more), that a generalization of the formalism of wave mechanics is necessary. It is also clear that this generalization should consist of the replacement of the wave function, in the role of the representation of a state, by some mathematical entity, not as explicitly dependent on the coordinates and therefore more abstract. But, of course, the essential features of wave mechanics, those which made it so successful in describing observed facts, must be preserved in that operation.

Now, if one investigates what is the most general—and therefore the most characteristic—attribute of the wave concept, one soon recognizes that it is the property of algebraic addition of the amplitudes: The sum of two amplitudes (i.e., of two functions which describe a possible pattern of the oscillating medium) is also, in general, a wave amplitude, that is, a function which also describes a possible pattern of the medium. This property is in fact nothing else than the superposition principle, and it is common to all possible kinds of waves, such as acoustic, hydraulic, and electromagnetic waves. It is independent of the number of dimensions involved in the problem. One is thus led to suspect that it is the essential property that should be kept, and to replace accordingly the concept of the wave function by a more general and abstract one, which emphasizes only this formal aspect. Quantum mechanics, in the form given to it by Dirac, is nothing but the realization of this program. It describes the possible states of a system, not by wave functions, but by abstract entities called state vectors, whose *defining* property is just that they obey the superposition principle.

Advances in theoretical physics often require that images should be abandoned, or should at least be recognized as being mere, imperfect images. Wave mechanics rightly pointed out that the concept of a point particle with definite position and momentum is, in that sense, just an image. But the concept of wave is also an image and in precisely the same sense. This is not to say that such images are useless representations; on the contrary, their value is obvious since they guide imagination. M. Gell-Mann and V. Telegdi once compared processes of this kind to a particular cooking recipe. Although, they say, this recipe is famous in some regions in France, we are not sure that we can reproduce it here with rigorous exactitude; the general idea, however, is this: Take a morsel of veal and cook it between two wings of peacock. When the aroma of the peacock has sufficiently impregnated the veal, throw the peacock wings away and serve what remains. In other words, let your friends enjoy the essence of the peacock wing flavor thus infused into the veal. This, it is claimed, is infinitely more satisfactory than eating the peacock itself. In a similar way, we shall find that the essence of the superposition principle, as concentrated in the ket vectors, is, to a properly refined sensibility, infinitely more palatable than any kind of wave function.

Because of its relevance to physics, the mathematical formalism of the linear abstract vector spaces is described in most textbooks on quantum mechanics (see, e.g., Ref. [1] or [2]). It is here assumed to be known, along with all the most common definitions and notations.

REFERENCES

[1] A. Messiah, *Mécanique Quantique*, Dunod, Paris, 1958.
[2] K. Gottfried, *Quantum Mechanics*, 1966, (4th printing, with corrections, 1974), W. A. Benjamin, Inc., Reading, Massachusetts.

CHAPTER 3

STATEMENT OF THE RULES

Before the rules are stated, two definitions should be given.

Definition 1: Physical system

Classical and quantum physics both make considerable use of the expressions "physical system" and "state of a physical system." Indeed, any sufficiently isolated thing—a voltmeter, an electron, or a molecule—is a physical system (the latter two, and, more generally, systems which are of molecular size or smaller, and for which a description in terms of classical physics is obviously insufficient, are often called "microscopic" systems). As a matter of fact, the concepts implied by the two expressions above appear so obvious to us that we are tempted to consider them as a priori notions, which are inherently meaningful. A word of caution is therefore appropriate at the start. As results from the discussion in the following chapters, the concepts of systems and states, which are also extremely useful in quantum mechanics, should nevertheless be handled with some care in that formalism. In particular, it should *not* be assumed that in all cases where we intuitively speak of a physical system (and even of a "completely isolated physical system") this system (supposing the word is used) is necessarily in some definite, known or unknown "state." In other words, we may not always assume that a system possesses at all times a constant number of known or unknown properties (momentum, spin component along some direction, etc.) that can properly be considered as pertaining to it alone. Counterexamples to such an assumption are given below.

In what follows, the expression *physical system* is taken in its familiar, intuitive sense as discussed above. For instance, a partially but sufficiently isolated atom (or electron, molecule, etc.) is a physical system or, for short, a system. As for the expressions *state* and *quantum state,* they are not always given the same definitions in the existing literature. At this stage, therefore, we avoid introducing them.

Definition 2: Ensembles

Quantum mechanics is essentially a statistical theory. Except in special cases it makes no prediction that bears on individual systems. Rather, it predicts statistical frequencies. In other words, it predicts, as a rule, the number n of times that a given event will be observed when a large number N of physical systems of the same type and satisfying specified conditions are subjected to a given measurement process. The statistical frequency or *probability* of the event is then the number n/N. The collection of N systems considered above is called a *statistical ensemble* or, for short, an *ensemble* of systems. Each individual system is an *element* of the ensemble. The elements of the ensemble are of course noninteracting. Ideally, it should be imagined that to each element corresponds one particular apparatus by means of which the measurement is made. Only in the cases $n/N = 1$ or 0 is it possible to say that the theory makes a prediction that bears on individual systems, since in these cases the prediction is that the given event either will or will not be observed on any individual element.

The rules of quantum mechanics can now be summarized as follows (comments and alternative choices of axioms are postponed to later chapters).

Rule 1. A vector space $\mathscr{H}$ which has the structure of a Hilbert space is introduced for describing ensembles of physical systems of the same type. There are then instances in which, for the purpose of predicting the results of definite experiments, such an ensemble can be associated with a definite element of $\mathscr{H}$. Such an element, or *ket*, is then called a *state vector*.

Rule 1 calls for some remarks.

Remark 1: As Rule 6 in this section makes apparent, two kets differing only by a phase factor give rise to the same predictions. Such a phase factor therefore has no physical meaning, although, in a superposition such as those considered in Rule 3, the relative phases are of course meaningful. The expression "definite element of $\mathscr{H}$" should be understood with this restriction.

Remark 2: A frequent practice is to associate individual systems with state vectors. Often, this may be done without inconvenience. From a theoretical standpoint, however, the use of ensembles is preferable, unless special precautions are taken (see Chapter 10). Rule 1 does not imply that every ensemble (or, in the loose language just alluded to, every system) can be described by a state vector. Indeed, the majority of systems are such that no state vector can be attributed to them, not even in a loose language (see Sections 7.2 and 8.1).

Rule 1a (Superposition Principle). Any ket of $\mathscr{H}$ can usually describe an ensemble of physical systems.

Exceptions exist. They are known as *superselection rules* (see Section 7.1).

When a ket $|\psi\rangle$ can be written as a linear combination of several kets $|\Phi_1\rangle$, . . . , $|\Phi_N\rangle$, one says that $|\psi\rangle$ is a superposition of $|\Phi_1\rangle$, . . ., $|\Phi_N\rangle$.

Rule 2. If two nonidentical, noninteracting systems S and S' are elements of ensembles that can be described by means of the kets $|\psi\rangle$ and $|\psi'\rangle$, respectively, the ensemble of the compound systems composed of S and S' can also be described by a ket noted as $|\psi\rangle|\psi'\rangle$. This ket is defined as an element of the tensor product $\mathcal{H} \otimes \mathcal{H}'$ of the vector spaces $\mathcal{H}$ and $\mathcal{H}'$ corresponding to S and S', respectively.

Rule 3. The evolution in time of a state vector $|\psi\rangle$ that describes an ensemble of systems subjected only to external forces is causal and linear. This means that $|\psi(t)\rangle$ uniquely determines $|\psi(t')\rangle$ (t and t' being two different times) and that if

$$|\psi(t)\rangle = \alpha_1|\psi_1(t)\rangle + \alpha_2|\psi_2(t)\rangle \tag{3-1}$$

then

$$|\psi(t')\rangle = \alpha_1|\psi_1(t')\rangle + \alpha_2|\psi_2(t')\rangle \tag{3-1a}$$

In the statement of this rule, the term "external forces" means, as usual, that the influence (often called the "reaction") of the system on the physical entities that create the forces can be neglected. If this is not the case, a larger system, including these entities, should be considered, and Rule 3 should be applied to the larger system.

It should also be noted that the usual Schrödinger equation

$$H|\psi(t)\rangle = i\hbar \frac{\partial}{\partial t}|\psi(t)\rangle \tag{3-2}$$

where H is the Hamiltonian, is just a particular realization of Rule 3. The point is that Rule 3 embodies all the general properties of the Schrödinger equation that are of interest in the subsequent discussion.

Rule 4. To every measurable physical quantity A there corresponds a Hermitean operator operating on the elements of the vector space defined in Rule 1.

For convenience we call "Hermitean" a Hermitian operator for which the eigenvalue problem can be solved (see von Neumann ref. [1], Section 2.9). Following Dirac, we henceforth call a measurable physical quantity an *observable*. Two observables are said to be *compatible* if the corresponding operators commute.

Rule 5. The measurement of an observable A necessarily yields one of the eigenvalues of the operator associated with A.

Henceforth the expression "eigenvalue (or eigenfunction) of observable A" is often used as an abbreviation for "eigenvalue (or eigenfunction) of the operator associated with A."

Rule 6. Let the symbols a_k and $|a_k, r\rangle$ denote, respectively, the eigenvalues and the normalized eigenvectors of the observable A (r is a degeneracy index), and let $|\psi\rangle$ be a normalized ket that describes an ensemble E of physical systems S immediately before a measurement of A on S. Then the probability that this measurement will yield the value a_k is given by

$$w_k = \sum_r |\langle a_k, r|\psi\rangle|^2 \qquad (3\text{--}3)$$

Again, what we mean here by "probability" is just the relative statistical frequency with which, according to our predictions, the result a_k will be obtained upon measurement of A.

Remark 1: The possibility of generalizing the concept of measurement and, in particular, of dissociating it from "the observer," or of reducing the observer to an ordinary physical system is discussed in later chapters. What we mean here by the word "measurement" is (together with the corresponding reaction) the action on system S of a generally complex entity S' incorporating, along with measuring apparatus and so forth, the observer himself. At this stage, we make no effort to describe the observer in the same way as an ordinary physical system, and for the time being the entity S' need not be analyzed further as to its constituents. In other words, we formulate prediction rules that are for the time being independent of the more subtle question of whether or not the observer himself (for whom the predictions are made) is to be identified with a "physical system" in the ordinary sense. At this stage we simply leave that question open.

Remark 4: The observables which have a continuous eigenvalue spectrum have eigenvectors that are not normalizable. As regards the measurement of these observables, some technical changes should therefore be made in expression (3–3). These are described in the textbooks. Then, when applied to a position measurement, Eq. (3–3) reproduces Eq. (2–5).

Remark 5: It is important to realize that the degeneracy index r can (and usually does) correspond to the existence of observables other than A and compatible with A. For instance, if B is such an observable, with eigenvalues b_l, and if the complete Hilbert space of the system is spanned by $|a_k, b_l\rangle$, the probability of obtaining the result a_k when A alone is measured is given by formula (3–3), where now b_l acts as a degeneracy index

$$w_k = \sum_l | \langle a_k, b_l | \psi \rangle |^2 \tag{3–3a}$$

Therefore, in this case, w_k is just equal to the value obtained by adding the probabilities that, upon measurements of A and B, a definite value a_k of A is found, together with any possible value of B.

When the kets are not normalized to unity, expressions (3–3) and (3–3a) must be modified in such a way that they are not affected by the arbitrariness in the normalization; expression (3–3) should then, for instance, be replaced by

$$w_k = \sum_r \frac{| \langle a_k, r | \psi \rangle |^2}{\langle a_k, r | a_k, r \rangle \langle \psi | \psi \rangle} \tag{3–3b}$$

Rule 7. For any observable, it is always possible, at least conceptually, to invent a measuring device that satisfies the following condition: if the observable is measured twice on a given system, and if the time interval between the two measurements is vanishingly small, then both measurements give the same result.

Such a measurement will be called an *ideal* one or a *measurement of first kind*. All other measurements are measurements of the second kind.*

Rule 8. If, on a given system, three ideal measurements are made in succession: one of A, one of B, and again one of A, then, if the observables A and B are compatible and if the time intervals between two successive measurements are vanishingly small, the result of the last measurement is the same as the result of the first one.

N.B. This statement is not really a rule but rather a generalization or idealization of experimental findings. It is incorporated in the present list because these experimental facts create the conditions that make the statement of the other rules meaningful.

It follows from Rule 8 that the order of succession of two ideal measurements made on compatible observables is irrelevant if the time interval between them is vanishingly small. Thus Rule 8 gives meaning to the notion of "simultaneous measurements" of two such observables. This notion can obviously be extended so as to cover the case of an arbitrary number of compatible observables. In particular, a *complete* measurement on a system S is a set of simultaneous (ideal) measurements of the observables corresponding to

---

*Note that some authors use the word "ideal" to designate what we call a "complete" measurement below.

a complete set of commuting Hermitean operators acting on the vector space associated with S.

Rule 9. Let $A, B, \ldots$ be a complete set of compatible observables pertaining to some type of systems. If for some reason the results of simultaneous measurements of $A, B, \ldots, F$ are known in advance on an ensemble E of such systems and are the same for all the elements, S, of E, then E can be described by a ket.

Let $|\psi\rangle$ be this ket. Let $a_k, b_l, \ldots$ be the (known) results of the measurement to be made on S. The probability of finding the results $a_k, b_l, \ldots$ is unity, so that by Eq. (3–3b)

$$1 = \frac{|\langle a_k, b_l, \ldots |\psi\rangle|^2}{\langle a_k, b_l, \ldots |a_k, b_l, \ldots\rangle \langle \psi|\psi\rangle}$$

This is precisely the Cauchy-Schwarz inequality in the case where it reduces to an equality: it follows that $|\psi\rangle$ is proportional to $|a_k, b_l, \ldots\rangle$. Since a numerical factor is just a normalization constant without physical significance, the following proposition holds.

Proposition
The ket considered in Rule 9 is necessarily

$$|\psi\rangle = |a_k, b_l, \ldots\rangle$$

A special case of Rule 9 is the following: the prior knowledge of the results of simultaneous measurements of $A, B, \ldots$ to be made at time t comes from a previous measurement of these observables made immediately before t. In this case, Rule 9 and the corresponding proposition are known as the *reduction of the wave packet*. For future reference we restate it as follows.

Rule 10 (Reduction of the Wave Packet). Let E_0 be an ensemble of systems S. Immediately after simultaneous (ideal) measurements have been performed on a complete set of observables pertaining to the systems S, every subensemble E of E_0 that is composed of systems S on which the measurements have produced identical sets of results can be described by a ket. This ket then necessarily is the eigenket common to the eigenvalues found as a result of the measurements.

Finally, we must also consider the case in which an incomplete measurement is made on a system. We then make the following postulate.

Rule 10a. Let E_0 be an ensemble of systems S that is described by a ket. Immediately after simultaneous (ideal) measurements have been performed on an incomplete set of observables A, B, . . . pertaining to the systems S, every subensemble E that is composed of systems S on which the measurements have produced identical sets of results can be described by a ket.

Let this ket be denoted by $|\psi\rangle$. Here, again, the probability law (3–3) gives some information about $|\psi\rangle$ once its existence is known. The information, however, is only partial in this case: what (3–3) shows is simply that (in the ideal case) the ket $|\psi\rangle$ is in the manifold M spanned by the eigenvectors that have in common the observed eigenvalues of the operators A, B, . . . corresponding to A, B, ...

A particular type of measurement, sometimes called *moral measurement,* is often considered. In a moral measurement $|\psi\rangle$ is simply (to within a renormalization factor, of course) the projection on M of the ket describing the system before the measurement was made.

Although the postulate described by Rule 10a is usually made, it is not the only possibility. Another one is considered by von Neumann (ref. [1], Section 4.3)

REFERENCES

[1]   J. von Neumann, *Mathematical Foundations of Quantum Mechanics,* Princeton University Press, Princeton, N. J., 1955, p. 348.

Further Reading

G. Lüders, *Ann. Phys.* **8**, 322 (1951)
F. Herbut, *Int. J. Theor. Phys.* **11**, 193 (1974).

CHAPTER 4

COMMENTS

We have stated the rules of quantum mechanics in a condensed way that leaves many questions open. Some important specifications not yet given will be developed in later sections. Even at this stage, however, a few comments are in order.

4.1 ON THE DESCRIPTION OF ENSEMBLES BY KETS

(*a*) It should be kept in mind that quite often an ensemble of systems can be associated with a (known or unknown) ket only with reference to a certain set K of possible planned experiments. In other words, in many circumstances (see Chapter 8) it is legitimate to assume that a statistical ensemble E of systems corresponds to a known or unknown ket *only for the limited purpose of predicting (statistically) the results of an experiment chosen from the set K,* whereas such an assumption may lead to false results if it is used in a calculation aimed at predicting the results of an experiment chosen outside this set.

For instance, if E is an ensemble of systems S, each of which has previously interacted with several other systems S_1, S_2, . . ., then K (see Chapter 8) can contain as a subset the set of all experiments in which no correlation is measured between physical quantities pertaining to system S and physical quantities pertaining to system S_1 or S_2, and so forth. Such correlation measurements, on the contrary, often fall outside K.

Under such circumstances a meaningful question of principle is: "In the context of Rule 9 is the meaning of the expression 'can be described by a ket' absolute or is it, on the contrary, similarly relative?" This question involves measurement theory, however. We must therefore postpone its investigation.

(*b*) In many instances Rule 10 (reduction of the wave packet) is the one that is used to determine the ket of an ensemble of systems; filtration, in particular, can be roughly assimilated to a measurement. For example, when we attribute a wave function (essentially a plane wave) to particles emerging from an accelerator, we, in fact, just apply Rule 10 to a kind of implicit measurement of the momentum. Under these circumstances the necessity of introducing the more general Rule 9 may a priori be questioned.

Let us therefore point out explicitly that Rule 9 is in fact necessary. Consider a scattering experiment of, for instance, muons by hydrogen atoms. In order for the calculation to be at all possible, it is necessary that some wave function be attributed not only to the incident particle (muon) but also to the target (hydrogen atom). The experimentalist has not really measured the energy of the particular atoms on which the scattering is going to take place. He knows precisely the energy of these atoms only because of his general knowledge of the quantization rules and of his particular knowledge that the temperature is small. Rule 9 is therefore necessary in order for this knowledge to be sufficient to justify the attribution of a definite ket (or wave function) to the target.*

4.2 ASSUMPTION Q

We have just observed that some prescription similar or equivalent to Rule 9 is necessary. But of course this does not mean that we cannot replace the latter by another, more general formulation that would imply Rule 9 as a particular case. The opportunity of trying to find such a generalization becomes apparent when it is observed that the ensembles considered in the formulation of Rule 9 are rather special ones indeed. As we shall see in Chapter 6, there exist more general formulations of the quantum rules that can be applied just as well to these special ensembles and to ensembles of a more general character (mixtures). Now, Rule 9 is in fact nothing else than a precise formulation of the general idea that the elementary rules of quantum mechanics are universally applicable to the hitherto considered ensembles. Therefore, the generalization we are seeking should somehow assert that quantum mechanics is universally valid for all the ensembles to which the more general formalism just alluded to is applicable. But indeed the only condition that this new formalism explicitly postulates as regard the ensembles with which it deals is that these should not be specified with reference to future observations bearing on their elements. Hence we are led in a very natural fashion to formulate in a

---

*The possibility of separately attributing here a ket to each atom (or ensemble of these) may—apart from the question of indiscernability, which requires separate arguments (see, e.g., Messiah, T., II)—be simply justified by the fact that the set K of experiments for which predictions are calculated does not include sophisticated correlations between the mu meson and *several* hydrogen atoms: the motivation is then the same as in (*a*). This does not account, however, for the choice of the ket as being, for instance, that which corresponds to the ground state, or, with some probability, that which corresponds to the first excited state, depending on the temperature.

precise way the hypothesis that quantum mechanics has a kind of universal validity by stating it in the form of the following assumption.

Assumption Q

Any ensemble of systems the specification of which involves no reference to the results of future measurements that are to be performed on these systems obeys the quantum rules.

The fact that Assumption Q is indeed a workable substitute for Rule 9 within the framework of the more general formalism will become clear in Chapter 7.

For later reference we define ensembles of systems that obey the quantum rules as "quantum ensembles."

Remark 1

The restriction that the specification of these ensembles should not refer to the results of future measurements excludes ensembles selected from larger ones through the requirement that *future* measurements of some observables $A, B, C, \ldots$ will give results $a_k, b_l, c_m, \ldots$. Since, as discussed below and in Section 7.3, any subensemble E' of an ensemble E described by some ket $|\psi\rangle$ is also described by $|\psi\rangle$, the exclusion just mentioned is indeed necessary for the internal consistency of the formalism. This is obvious since, if E' is the subensemble constituted of all the systems upon which a measurement of A immediately *after t* will give result a_k, then in general E' does not obey Rule 6 at time t (indeed this rule would, in general, give $w_k \neq 1$).

Remark 2

In Assumption Q, as well as in Remark 1, the word "measurement" should be understood in a very general sense. For instance, if an ensemble of particles in some definite spin state is deflected by a Stern-Gerlach device D in such a way that several beams are produced, and if all the beams except one are intercepted, the surviving particles must be considered to have undergone a measurement of their spin component along the direction defined by D.

Remark 3

One of the views most often expressed by the supporters of the "Copenhagen interpretation" is that quantum mechanics is a *complete* theory. In fact, several important features of that interpretation depend on the validity of such a view (this is well known and will also emerge from the contents of several of the following chapters). Our formulation of Assumption Q is an attempt to make the view in question more definite. Indeed Assumption Q is obviously equivalent to the assumption that there are no nonquantum ensembles (except those defined by referring to the results of later measurements). In particular, it implies therefore that there are no dispersion-free states, as defined by von Neumann [1] and others. Since these dispersion-free states, if they existed, would be specified not only by the state vector but also by additional "hidden variables," Assumption Q is a somewhat strengthened version of the usual hypothesis that "hidden variables do not exist." In fact, it is the least restrictive positive counterpart to that hypothesis that we have found to be workable. The notion of hidden variables is discussed in Section 7.5 and further in Chapter 11.

4.3 INDETERMINISM

The assumption is usually made that, if an ensemble can be described by a ket $|\psi\rangle$, all its subensembles can be described by the same ket. This assumption implies that it is not possible to split an ensemble E corresponding to a given ket into two subensembles for which the various expectations would (in percentage, of course) be different from what they are for E. Clearly, it also implies that such a splitting already realized can never be found.

Of course, the main motivation for the assumption is that such splittings cannot be made in any straightforward manner. Let us consider, for example, the two-hole experiment discussed in Chapter 2. When both holes are open, the density distribution of the impact points on the screen is not the sum of the two distributions produced when one or the other hole is open alone. For this reason it is very difficult to consider the ensemble of particles under study as composed of two distinct ensembles, each having as its elements the particles that go through one specific hole. As a matter of fact, this is even impossible, unless it is assumed that a particle that goes through one hole is somehow influenced by the conditions prevailing at the place where the other hole is located. However, this assumption implies the existence of particular, hitherto unknown forces, and it is therefore quite generally considered as unattractive.

Thus the probabilities, or statistical distributions, that conventional quantum mechanics introduces into physics are inherently different from those found in classical mechanics, for example, in the kinetic theory of gases. The appearance of the latter is but a reflection of the physicist's ignorance of the finer details of the systems involved, whereas no simple interpretation of this kind can be given to the former.

Formally, the same conclusion results from a somewhat more abstract discussion of the quantum rules. Let us, for simplicity, consider a vector space $\mathcal{H}$ with two dimensions only. Let A and B be the two noncommuting observables whose operators operate in $\mathcal{H}$. Let their eigenvalue equations be

$$A|a_i\rangle = a_i|a_i\rangle \tag{4-1}$$

and

$$B|b_j\rangle = b_j|b_j\rangle \tag{4-2}$$

respectively ($i, j = 1, 2$). On the other hand, let

$$|\psi\rangle = c_1|a_1\rangle + c_2|a_2\rangle \tag{4-3}$$

with

$$|c_1|^2 + |c_2|^2 = 1$$

describe an ensemble E of N systems. The number of cases in which the value a_i is found upon measurement of A on these systems is approximately

$$N|\langle a_i|\psi\rangle|^2 = N|c_i|^2 \tag{4-4}$$

Again, one could try to interpret this result by the following assumption: $N|c_i|^2$ systems in E have, before measurement, a value $A = a_i$ of A or, in other words, compose an ensemble described by $|a_i\rangle$. If this were the case, however, the number of cases in which a value b_j of B would be found upon measurement of B would be approximately

$$N_j = \sum_i N|c_i|^2 p_{ji} \tag{4-5}$$

where p_{ji} is the probability that such a measurement, performed on an ensemble described by $|a_i\rangle$, results in the value b_j. Now, from the rules of quantum mechanics,

$$p_{ji} = |\langle b_j|a_i\rangle|^2 \tag{4-6}$$

so that Eq. (4–5) reads

$$N_j = N[|c_1|^2|\langle b_j|a_1\rangle|^2 + |c_2|^2|\langle b_j|a_2\rangle|^2] \tag{4-7}$$

When directly applied to $|\psi\rangle$, these same rules, however, immediately give

$$N_j = |c_1\langle b_j|a_1\rangle + c_2\langle b_j|a_2\rangle|^2 \tag{4-8}$$

As we see, Eq. (4–7) is, in general, different from the correct quantum-mechanical prediction in Eq. (4–8), so that the hypotheses on which Eq. (4–7) rests cannot be correct. In other words, the two assumptions that (a) the quantum rules hold for any ensemble (Assumption Q, Section 4.2) and (b) some elements of E have $A = a_1$, and the others have $A = a_2$, are not mutually consistent. More generally, it is shown in Chapter 7 that, if an ensemble E of microsystems is described by a ket, if Assumption Q is made, and if it is assumed that every Hermitean operator corresponds to an observable, then no

splitting of that ensemble into two or more ensembles having different statistical properties can be consistently imagined. For this reason, it is convenient
to say that under the conditions just stated all the elements of an ensemble
that is described by a ket are identical to one another (in the case in which
not all Hermitean operators correspond to observables such a convention
loses a part of its attractiveness, as we shall see).

It should, however, be noticed at this point that, if Assumption Q is relaxed,
the conclusion no longer holds, and that, moreover, Assumption Q *can* indeed be relaxed without altering in practice the predictions of conventional
quantum mechanics. This has been demonstrated by Bohm [2], who has explicitly constructed a fully deterministic version of quantum mechanics. In
that theory, whenever individual observations performed on the elements of
any ensemble E lead to different results, these elements are indeed different,
even before the time when the observations are actually made. They differ by
the values of some variables, usually called "hidden variables" because they
should not be confused with the observables of the theory. In such a theory,
any individual electron goes through one particular hole of the device used in
the two-hole experiment. However, it experiences the influence of hitherto
unknown forces which are different according to whether the other hole is
open or closed.

One basic feature of Bohm's theory is that *the quantum-mechanical predictions are exactly reproduced for any ensemble E in which the hidden variables
are distributed at random.* This suffices to account for the agreement between
quantum mechanics and experience. Let us, however, stress the fact that such
an assumption is far less stringent than Assumption Q. In Bohm's theory
the quantum rules, for instance, are not valid for ensembles of systems which
have $A = a_i$ ($i = 1, 2$) in the example studied above; the hidden variables
should therefore not be considered as randomly distributed in these subensembles. Thus Bohm's theory does not contradict any of the arguments
previously given, and we should not wonder that it could be constructed.

For any discussion on the conceptual foundations of quantum mechanics
the mere fact that a deterministic theory can be exhibited is of great relevance.
However, it is shown below that any hidden variables theory which reproduces
the quantum predictions in the sense just defined necessarily has very strange
nonlocal features. Moreover, the deterministic description leads, in every
known case, to a cumbersome formalism that is not easily generalizable outside the realm of elementary nonrelativistic quantum mechanics. For these
reasons it is not popular with physicists.

Unless otherwise stated, the assumption that hidden variables exist is not
made here. And whenever it is *not* made, we hold fast to Assumption Q.

If hidden variables are assumed not to exist, the theory obviously has an intrinsic indeterminism built in, since identical measurements made on identical
systems do not give identical results. This "elementary" indeterminism is

balanced, however, by a kind of statistical determinism, which is a direct consequence of the rules given in Chapter 3. Indeed, this result is so important that it deserves an explicit statement in the form of a principle.

Principle of Statistical Determinism

If two statistical ensembles of physical systems are submitted to identical treatments, and if subsequent observation reveals significant statistical difference between them, the implication is that the two ensembles were not identical at the start.

4.4 MISCELLANEOUS REMARKS

(a) Relativistic Formulations

In Chapter 3 the rules of quantum mechanics were stated without taking care of Lorentz invariance. If we wanted the theory to be relativistically invariant, we should therefore make further requirements about it. However, relativistic quantum physics is still a subject of controversy among the experts. A complete, self-consistent, and useful set of axioms in this field has not yet been developed. Hence it seems worthwhile to focus one's effort of comprehension, as we shall do, on the general principles stated above, which are well defined and have a well-defined field of application, namely, nonrelativistic quantum mechanics.

On the other hand, the very general ideas on which the analyses that follow are grounded (superposition and linearity of evolution, for instance) are features that are shared by all the recent theoretical attempts to develop relativistic quantum theories at a sufficiently basic level, including, of course, quantum field theory and S matrix theory. Hence a discussion of the general axioms of quantum mechanics and of their broad implications necessarily has a bearing on the possible interpretation of these more recent theories.

(b) Undiscernability

As experiment shows, indistinguishable particles exist; and when two or more such particles are present in a physical system or participate in a reaction, the rules described in Chapter 3 must be changed somewhat.

There are several ways of changing the rules so as to bring them into agreement with experiment, but the most conservative is just to add a new one to the list. This supplementary rule is the well-known symmetrization (antisymmetrization) principle, which states that the ket describing the whole system should be made completely symmetrical (antisymmetrical) with respect to the boson (fermion) variables. [As is well-known, there are two kinds of particles, bosons and fermions; the former (latter) have integer (half-integer) spins.] More explicitly, let us consider, for instance, two bosons of a given type. Let $q^{(1)}$ and $q^{(2)}$ each represent the complete set of observables for a boson of that type (the duplication corresponds to the presence of two bosons), and let $|q_\alpha\rangle^{(1)}$, $|q_\alpha\rangle^{(2)}$ be normalized eigenvectors pertaining to the operators associated with $q^{(1)}$ and $q^{(2)}$ and corresponding to the eigenvalue q_α. The indices (1) and (2) that are attached to these kets mean that the first (second) one is in Hilbert space $\mathscr{H}_1$ ($\mathscr{H}_2$). Then $|\psi\rangle \in \mathscr{H}_1 \otimes \mathscr{H}_2$ is a possible ket describing a system of two bosons only if

$$\langle \psi | q_\alpha \rangle^{(1)} | q_\beta \rangle^{(2)} = \langle \psi | q_\beta \rangle^{(1)} | q_\alpha \rangle^{(2)} \tag{4–9}$$

for any α, β. If $|\psi\rangle$ has the form

$$|\psi\rangle = \sum_{ij} C_{ij} |u_i\rangle^{(1)} |u_j\rangle^{(2)} \tag{4–10}$$

where the C_{ij} are numbers, the condition implies that

$$C_{ij} = C_{ji} \tag{4–11}$$

Corresponding statements can be made for fermions (in which case $C_{ij} = -C_{ji}$).

(c) Second quantization

A formalism exists, often called "second quantization," which is somewhat different from that described in the foregoing section and has two remarkable advantages over it. The first is that it treats coordinates x and t of space and time on the same level; this condition is a necessary requirement for relativistic covariance and is obviously not satisfied in the formalism we have described, since x corresponds to an operator whereas t does not (in the second quantization picture both x and t are just parameters, not corresponding to any operator). Another virtue of the second quantization formalism is as follows. A situation in which a complete set of compatible observables pertaining to a system of two indistinguishable particles is entirely defined is automatically described, in this formalism, by at most one ket—never by two possible kets. Consequently, for identical particles Boltzmann statistics is automatically excluded as soon as all states are considered as equally probable. This result is a considerable step toward the symmetrization (antisymmetrization) principle. The fact that it simply follows from the second quantization formalism without having to be added to it as a supplementary requirement is surely to be considered as indicating that the second quantization formalism is a more faithful description of reality (whatever this word means) than the elementary formalism of Chapter 3.

Nevertheless, the formalism of Chapter 3, supplemented by the symmetrization-antisymmetrization principle, is the one that is discussed most extensively in this book. The reasons are that (*a*) it is in many respects simpler and more easily visualized, and (*b*) although the characters of systems are appreciably different in the two formalisms, all the other conceptual and mathematical structures are so similar that in most cases the results of a discussion concerning one formalism apply also to the other.

REFERENCES

[1] J. von Neuman, *Mathematical Foundations of Quantum Mechanics,* Princeton University Press, Princeton, N. J., 1955.

[2] D. Bohm, *Phys. Rev.* **85,** 166 (1952)

CHAPTER 5

ALTERNATIVE FORMULATIONS

The formulation of the rules of quantum mechanics given in Chapter 3 is not, of course, the only acceptable one. Some other possibilities are briefly reviewed here.

5.1 THE MEAN VALUE RULE

Rule 6, which gives the probability [Eq. (3–3)] of obtaining the value a_k as a result of a measurement of observable A on an ensemble E of systems described by the normalized ket $|\psi\rangle$, makes it possible to calculate the mean value $\bar{A}$, that is, by definition, the quantity

$$\bar{A} = \lim_{N \to \infty} \frac{n_k a_k}{N} \tag{5-1}$$

Here N is the number of systems in E, and n_k is the number of those systems for which the measurement gives the value a_k. Since

$$\lim n_k/N = w_k \tag{5-2}$$

we have

$$\begin{aligned}
\bar{A} &= \sum_k w_k a_k \\
&= \sum_{k,r} a_k \langle \psi | a_k, r \rangle \langle a_k, r | \psi \rangle \\
&= \sum_{k,r} \langle \psi | a_k | a_k, r \rangle \langle a_k, r | \psi \rangle
\end{aligned} \tag{5-3}$$

Then, using the same symbol A for the observable and for the corresponding operator and applying successively (*a*) the eigenvalue equation

$$a_k | a_k, r \rangle = A | a_k, r \rangle \tag{5–4}$$

and (b) the completeness relation, we have

$$\bar{A} = \sum_{k,r} \langle \psi | A | a_k, r \rangle \langle a_k, r | \psi \rangle \tag{5–5}$$

$$= \langle \psi | A | \psi \rangle \tag{5–6}$$

Similarly,

$$\overline{A^n} = \langle \psi | A^n | \psi \rangle \tag{5–7}$$

When $| \psi \rangle$ is *not* normalized to unity, then, of course, the expressions should be changed to

$$w_k = \frac{\sum_r |\langle a_k, r | \psi \rangle|^2}{\langle \psi | \psi \rangle} \tag{3–3a}$$

$$\bar{A} = \frac{\langle \psi | A | \psi \rangle}{\langle \psi | \psi \rangle} \tag{5–6a}$$

$$\overline{A^n} = \frac{\langle \psi | A^n | \psi \rangle}{\langle \psi | \psi \rangle} \tag{5–7a}$$

Expression (5–6a) has thus been deduced from Rule 5 (possible results of a measurement) and Rule 6 (statistical frequencies). Conversely, (5–6a) *can be taken as an axiom.* Rules 5 and 6 are then merely theorems; indeed, they follow from (5–6a), as we presently show.

The most straightforward proof is based on the use of the so-called *characteristic function.* In classical probability theory let

$$\rho(x) \, dx \tag{5–8}$$

be the frequency function describing the probability that a mesurement of quantity X will give a value lying in the interval $(x, x + dx)$. The case in which X can take only discrete values x_k, with probabilities w_k, can be incorporated into this description by choosing

$$\rho(x) = \sum_k w_k \, \delta(x - x_k) \tag{5–9}$$

The so-called characteristic function corresponding to (5–8) is by definition

$$\phi(t) = \int e^{itx} \rho(x) \, dx \tag{5–10}$$

In other words it is, up to a constant factor, the Fourier transform of $\rho(x)$.

In particular, if $\rho(x)$ is described by Eq. (5–9), then Eq. (5–10) gives immediately

$$\phi(t) = \sum_k w_k \, e^{itx_k} \qquad (5\text{–}11)$$

It is important to notice that, conversely, Eq. (5–11) uniquely determines the frequency function as being given by Eq. (5–9); if some (other) frequency function $\rho'(x)$ also had expression (5–11) as its characteristic function, the difference would give

$$\int e^{itx} \left[\rho'(x) - \sum_j w_j \, \delta(x - x_j)\right] dx = 0 \qquad (5\text{–}12)$$

for any t, so that the expression in brackets would necessarily be zero:* $\rho'(x)$ cannot therefore be different from (5–9).

On the other hand, let us expand e^{itx} in Eq. (5–10). This gives

$$\phi(t) = 1 + i\bar{x}t + \frac{i^2}{2}\overline{x^2}\, t^2 + \cdots + \frac{i^n}{n!}\overline{x^n}\, t^n + \cdots \qquad (5\text{–}13)$$

where

$$\overline{x^n} = \int \rho(x)\, x^n \, dx \qquad (5\text{–}14)$$

is the mean value of the quantity x^n.

Returning now to the quantum problem, we see that Eqs. (5–7a) and (5–13) make it possible to expand in powers of t the characteristic function corresponding to the frequency function for A. This expansion is

$$\phi(t) = \frac{1}{\langle\psi|\psi\rangle}\left[\langle\psi|\psi\rangle + i\langle\psi|A|\psi\rangle\, t\right.$$
$$\left. + \cdots + \frac{i^n}{n!}\langle\psi|A^n|\psi\rangle\, t^n + \cdots\right] \qquad (5\text{–}15)$$

$$= \frac{1}{\langle\psi|\psi\rangle}\langle\psi|1 + iAt + \cdots + \frac{i^n}{n!}A^n t^n \cdots |\psi\rangle \qquad (5\text{–}16)$$

$$= \frac{1}{\langle\psi|\psi\rangle}\langle\psi|e^{(itA)}|\psi\rangle \qquad (5\text{–}17)$$

where the exponential is defined formally by its series expression.

---

*Except perhaps on a set of zero measure, where, however, it could *not* behave as a "delta function," but should behave as a function whose Fourier transform is zero. For our purpose, this is equivalent to its being identically zero everywhere.

Let A have discrete eigenvalues, as above:

$$A|a_k, r\rangle = a_k|a_k, r\rangle \qquad (5\text{--}18)$$

Then

$$e^{(itA)}|a_k, r\rangle = e^{ita_k}|a_k, r\rangle \qquad (5\text{--}19)$$

as the series expansion clearly shows. On the other hand, from the decomposition of the unit operator, expression (5–17) can be written as

$$\phi(t) = \frac{1}{\langle\psi|\psi\rangle} \sum_{k,r} \langle\psi|e^{itA}|a_k, r\rangle \langle a_k, r|\psi\rangle \qquad (5\text{--}20)$$

that is, from (5–19)

$$\phi(t) = \frac{1}{\langle\psi|\psi\rangle} \sum_{k,r} e^{ita_k} \langle\psi|a_k, r\rangle \langle a_k, r|\psi\rangle \qquad (5\text{--}21)$$

$$= \sum_{k,r} e^{ita_k} \frac{|\langle a_k, r|\psi\rangle|^2}{\langle\psi|\psi\rangle} \qquad (5\text{--}22)$$

This is just an expression of the form (5–11). As pointed out above, this expression for $\phi(t)$ uniquely determines the frequency function to be

$$\rho(A) = \sum_k w_k \, \delta(x - a_k) \qquad (5\text{--}23)$$

with

$$w_k = \frac{\sum_r |\langle a_k, r|\psi\rangle|^2}{\langle\psi|\psi\rangle} \qquad (5\text{--}24)$$

Now Eq. (5–23) is just another expression of Rule 5, since the presence of the delta function implies that the probability for a measurement of A to give a result different from one of the eigenvalues a_k is zero. Similarly, Eq. (5–24) is obviously just Rule 6. The cases in which A has a continuous or partly continuous spectrum can be handled in the same way.

In summary, it has been shown that Rule 5 (possible values of A on a measurement of this quantity) and Rule 6 (probabilities of these values) can be replaced by the unique rule that follows.

Mean Value Rule

Let E be an ensemble of systems that can be described by a ket $|\psi\rangle$. Then, if a measurement of an observable A is made on each system, the mean result is given by Eq. (5–6a).

The formulation of Chapter 3 that treats Rules 5 and 6 as basic and subsequently derives the mean value rule is equivalent to the present formulation, which does the reverse. It is therefore purely a matter of taste to prefer one over the other.

Remark

Let it be pointed out that, in the language we have used, *both* formulations are expressed in terms of operational rules, that is, rules of the type "if such and such measurements are made, such and such results are obtained." The question may be asked whether such a wary formulation can be replaced by a more ambitious one that would claim to give information on the physical properties really possessed by systems or by ensembles of systems, whether or not we observe them. For some physicists, this question makes sense; for others, it does not. The point we now want to make, however, is that, if it makes sense, the problems it raises are clearly of the same nature in both formulations, since the two are completely equivalent. It would, in particular, be erroneous to believe that replacement of Rules 5 and 6 by the mean value rule is by itself sufficient to eliminate from the theory every reference to the community of observers. Since expression (5–23) is a consequence of the mean value rule, such a standpoint would imply, for example, that in the ensemble E considered above there *are* $n_k = N w_k$ systems for which the physical quantity A has the value a_k even when no measurement is performed. Such a statement, however, is wrong, as already pointed out.

5.2   THE HEISENBERG PICTURE FOR TIME EVOLUTION

The description of the time evolution of systems given in Chapter 3, Rule 3, is technically known as the "Schrödinger picture." It is not the only possible one. An important alternative description is the "Heisenberg picture," which indeed was the one Heisenberg used when he first formulated quantum mechanics. The Schrödinger and Heisenberg pictures represent two different but equivalent sets of rules. In other words, when applied to a given physical situation they lead to the same observable predictions. In some respects the Heisenberg picture presents definite technical advantages. It is particularly convenient, for instance, in quantum field theory. In this book, however, we find it more practical to work consistently with the equivalent Schrödinger set of rules, described above. For this reason the Heisenberg formulation is only briefly sketched.

Let us, for simplicity (this assumption is not essential), consider a time-independent Hamiltonian. Then the Schrödinger equation (3–2), together with the initial condition that $|\psi(t_0)\rangle$ is known, can be rewritten as

$$|\psi(t)\rangle = U(t, t_0)|\psi(t_0)\rangle \tag{5–25}$$

where

$$U(t_0, t_0) = 1 \tag{5–26}$$

$$U^+(t, t_0)\, U(t, t_0) = U(t, t_0)\, U^+(t, t_0)$$
$$= 1 \tag{5–27}$$

and

$$i\hbar \frac{dU(t, t_0)}{dt} = HU(t, t_0) \tag{5–28}$$

A solution to this operator equation that satisfies the initial condition (5–26) is

$$U(t, t_0) = \exp\left[-i\hbar^{-1}H(t - t_0)\right] \tag{5–29}$$

where the operator e^A is defined by the formal expansion of the exponential.

Let us now describe the ensemble E under consideration, not by $|\psi(t)\rangle$, but by the ket

$$|\phi(t)\rangle = V(t)|\psi(t)\rangle \tag{5–30}$$

where $V(t)$ is any given unitary operator (the norms of ψ and ϕ are thus the same). As shown above, nothing is changed in the physical content of the formalism as long as the predictions for the mean values of observables to be measured at time t are not changed. Since in the old description these mean values are given by

$$\bar{A} = \frac{\langle\psi(t)|A|\psi(t)\rangle}{\langle\psi(t)|\psi(t)\rangle} \tag{5–31}$$

in the new description they should be given by the equation

$$\bar{A} = \frac{\langle\phi(t)|V(t)\,A[V(t)]^{-1}|\phi(t)\rangle}{\langle\phi(t)|\phi(t)\rangle} \tag{5–32}$$

that is, by

$$\bar{A} = \frac{\langle\phi(t)|A'(t)|\phi(t)\rangle}{\langle\phi(t)|\phi(t)\rangle} \tag{5–33}$$

with

$$A'(t) = V(t)A[V(t)]^{-1} \tag{5–34}$$

The Heisenberg picture corresponds to the particular choice

$$V(t) = U(t, t_0)^{-1} \tag{5-35}$$

whence, from Eqs. (5–25), (5–26), and (5–30),

$$|\phi(t)\rangle = |\psi(t_0)\rangle = |\phi(t_0)\rangle \equiv |\phi\rangle \tag{5-36}$$

In other words, in the Heisenberg picture the ket $|\phi\rangle$ that describes E and is defined by Eq. (5–36) does not change with time. On the other hand, in the Heisenberg picture the observables correspond to the operators [Eq. (5–34)]

$$A'(t) = U^{-1}(t)AU(t) \tag{5-37}$$

In other words, the operators are time dependent. The differential equation that is satisfied by these operators is easily derived from Eq. (5–37) (for simplicity, the assumption is made that A is independent of time in the Schrödinger picture):

$$i\hbar \frac{dA'(t)}{dt} = U^{-1}[A, H]U \tag{5-38}$$

$$= [A', H]$$

This equation, in particular, shows in a direct way that an observable is a constant of the motion if it commutes with the Hamiltonian of the system.

As regards Rule 6 (the probability that a measurement of A will give a result a_k), it reads as

$$w_k = \sum_r \frac{|\langle a_k, r | \psi(t)\rangle|^2}{\langle \psi(t)|\psi(t)\rangle}$$

$$= \sum_r \frac{|\langle a_k, r | U(t, t_0)|\psi(t_0)\rangle|^2}{\langle \psi(t_0)|\psi(t_0)\rangle}$$

in the old formulation; therefore it reads, because of (5–36), as

$$w_k = \sum_r \frac{|\langle a_k, r, t | \phi\rangle|^2}{\langle \phi | \phi\rangle} \tag{5-39}$$

with

$$|a_k, r, t\rangle \equiv U^{-1}(t, t_0)|a_k, r\rangle \tag{5-40}$$

in the new formalism. Equations (5–33) and (5–39) show that the expressions used in the Schrödinger picture can be carried over to the Heisenberg picture. This implies, however, that now the kets associated with the possible values that can be found upon measurement of an observable are themselves time-dependent kets, $|a_k, r, t\rangle$. Moreover, the time dependence of these kets is determined by the Hamiltonian H representative of the forces to which one particular system is subjected. These features were to be expected since $|a_k, r, t\rangle$ is an eigenvector of a complete set of observables *at a given time t*.

PART TWO

DENSITY MATRICES AND MIXTURES

In quantum physics, we often need to consider ensembles that cannot be described by state vectors. This is so, for instance, when an ensemble is composed of several subensembles that are described by nonequivalent state vectors. It also happens in most cases in which we consider the ensemble of all the subsystems of some larger systems. The density matrix is a very convenient algorithm that makes it possible to deal with such situations swiftly and efficiently.

This part consists of two chapters. The first one (Chapter 6) is an introduction to the mathematical formalism. Its purpose is not to give a systematic review of all the properties of density matrices. Rather, it describes *those* properties that are most useful for the particular applications we have in mind, that is, the description of mixtures, the various kinds of mixtures, and the problems associated with nonseparability and with the theory of measurement. This is the reason why some known properties of density matrices are not recalled in Chapter 6 and why, on the contrary, some propositions are shown there that are not mentioned in the textbooks.

The second chapter in this part (Chapter 7) provides a description and an examination of the physical concept of mixtures. Along with more conventional aspects of the theory of mixtures, it introduces in particular the notion of proper and of improper mixtures and shows the physical difference between them. It also investigates in what cases an observable can be said to *have* a property in a system, and what consequences this has—a question that is often left with only implicit and therefore vague answers in more formal descriptions of the theory.

CHAPTER 6

THE DENSITY MATRIX FORMALISM

Along with an elementary description of the well-known density matrix formalism and of its application to the description of quantum-statistical ensembles, this chapter contains the proofs of a few propositions that are especially useful for understanding most of the further developments in the book. These propositions are collected here for the sake of convenience. However, although some of them will turn out to be of constant use below, others will be applied only episodically. The latter are dealt with in Sections 6.5, and 6.6, which can be bypassed on the first reading.

6.1 THE DENSITY MATRIX

Definition

Let $E_1 \ldots E_\alpha \ldots E_\mu$ be μ ensembles of physical systems of the same type, let N_α be the number of elements of E_α, and let $\hat{E}$ be the ensemble of all the $N = N_1 + \cdots + N_\alpha + \cdots + N_\mu$ elements of the various E_α. Let us assume, moreover, that each E_α can be described by a normalized ket $|\phi_\alpha\rangle$ (the kets $|\phi_\alpha\rangle$ are not necessarily mutually orthogonal). Then the operator

$$\rho = \sum_{\alpha=1}^{\mu} |\phi_\alpha\rangle \frac{N_\alpha}{N} \langle \phi_\alpha| \qquad (6\text{--}1)$$

is called the "statistical operator of ensemble $\hat{E}$." Although, properly speaking, ρ is an operator, not a matrix, it is quite often also called the *density matrix* associated with $\hat{E}$.

Some mathematical properties

The basic mathematical properties of ρ are as follows.

(i) ρ is Hermitian.
(ii) ρ is positive definite, that is,

$$\langle u|\rho|u\rangle \geqslant 0 \qquad (6\text{-}2)$$

for any $|u\rangle$.
(iii) Let us define the trace of an operator as the trace of the corresponding matrix in any matrix representation (it is easily verified that the value of the trace is independent from the choice of the representation). Then

$$T_r(\rho) = 1 \qquad (6\text{-}3)$$

This follows from Eq. (6–1), since

$$
\begin{aligned}
T_r(\rho) &= \sum_{\alpha,n}\left(\frac{N_\alpha}{N}\right)\langle\phi_\alpha|n\rangle\langle n|\phi_\alpha\rangle \\
&= \sum_\alpha\left(\frac{N_\alpha}{N}\right)\langle\phi_\alpha|\phi_\alpha\rangle = 1
\end{aligned}
\qquad (6\text{-}4)
$$

Properties (i), (ii), and (iii) have the following consequences.
(iv) Every diagonal element of ρ in any matrix representation is nonnegative. In particular, the eigenvalues of ρ are nonnegative.
This is an immediate consequence of (ii).

(v) The eigenvalues p_n of ρ satisfy

$$0 \leqslant p_n \leqslant 1 \qquad (6\text{-}5)$$

This follows from properties (iii) and (iv).

(vi) If ρ is a projection operator, it projects into a one-dimensional subspace. If ρ is a projection operator, then

$$\rho^2 = \rho. \qquad (6\text{-}6)$$

The same algebraic equality holds in regard to any given eigenvalue:

$$p_n{}^2 = p_n \qquad (6\text{-}7)$$

so that $p_n = 0$ or $p_n = 1$. Property (iii), however, entails that

$$\sum_n p_n = 1 \qquad (6\text{-}8)$$

so that only one eigenvalue is equal to one, all the other ones being zero. The general expression $\rho = \sum_n |n\rangle p_n \langle n|$ thus reduces in this case to one term only.   Q.E.D.

(vii)

$$\text{Tr}(\rho^2) \leqslant 1 \tag{6-9}$$

and the equality sign can hold only if ρ is a projection operator.

This inequality is a consequence of the fact that no point of the hyperplane (6–8) with nonnegative coordinates p_n lies outside the hypersphere

$$\text{Tr}(\rho^2) \equiv \sum_n p_n{}^2 = 1 \tag{6-10}$$

Concerning the equality case, note that when the numbers $p_1 \ldots p_n$ satisfy both Eqs. (6–8) and (6–10), they also satisfy

$$\sum_n p_n(1 - p_n) = 0 \tag{6-11}$$

Since none of the individual terms in the sum can be negative because of inequality (6–5), all of them are zero. Equation (6–8) then entails that only one p_n is equal to one, all the others being zero.   Q.E.D.

(viii) *When ρ is written as in Eq. (6–1), a necessary and sufficient condition for ρ to be a projection operator is that all the ϕ_α be identical up to phase factors, or, in other words, that the sum in Eq. (6–1) reduce to one term.*

The condition is obviously sufficient. To show that it is necessary, let us use Eq. (6–1) to write

$$\text{Tr}(\rho^2) = \sum_{\alpha,\beta,n} \frac{N_\alpha N_\beta}{N^2} \langle n|\phi_\alpha\rangle\langle\phi_\alpha|\phi_\beta\rangle\langle\phi_\beta|n\rangle$$

$$= \sum_{\alpha,\beta,n} \frac{N_\alpha N_\beta}{N^2} \langle\phi_\alpha|\phi_\beta\rangle \langle\phi_\beta|n\rangle \langle n|\phi_\alpha\rangle$$

$$= \sum_{\alpha,\beta} \frac{N_\alpha N_\beta}{N^2} |\langle\phi_\alpha|\phi_\beta\rangle|^2$$

The condition $\text{Tr}(\rho^2) = 1$ thus gives

$$\sum_{\alpha,\beta} N_\alpha N_\beta |\langle\phi_\alpha|\phi_\beta\rangle|^2 = N^2 = (\sum_\alpha N_\alpha)^2$$

$$= \sum_{\alpha,\beta} N_\alpha N_\beta$$

or

$$\sum_{\alpha,\beta} N_\alpha N_\beta(1 - |\langle\phi_\alpha|\phi_\beta\rangle|^2) = 0$$

since all the $|\phi_a\rangle$ are normalized, none of the numbers of the sum can be negative. Each one of them must therefore be zero. Thus, since both N_a and N_β can be assumed to be different from zero,

$$|\langle\phi_a|\phi_\beta\rangle|^2 = 1 \equiv \langle\phi_a|\phi_a\rangle\langle\phi_\beta|\phi_\beta\rangle$$

This is the case in which the Cauchy-Schwarz inequality reduces to an equality. The implication of this equality is, therefore, that for any α, β

$$|\phi_a\rangle = \lambda_{a\beta}|\phi_\beta\rangle \qquad\qquad (6\text{--}12)$$

where $\lambda_{a\beta}$ is a number whose modulus is, moreover, equal to one since both $|\phi_a\rangle$ and $|\phi_\beta\rangle$ are normalized.   Q.E.D.

In all discussions about the conceptual implications of quantum mechanics, the role of each proposition listed above, and particularly of proposition (viii), is essential.

6.2 PURE CASES AND MIXTURES

Pure Cases*

An ensemble E which can be described by a state vector $|\psi\rangle$ is called a *pure case*.

Instead of describing E by means of the (normalized) ket $|\psi\rangle$, we can equally well describe it by means of the density matrix

$$\rho = |\psi\rangle\langle\psi| \qquad\qquad (6\text{--}13)$$

which is a projection operator. Let A be an observable. Then

$$\mathrm{Tr}(\rho A) = \sum_n \langle n|\psi\rangle\langle\psi|A|n\rangle$$
$$= \sum_n \langle\psi|A|n\rangle\langle n|\psi\rangle = \langle\psi|A|\psi\rangle$$

In this formalism, therefore, the mean value of an observable A is given by the expression

$$\bar{A} = \mathrm{Tr}(\rho A) \qquad\qquad (6\text{--}14)$$

Similarly, let

$$P(a_k) = \sum_r |a_k, r\rangle\langle a_k, r|$$

---

*This is not the original definition of a *pure case* by von Neumann (op. cit.). But the two definitions are equivalent in the case in which all Hermitian operators correspond to observables. See Chapter 7 for details.

be the projection operator into the subspace H_k spanned by the eigenvectors of A corresponding to the eigenvalue a_k. Then

$$\text{Tr}\,[\rho P(a_k)] = \sum_{n,r} \langle n|\psi\rangle \langle\psi|a_k,r\rangle\langle a_k,r|n\rangle$$

$$= \sum_r |\langle a_k,r|\psi\rangle|^2$$

Thus the statistical frequency with which a measurement of A is predicted to result in a value a_k is just

$$w_k = \text{Tr}\,[\rho P(a_k)] \tag{6-15}$$

Mixtures

An ensemble $\hat{E}$ obtained by combining all the elements of several subensembles E_α is a *mixture,* although, as we shall see, this statement is not a general definition of the concept. If every E_α can be described by a state vector $|\phi_\alpha\rangle$, and if these $|\phi_\alpha\rangle$ are not all identical up to phase factors, the mixture obtained is not a pure case. We make the convention of using the word "mixture" in the restrictive sense that excludes the pure cases.

It is very convenient to describe $\hat{E}$ by means of the density matrix ρ defined by expression (6–1). Then the mean value of any observable A of $\hat{E}$ is

$$\bar{A} = \frac{1}{N} \sum_\alpha N_\alpha \bar{A}_\alpha$$

where $\bar{A}_\alpha$ is the mean value of A on E_α. Equation (6–14) then shows that

$$\bar{A} = \frac{1}{N} \sum_\alpha N_\alpha \,\text{Tr}(|\phi_\alpha\rangle \langle\phi_\alpha|A)$$

$$= \text{Tr}\,(\rho A)$$

Thus expression (6–14) for the mean values, derived above for the pure cases, applies equally well to the mixtures considered here. A similar argument shows that the same is true in regard to expression (6–15), which gives the probabilities of observation. What differentiates such mixtures from pure states is that the density matrices that describe mixtures are not projection operators:

$$\rho^2 \neq \rho$$

This follows from proposition (viii) above, which shows that, if the density matrix ρ expressed by (6–1) were a projection operator, all the operators

$|\varphi_\alpha\rangle\langle\varphi_\alpha|$ would be identical to each other. Under such conditions $\hat{E}$ would be the pure case described by $|\varphi_\alpha\rangle\langle\varphi_\alpha|$ or by the corresponding ket $|\varphi_\alpha\rangle$.

Finally, it is interesting to note that, if an ensemble $\hat{E}$ is not a pure case, there are several ways in which its density matrix ρ can be expressed in the form (6–1). If, for instance, the $|\varphi_\alpha\rangle$ are not mutually orthogonal, it is always possible to consider a representation $|n, r\rangle$ in which the matrix ρ is diagonal. Then ρ can also be expressed as

$$\rho = \sum_{n,r} |n, r\rangle\, p_n \,\langle n, r| \tag{6–16}$$

In Eq. (6–16) the eigenvalues p_n of ρ are, of course, the statistical frequencies with which a complete measurement of a set of compatible observables having $\{|n, r\rangle\}$ as eigenvectors would give the definite set of values labeled n.

Remark

Let us show an interesting property of the density matrices. It concerns the case in which it is known with certainty that a measurement of an observable A will yield one particular result, a_n. Formulas (6–1) and (6–15) then give

$$\sum_{\alpha,r} \frac{N_\alpha}{N} |\langle a_k, r | \varphi_\alpha\rangle|^2 = \delta_{k,n}$$

Since the N_α are all different from zero by assumption, this shows that

$$\langle a_{k,r} | \varphi_\alpha\rangle = 0$$

for any k different from n, in other words, it shows that every $|\varphi_\alpha\rangle$ belongs to the subspace H_n spanned by the $|a_{n,r}\rangle$. In the case in which it is known with certainty that simultaneous measurements of the compatible observables $A, B, C \ldots$ would give the results $a_n, b_m, c_l \ldots$, the foregoing argument shows that the $|\varphi_\alpha\rangle$ are all contained in the intersection of the subspaces $H_n, H_m, H_l \ldots$. When this intersection is one dimensional, $\hat{E}$ reduces to the pure case described by the eigenvector common to $a_n, b_m, c_l \ldots$. Because of property (iv) and of the possibility of writing down any density matrix in the form (6–16), the latter result holds good for any density matrix ρ, irrespective of whether or not it is given explicitly by a formula such as (6–1).

6.3 TIME DEPENDENCE OF DENSITY MATRICES

Let ρ be a density matrix that describes a pure case or a mixture of the kind we have just defined; ρ is given by the general formula (6–1). The time evolution of the $|\phi_\alpha\rangle$ in the Schrödinger representation is given by (5–25), so that

$$\rho(t) = U(t, t_0)\rho(t_0) \ U^+(t, t_0) \tag{6–17}$$

Because of Eq. (5–28) for the time variation of $U(t, t_0)$ and of the corresponding equation concerning U^+, (6–17) gives

$$i\hbar \frac{d\rho(t)}{dt} = [H, \rho] \tag{6–18}$$

Formula (6–18) is called the "Schrödinger equation for the density matrix." In spite of a formal similarity (up to a sign) with Eq. (5–38), it holds in the Schrödinger picture.

It is important to observe that

$$\rho^2(t) = U(t, t_0)\rho(t_0) \ U^+(t, t_0) \ U(t, t_0) \ \rho(t_0) \ U^+(t, t_0)$$

$$= U(t, t_0) \ \rho^2(t_0) \ U^+(t, t_0)$$

so that, if $\rho^2(t_0) = \rho(t_0)$, then

$$\rho^2(t) = \rho(t)$$

In other words, if an ensemble of systems is a pure case, it cannot evolve into a mixture if the time evolution is governed by the Schrödinger equation (3–2).

6.4 ALTERNATIVE FORMULATION OF QUANTUM RULES

The rules given in Chapter 3 apply only to ensembles that can be described by state vectors. As shown above, these rules entail the density matrix rules described in the present Chapter as consequences. However, it is also possible to formulate the density matrix rules directly instead. These rules are then chosen as basic axioms, and the rules described in Chapter 3 become but special consequences of these axioms. In the new formulation, ensembles (be they pure cases or mixtures) are directly described by density matrices, observables are still described by Hermitean operators, and the rules giving the statistical distribution and the mean values of the predicted measurement results are, respectively, expressions (6–15) and (6–14).

When stated in this manner, the rules of quantum mechanics apply directly not only to pure states but also to mixtures. Formally, this is a distinct advantage. In the case of mixtures defined by their density matrices, the new formulation also has another merit. Since such a mixture is, as shown above, equivalent not to *one* but to *several* physical mixtures such as $\hat{E}$, the old formulation, in terms of state vectors, necessarily implies some arbitrariness in the choice of the latter. The new formulation has no such defect.

Finally, it should be mentioned that such considerations have opened the way to other axiomatic formulations of quantum mechanics that are both concise and very general. One of them goes back to Birkhoff and von Neumann [1] and was further developed by von Weizsaecker [2], Mackey [3], Jauch and Piron [4,5] and others. The basic structure in this approach is the set of all the "yes"–"no" questions. Mathematically, this set constitutes a lattice. Supplementary axioms are necessary to recover all the usual principles of quantum mechanics, so that this approach has a certain degree of adaptability that may offer some advantages. Starting from it, Jauch [6], for instance, has suggested some modifications of quantum mechanics that would eliminate a few of the difficulties described below. At the same time, these modifications change some of the predictions of quantum theory. They thus originate a *new* theory and hence fall outside the scope of this study.

Another axiomatic formulation was initiated by Jordan, Wigner, and von Neumann [7] and developed by Segal [8], Haag and Kastler [9], and others. Its basic idea is to give axiomatically the structure of an algebra to the set of all bounded observables and to relate formalism and experiment by means of a mean value rule, which reads as follows: *The mean value of the results of measurements of any observable A on an ensemble of systems of the same type is a real, positive linear functional m(A) of the element A of the algebra that corresponds to that observable.* Again, it can be shown that the usual formulation in terms of Hilbert spaces, density matrices, and so forth can be recovered at the price of some supplementary assumptions of a quite general nature. Nowadays the algebra is extended so as to include non-Hermitian elements, not related to observables.

These axiomatics have distinct formal advantages and constitute a very inspiring field of study [9]. However, it should be remembered that in the domain of validity of elementary quantum mechanics, they are finally entirely equivalent to the formalism that is based on density matrices and/or state vectors. For that reason, the latter one remains the main subject of our investigations. Henceforth, we understand the expression "the quantum rules" as meaning "the rules dealing with density matrices and/or state vectors." We can therefore make Assumption Q more precise by restating it as follows.

Assumption Q'. For the purpose of making statistical predictions about measurements bearing on the observables that belong properly to the elements of an ensemble E of systems, where E is specified without reference to future observations on these S, E can always be described by some density matrix. By "belong properly," it is meant that the said observables can be measured by operating exclusively on S. In other words, the correlations with other systems are excluded.

The study of "elementary quantum mechanics" thus understood is the main subject of this book. A significant exception to this restriction in the scope of our studies is represented, however, by the content of Chapter 12, which is self sufficient and is based on assumptions of an even greater generality than those of the axiomatics grounded on the lattice of the propositions.

6.5   MIXTURES DEFINED BY ONE AND THE SAME DENSITY MATRIX*

It was observed at the end of Section 6.2 that a given density matrix usually describes not just one ensemble but a great number—usually an infinity—of ensembles of the type of the ensembles $\hat{E}$ (as defined in that section) that are all different from one another. The purpose of this section is to present some more information concerning this phenomenon.

If a density matrix has a degenerate eigenvalue p_n, then, of course, there are infinitely many ways of choosing the corresponding orthogonal eigenvectors. In other words, there are infinitely many mixtures that have this density matrix. Let $\{|n, r\rangle\}$ and $\{|n, s\rangle\}$, where r and s are degeneracy indices, be two sets of orthonormal eigenvectors corresponding to the eigenvalue p_n. Then the $|n, s\rangle$ can be written as linear combinations of the $|n, r\rangle$, and conversely. This holds good, of course, for any degenerate eigenvalue. Now, if a density matrix can be expressed as

$$\rho = \sum_{n,r} |n, r\rangle p_n \langle n, r| \tag{6-16}$$

where the $|n, r\rangle$ are orthonormal kets, then the p_n are the nonzero eigenvalues of ρ and the $|n, r\rangle$ are the corresponding eigenvectors. It follows that, if ρ can be expanded both according to Eq. (6–16) and according to

$$\rho = \sum_{m,s} |m, s\rangle q_m \langle m, s| \tag{6-16a}$$

where the $|m, s\rangle$ are also orthonormal kets, then the $\{q_m\}$ are identical to the $\{p_n\}$ and the $|m, s\rangle$ are linear combinations of the $|n, r\rangle$. The simplest example of such a situation is provided by an unpolarized beam of spin-$\frac{1}{2}$ particles. The corresponding density matrix is (in any orthonormal basis)

$$\rho = \tfrac{1}{2} \begin{pmatrix} 1 & 0 \\ 0 & 1 \end{pmatrix}$$

It can be written either as

$$\rho = \frac{1}{2} \left[|u_+\rangle \langle u_+| + |u_-\rangle \langle u_-| \right]$$

---

*The content of this section and the following one in this chapter can be bypassed on the first reading.

or as

$$\rho = \frac{1}{2}\left[\,|\,v_+\,\rangle\langle\,v_+\,| + |\,v_-\,\rangle\langle\,v_-|\,\right]$$

or as

$$\rho = \frac{1}{2}\left[\,|\,w_+\,\rangle\langle\,w_+\,| + |\,w_-\,\rangle\langle\,w_-|\,\right]$$

and so on, where

$$|\,u_\pm\,\rangle$$
$$|\,v_\pm\,\rangle = 2^{-1/2}\,(\,|\,u_+\,\rangle \pm |\,u_-\,\rangle)$$

and

$$|\,w_\pm\,\rangle = 2^{-1/2}\,(\,|\,u_+\,\rangle \pm i\,|\,u_-\,\rangle)$$

are the eigenvectors of the spin components along three mutually orthogonal axes.

Let us now consider the general problem, which we can state as follows. Given a mixture of N systems, N_α of which constitute a pure case described by the normalized ket $|\,\phi_\alpha\,\rangle\,(\Sigma_\alpha N_\alpha = N)$; given another mixture of N systems of the same type as the one above, but N_i of which constitute now a pure case described by the normalized ket $|\,\psi_i\,\rangle\,(\Sigma_i N_i = N)$; and assuming that these two mixtures have the same density matrix ρ_M what can we say about the relations between the $|\,\psi_i\,\rangle$ and the $|\,\phi_\alpha\,\rangle$? The difference between this problem and the one previously investigated is that here we do not assume that the $|\,\psi_i\,\rangle$ are mutually orthogonal. Indeed, we do not even assume that they are linearly independent. Nor is any such assumption made regarding the $|\,\phi_\alpha\,\rangle$ either.

Then, let $\{|\,n\,\rangle\}$ be a complete set of eigenvectors of ρ. Moreover, let $\{|\,m\,\rangle\}$ be the subset of $\{|\,n\,\rangle\}$ that corresponds to nonzero eigenvalues of ρ, and let $\{|\,t\,\rangle\}$ be the complementary subset, which corresponds to eigenvalues all equal to zero. By assumption

$$\rho = \sum_\alpha |\,\phi_\alpha\,\rangle\,\frac{N_\alpha}{N}\,\langle\,\phi_\alpha| = \sum_i |\,\psi_i\,\rangle\,\frac{N_i}{N}\,\langle\,\psi_i| \qquad (6\text{–}19)$$

where all the numbers N, N_α, N_i are greater than zero. The eigenvalue that corresponds to $|\,n\,\rangle$ is given by

$$p_n = \rho_{nn} = \langle\,n|\rho|n\,\rangle = \sum_\alpha \frac{N_\alpha}{N}\,|\,\langle\,n|\phi_\alpha\,\rangle\,|^2$$

$$= \sum_i \frac{N_i}{N}\,|\,\langle\,n|\psi_i\,\rangle\,|^2 \qquad (6\text{–}20)$$

Now, in each of these two sums the coefficients N_α/N, N_i/N are nonzero and positive. The condition $p_t = 0$ entails, therefore,

$$\langle t|\phi_\alpha\rangle = \langle t|\psi_i\rangle = 0 \tag{6-21}$$

so that the expansions

$$|\phi_\alpha\rangle = \sum_\alpha |n\rangle\langle n|\phi_n\rangle \tag{6-22}$$

and

$$|\psi_i\rangle = \sum_n |n\rangle\langle n|\psi_i\rangle \tag{6-23}$$

reduce to

$$|\phi_\alpha\rangle = \sum_m |m\rangle\langle m|\phi_\alpha\rangle \tag{6-24}$$

and

$$|\psi_i\rangle = \sum_m |m\rangle\langle m|\psi_i\rangle \tag{6-25}$$

respectively. On the other hand, ρ is diagonal in the representation $\{|n\rangle\}$, with eigenvalues p_n. Hence, multiplying Eq. (6–19) by $|n\rangle$ on both sides gives

$$p_n|n\rangle = \sum_\alpha \frac{N_\alpha}{N} |\phi_\alpha\rangle\langle\phi_\alpha|n\rangle \tag{6-26}$$

since, for $n = m$, $p_n \neq 0$, Eq. (6–26) shows that any $|m\rangle$ is a linear combination of the $|\phi_\alpha\rangle$. Since any $|\psi_i\rangle$ is a linear combination of vectors $|m\rangle$ [see Eq. (6–25)], it follows that any $|\psi_i\rangle$ is a linear combination of the $|\phi_\alpha\rangle$. The same argument also shows, of course, that any $|\phi_\alpha\rangle$ is a linear combination of the $|\psi_i\rangle$. In other words, we have shown the validity of the following proposition.

Proposition

If two proper mixtures are described by the same density matrix, any state vector representing a subensemble of one of these mixtures can be expressed as a linear combination of the state vectors representing the subensembles of the other mixture.

6.6 A PROPOSITION BEARING ON CORRELATIONS*

Let G be the tensor product of two Hilbert spaces $H^{(u)}$ and $H^{(v)}$ that have

*This proposition (Proposition 1) essentially serves in proving a point discussed in Chapter 16. The reader who does not want to be distracted from the general line of the argument developed in the book is urged to postpone this section for a later reading.

$\{|u_i\rangle\}$ and $\{|v_r\rangle\}$, respectively, as orthonormal bases. Let A be an operator in G. A useful concept for what follows is that of the *partial traces* $\mathrm{Tr}^{(u)}A$ and $\mathrm{Tr}^{(v)}\,A$ with respect to $H^{(u)}$ and $H^{(v)}$, respectively. We recall that, if A is expanded according to the formula

$$A = \sum_{i,j;s,r} A_{i,r;j,s}\,|u_i\rangle\langle u_j| \otimes v_r\rangle\langle v_s|$$

then by definition we have

$$\mathrm{Tr}^{(u)}\,A = \sum_{k}\langle u_k|A|u_k\rangle = \sum_{k,r,s} A_{k,r;k,s}|v_r\rangle\langle v_s|$$

and a similar definition holds for $\mathrm{Tr}^{(v)}A$.

With this convention let U and V be two systems that have once interacted, let $H^{(u)}$ and $H^{(v)}$ be two finite-dimensional Hilbert spaces corresponding to U and V, respectively, and let ρ_1 be a density matrix describing an ensemble of composite systems $\Sigma = U + V$. Then ρ_1 operates in the Hilbert space $G = H^{(u)} \otimes H^{(v)}$.

Let us define

$$\rho = \mathrm{Tr}^{(u)}\rho_1 \qquad\qquad (6\text{--}27)$$

and, similarly,

$$\rho' = \mathrm{Tr}^{(v)}\rho_1 \qquad\qquad (6\text{--}28)$$

as the partial traces of ρ_1 over $H^{(u)}$ and $H^{(v)}$, respectively. Let $\{|u_i\rangle\}$ be an orthonormal basis of $H^{(u)}$ over which ρ' is diagonal; ρ_1 can always be written as

$$\rho_1 = \sum_{n} p_n\,|\psi_n\rangle\langle\psi_n| \qquad\qquad (6\text{--}29)$$

with $p_n \rangle 0\,(\Sigma_n p_n = 1)$ and with

$$|\psi_n\rangle = \sum_{i} |u_i\rangle\,|\Phi_i{}^{(n)}\rangle \qquad\qquad (6\text{--}30)$$

the non-orthogonal kets$|\Phi_i{}^{(n)}\rangle$ thus defined being vectors of $H^{(v)}$ and the p_n being probabilities. Since ρ' is diagonal in $\{|u_i\rangle\}$, it can be written as

$$\rho' = \sum_{i} r_i\,|u_i\rangle\langle u_i| \qquad\qquad (6\text{--}31)$$

On the other hand,

$$\rho_1 = \sum_{nij} p_n|u_i\rangle\langle u_j|\otimes|\Phi_i{}^{(n)}\rangle\langle\Phi_j{}^{(n)}| \qquad\qquad (6\text{--}32)$$

and, if $\{|v_s\rangle\}$ is a basis in $H^{(v)}$, then from Eq. (6–28)

$$\rho' = \sum_{nij} p_n |u_i\rangle \langle u_j| \sum_s \langle v_s|\Phi_i^{(n)}\rangle\langle\Phi_j^{(n)}|v_s\rangle$$

$$= \sum_{ij} |u_i\rangle\langle u_j| \sum_n p_n \langle\Phi_j^{(n)}|\Phi_i^{(n)}\rangle \tag{6-33}$$

Inserting the two expressions (6–31) and (6–33) between $\langle u_k|$ and $|u_l\rangle$, we obtain

$$r_k \delta_{kl} = \sum_n p_n \langle\Phi_l^{(n)}|\Phi_k^{(n)}\rangle \tag{6-34}$$

In particular, if $r_k = 0$, then

$$\sum_n p_n \|\Phi_k^{(n)}\|^2 = 0$$

Hence, because of the positivity of the p_n,

$$r_k = 0 \Rightarrow |\Phi_k^{(n)}\rangle = 0, \qquad \forall n \tag{6-35}$$

A consequence of (6–35) is the following lemma.

Lemma

Expression (6–30) for $|\psi_n\rangle$ involves only vectors $|u_i\rangle$ corresponding to nonzero eigenvalues r_i of ρ'.

This lemma has interesting consequences.

Let us consider the case in which the ensemble of the systems U is a pure case. Then only one eigenvalue of ρ'—let us label it by the index 1—is nonzero (and, of course, equal to 1). The lemma then shows that ρ_1 can be rewritten as

$$\rho_1 = |u_1\rangle\langle u_1|\otimes\rho \tag{6-36}$$

or

$$\rho_1 = \rho' \otimes \rho \tag{6-37}$$

Now, whenever ρ_1 can be written as a tensor product of ρ and ρ', the joint probability of $A = a_n$ and $B = b_m$ (where A and B are observables pertaining to U and V, respectively) is equal to the product of the probability for $A = a_n$ times the probability for $B = b_m$, as is easily verified; therefore, as is well known, this case is one in which there are no correlations between U and V. Hence (6–37) leads to the following proposition [10, 11].

Proposition 1

Whenever U is in a pure case, no correlation exists between U and V.

It is not without interest to note that this important proposition can be generalized as follows.

In the general case it is easily found that

$$\rho = \sum_{nk} p_n |\Phi_k{}^{(n)}\rangle\langle \Phi_k{}^{(n)}| \tag{6-38}$$

and from (6–31) and (6–34) that

$$\rho' = \sum_{ni} p_n |u_i\rangle\langle u_i| \; \|\Phi_i{}^{(n)}\|^2 \tag{6-39}$$

so that

$$1 = \mathrm{Tr}(\rho') = \sum_{ni} p_n \|\Phi_i{}^{(n)}\|^2 \tag{6-40}$$

Again let A be an observable of U, B an observable of V, and $\bar{A}$, $\bar{B}$ their mean values. Then

$$\bar{A} = \mathrm{Tr}(\rho'A) = \sum_{ni} p_n A_{ii} \|\Phi_i{}^{(n)}\|^2 \tag{6-41}$$

$$\bar{B} = \mathrm{Tr}(\rho B) = \sum_{nk} p_n \langle \Phi_k{}^{(n)}|B|\Phi_k{}^{(n)}\rangle \tag{6-42}$$

and

$$\overline{AB} = \mathrm{Tr}\,(\rho_1 AB) = \sum_{nij} p_n A_{ij} \langle \Phi_i{}^{(n)}|B|\Phi_j{}^{(n)}\rangle \tag{6-43}$$

Let us now consider a case in which some eigenvalues of ρ'—that without any loss of generality we can assume to be $r_{p+1} \ldots r_N$, where N is the dimension of $H^{(u)}$—are zero. Let us then choose for A an observable that, in the basis $\{|u_i\rangle\}$, is represented by a matrix satisfying

$$A_{ij} = \lambda \delta_{ij} \quad \text{for} \quad i < p \quad \text{and} \quad j < p \tag{6-44}$$

Then, because of implications (6–35),

$$\bar{A} = \lambda$$

and

$$\overline{AB} = \lambda \sum_{n,i} p_n \langle \Phi_i{}^{(n)}|B|\Phi_i{}^{(n)}\rangle \tag{6-45}$$

so that

$$\overline{AB} = \bar{A}\,\bar{B} \tag{6-46}$$

Let

$$A_1 = A - \bar{A} \tag{6-47}$$

and

$$B_1 = B - \bar{B}$$

We have

$$\overline{A_1 B_1} = \overline{AB} - \overline{A}\ \overline{B} - \overline{A}\ \overline{B} + \overline{A}\ \overline{B} = \overline{AB} - \overline{A}\overline{B}$$

Hence, because of (6–46),

$$l = \overline{A_1 B_1}\,(\overline{A_1{}^2}\ \overline{B_1{}^2})^{-1/2} = 0 \qquad\qquad (6\text{–}49)$$

The experimentally accessible quantity l is what is called the *coefficient of correlation* between A and B in conventional probability calculus. We have therefore proved the validity of the following assertion [11].

Proposition 2

If the eigenvalues $r_{p+1} \ldots r_N$ of ρ' are zero, and if A is an observable of U whose matrix satisfies (6–44) in the representation in which ρ' is diagonal, then *the coefficient of correlation between A and any observable of system V is zero.*

Remark 1

In the proofs of Propositions 1 and 2 the implicit assumption is made that there exists only one density matrix, namely ρ', that is fit for describing the ensemble of the systems U; there would be no necessity for requiring that ρ' have one only of its eigenvalues differing from zero if the said ensemble could be described by two distinct density matrices [one of them then being ρ' as given by Eq. (6–28), and the *other* one having one only of its eigenvalues differing from zero). As shown in Section 7.1, a possibility of describing the same ensemble by two (equivalent) density matrices can occur in the case in which not all the Hermitean operators of the system under consideration correspond to observables. Hence Propositions 1 and 2 are proved above only for the case in which all the Hermitean operators attached to the system U correspond to observables.

Remark 2

In regard to the word "correlation," it may be appropriate to recall here the fact that its meaning in the theory of probabilities—and also above—does not coincide completely with its usual meaning in the ordinary language. For example, let us consider a collection of N spin-$\frac{1}{2}$ particles U all having spin "up," each U being associated with a spin-$\frac{1}{2}$ particle V having spin "down." We might be tempted to say that the spin components of the U and the V along O_z are correlated. However, in the language of the theory of probability they are not, since the probabilities of finding the four possible results (up, up; up, down; down, up; down, down) are equal in each case to the products of the probalities bearing respectively on U alone and V alone.

REFERENCES

[1] G. Birkhoff and J. von Neumann, *Ann. Math. 37*, 823 (1936).
[2] C. F. von Weizsaecker, *Naturwissenschaften 42*, 521 (1955)
[3] G. Mackey, *Mathematical Foundations of Quantum Mechanics*, W. A. Benjamin, New York.
[4] C. Piron, *Helv. Phys. Acta 37*, 439 (1963).
[5] J. M. Jauch and C. Piron, *Helv. Phys. Acta 42*, 842 (1969).
[6] J. M. Jauch in *Foundation of Quantum Mechanics, Proceedings of the Enrico Fermi International Summer School, Course IL*, Academic Press, New York, 1971.
[7] J. von Neumann, P. Jordan, and E. P. Wigner, *Ann. Math. 35*, 29 (1934).
[8] I. B. Segal, *Ann. Math. 48*, 930 (1947).
[9] R. Haag and D. Kastler, *J. Math. Phys. 5*, 848 (1964). See also: A. M. Gleason, *Math. Mech. 6*, 885 (1957); H. Margenau and J. L. Park, *Int. J. Theor. Phys. 1*, 211 (1968).
[10] M. G. Doncel, L. Michel, and F. Minnaert, in *Matrices-Densités de Polarisation*, Summer School of Gif-sur-Yvette, 1970, R. Salmeron, Ed., Laboratoire de Physique de l' École Polytechnique, Paris.
[11] B. d'Espagnat, *Lett. Nuovo Cimento, 2* (16), 823 (1971).

Further Reading on Density Matrices U. Fano, *Rev. Mod. Phys. 29*, 74 (1957).

Further Reading on Axiomatics J. M. Jauch, *Foundations of Quantum Mechanics*, Addison-Wesley, Reading, Mass., 1968.

CHAPTER 7

MIXTURES

In regard to mixtures and their description, a few rather elementary aspects of the relevant mathematical theory were developed in Chapter 6. That theory could of course be expanded to a considerable degree. However, an abundant literature exists already on that subject. What is of greater interest to us (and what, also, has been explored in less detail by the various authors who have written about such matters) is the question of the possible physical interpretations of the formalism that we have at our disposal. This chapter is an approach to such problems. As a first, elementary way of clarifying them it introduces an unconventional but useful distinction between *improper* mixtures and *proper* mixtures. It then investigates the question of the homogeneity or heterogeneity of ensembles, as well as two related questions, which bear on the measurability of the density matrix and on the relation between Hermitean operators and observables, respectively. It also includes a discussion of the meaning that can be given to this sentence: "Such and such an observable *has* such and such a value on a system."

7.1 OPERATORS AND OBSERVABLES

Observables are described by Hermitean operators, and in the early days of quantum mechanics it was believed that this was a one-to-one correspondence. More precisely, the assumptions were made that (i) every observable can be related to some Hermitean operator, and (ii) every Hermitean operator corresponds to one observable.

It seems that no convincing evidence against the first assumption has yet been found (for a statement of the opposite opinion see Ref. [1]). Assumption (ii), on the other hand, is nowadays known to be wrong. Indeed Wick, Wightman, and Wigner [2] have shown that *superselection rules* exist in physics, according to which some linear combinations of well-known states (such as the states of an isotopic multiplet having different total charges) can never be physically realized.* As a consequence, the associated projection operators correspond to no observables. Thus the superselection rules provide a counterexample to assumption (ii), which shows that the latter is not correct. More generally, many Hermitean operators can be considered which are functions of operators related to well-known observables but for which no observable (i.e., no measuring device) has ever been found.†

The question of determining which operators correspond to observables and which do not is a very difficult one. At the present time, no satisfactory answer appears to be known. Nevertheless, it is interesting to investigate the relationship of this question to another, similar one: "What are the systems whose density matrices are measurable?" Should we, for instance, say that, if a given type of systems corresponding to a given Hilbert space has a measurable density matrix, then all the Hermitean operators defined on that space are measurable? And is the reverse proposition true?

What do we mean when we say that the density matrix corresponding to a given type of systems is measurable? Let an ensemble E of a sufficiently large number of systems of this type be given. Let us first separate it into subensembles E_λ, the elements of which are chosen at random in E. If, from the results of appropriate measurements on the E_λ, we can derive the value of every element of the matrix ρ that describes E in some fixed representation, we say that ρ is measurable.

Remark

If all the Hermitean operators Q of H are observables, this means that on a given ensemble E we can measure the quantities

$$\text{Tr}\,(\rho Q)$$

corresponding to any Hermitean operator Q (this is simply done by splitting E into a sufficient number of subensembles and by measuring mean values). In particular, we can measure the quantities $\text{Tr}(\rho Q)$ corresponding to the Q operators given by

$$|i\rangle\langle j| + |j\rangle\langle i| \qquad \text{and} \qquad i(|i\rangle\langle j| - |j\rangle\langle i|)$$

where $\{|i\rangle\}$ defines the representation. These quantities give the elements ρ_{ij} of the density matrix so that these elements can be measured.

---

*The observables whose eigenvectors corresponding to different eigenvalues cannot be physically superposed are often called "superselection charges."
†See, however, the investigations carried out by Lamb [3]. They show that more Hermitean operators are observable in principle than was thought to be the case.

7.2 PROPER AND IMPROPER MIXTURES

Let the normalized ket

$$|\psi\rangle = \sum_{ij} c_{ij} |v_i\rangle |u_j\rangle \tag{7-1}$$

describe, at a given time t, an ensemble $\mathscr{E}$ of N physical systems Σ, each composed of two subsystems, U and V. A Hilbert space $H^{(u)}$ $(H^{(v)})$ is attached to the $U(V)$ systems, and $\{|u_i\rangle\}$ $(\{|v_i\rangle\})$ is a complete system of orthonormal kets in $H^{(u)}$ $(H^{(v)})$. Such a situation frequently occurs (e.g., as a result of the Schrödinger time evolution) when U and V have interacted in the past. Here we are interested *in the ensemble of the V systems*, which we call E. Let A be an observable that pertains to systems V.

The mean value of A on $\mathscr{E}$ is

$$\bar{A} = \langle\psi|A|\psi\rangle = \sum_{i,j,r,s} c^*_{i,j}\, c_{r,s} \langle v_i|\langle u_j|A|v_r\rangle |u_s\rangle \tag{7-2}$$

Of course $\bar{A}$ is also the mean value of A on E. The fact that A belongs to systems V means that the operator A has no effect on the vectors in $H^{(u)}$, so that Eq. (7-2) can be written as

$$\bar{A} = \sum_{i,j,r} c^*_{i,j}\, c_{r,j} \langle v_i|A|v_r\rangle$$
$$= \sum \rho_{ri} A_{ir} = \mathrm{Tr}(\rho A) \tag{7-3}$$

with

$$A_{ir} = \langle v_i|A|v_r\rangle \tag{7-4}$$

$$\rho_{ri} = \sum_j c_{r,j}\, c^*_{i,j} \tag{7-5}$$

and

$$\rho = \sum_{r,i} |v_r\rangle \rho_{ri} \langle v_i| \tag{7-6}$$

Since the quantities

$$\rho_{rs;ij} = c_{r,s}\, c^*_{i,j} \tag{7-7}$$

are the elements of the density matrix ρ_1 which describes Σ in the representation $|v_i\rangle|u_j\rangle$, Eq. (7-5) can also be written as

$$\rho_{ri} = \sum_j \rho_{rj;ij} \tag{7-8}$$

and Eq. (7-6) as

$$\rho = \mathrm{Tr}^{(u)}\, \rho_1 \tag{7-9}$$

where

$$\rho_1 = \sum_{\substack{rs \\ ij}} |v_r\rangle |u_s\rangle \rho_{rs;ij} \langle v_i| \langle u_j| \qquad (7\text{--}10)$$

and where the symbol $\text{Tr}^{(u)}$ (*partial trace*) means, as we know, that the usual operation of taking the trace of the operator is carried over only in $H^{(u)}$:

$$\text{Tr}^{(u)} \rho_1 = \sum_t \langle u_t | \rho_1 | u_t \rangle \qquad (7\text{--}11)$$

Equation (7–3) shows that expression (6–14), which gives the mean value of an observable when E is a mixture of ensembles E_α that can be described by state vectors, can formally be extended to the ensemble E of V systems now under consideration. In a similar way, expression (6–15) for the statistical frequencies can be extended to the ensemble of the V systems. These extensions simply require that the density matrix of the ensemble of the V systems be defined by expressions (7–5) and (7–6). It is, moreover, easily verified that the operator ρ defined by these expressions has properties (i), (ii), and (iii) of Section 6.1. It therefore also has properties (iv), (v), and (vii). Finally, it can be verified that

$$\text{Tr}(\rho^2) \leq 1 \qquad (7\text{--}12)$$

In all these respects the ensembles E considered here do not differ from the mixtures $\hat{E}$ already encountered in Sections 6.1 and 6.2. This fact is quite remarkable. It motivates the usual convention which extends the generic name "mixtures" to ensembles of type E. Such a convention cannot, however, be fully accepted—that is, without reservations or qualifications—until it has been verified that any ensemble of type E can be identified with an ensemble of type $\hat{E}$. Let us therefore investigate whether or not this is the case, by considering every available observation.

At first sight, the arguments for identifying the two concepts look impressive. Indeed, it is well known that any measurement on an ensemble of systems can always be reduced to measurements of mean values on that ensemble. As a consequence, if E is an ensemble of N subsystems such as V, and if p_n and $|n, r\rangle$ are, respectively, the eigenvalues and eigenvectors of the density matrix (7–6) attached to E, a mixture $\hat{E}$ exists which is composed of subensembles of the E_α type (i.e., each E_α is describable by a ket) and which, on the other hand, cannot be distinguished from E by any measurement that bears on systems V alone.* This $\hat{E}$ is made up of the ensembles $\hat{E}_n$ that are described by the kets $|n, r\rangle$ and have the populations Np_n. If we could make measure-

---

*We exclude also measurements of correlations between elements of the same ensemble: see Section 10.2.

ments *only* on the systems V and not on the systems U, we should not be able to differentiate such an ensemble from a mixture of type E. The same is true, as is easily verified, if we are able to measure, along with the quantities pertaining to V, only a limited number of suitably chosen correlations between the U and V systems. Since the latter case often occurs in practice, the use of the generic name "mixture" for describing ensembles that are either of the $\hat{E}$ or of the E type is quite convenient for applications.

When fundamental questions are investigated, however, the situation is altogether different. Indeed, let us consider Assumption Q (Section 4.2). Let T be a certain type of physical systems S, and let D be the set of observables that refer to one system of type T alone (thus excluding observables that belong both to one S system and to other systems, such as correlations between S and other systems with which S interacted in the past). Then what Assumption Q precisely means is that, in regard to the predictions of measurements bearing on any observable belonging to set D, any ensemble E of such systems S specified without reference to future observations bearing on these S is describable by at least one density matrix ρ having the properties described in Chapter 6. When this is the case, we say for short that E is describable by ρ.

With the help of Assumption Q it is then possible to prove what follows. At least in respect to systems $\Sigma = U + V$, for which every Hermitean operator corresponds to an observable, the assumption (referred to as a below) that the ensemble of the systems U and the ensemble of the systems V are both separately mixtures of the type already encountered in Section 6.2 (type $\hat{E}$) is, in general, not consistent with Eq. (7–1).

To show this, let us remember that the mixtures defined (in physical terms) in Section 6.2 and described mathematically by Eq. (6–1) are (by definition) such that every element of such an ensemble is a member of a subensemble E_α which is describable by a ket $|\varphi_\alpha\rangle$ of $H^{(u)}$ (in regard to the predictions bearing on the observables belonging entirely to systems of this type). If assumption a were true, every U would belong to some E_α, and, similarly, every V would belong to a subensemble F_m of all the V's, which would also be describable by some ket $|f_m\rangle$ of $H^{(v)}$. We could then consider the subensemble $\mathscr{E}_{\alpha,m}$ of all the systems Σ whose subsystem U would belong to E_α *and* whose subsystem V would belong to F_m. The ensemble $\mathscr{E}$ of all the systems Σ would then be just the union of all these $\mathscr{E}_{\alpha,m}$. Now, because of Assumption Q, we know that each $\mathscr{E}_{\alpha,m}$ is a quantum ensemble and is therefore describable by a density matrix $\rho_{\alpha,m}$. In other words, we know there exists at least one density matrix $\rho_{\alpha,m}$ that gives the correct predictions of measurement results, in regard to all the observables belonging to systems of type Σ. Such a $\rho_{\alpha,m}$ must then, in particular, give the correct predictions, in regard to both the observables belonging entirely to U *and* the observables belonging entirely to V. But the only density matrix in $H^{(u)} \otimes H^{(v)}$ that gives these predictions (i.e., that gives, in particular, a probability $\delta_{\alpha,\beta}$ for finding a value 1 for the observable

$|\varphi_\beta\rangle\langle\varphi_\beta|$ of U and a probability $\delta_{m,n}$ for finding a value 1 for the observable $|f_n\rangle\langle f_n|$ of V) is

$$\rho_{\alpha,m} = |\varphi_\alpha\rangle\,|f_m\rangle\langle\varphi_\alpha|\langle f_m| \qquad (7\text{--}13)$$

as follows from the remark at the end of Section 6.2. Hence $\mathscr{E}_{\alpha,m}$ is a pure case describable by $\rho_{\alpha,m}$ (or, alternatively, by the ket $|\varphi_\alpha\rangle|f_m\rangle$).

Under these conditions, the fact that $\mathscr{E}$ should be describable as the union of all the $\mathscr{E}_{\alpha,m}$ implies that it should be describable also by a weighted sum ρ' of all the $\rho_{\alpha,m}$:

$$\rho' = \sum_{\alpha,m} p_{\alpha,m}\,\rho_{\alpha,m} \qquad (7\text{--}13a)$$

so that $\mathscr{E}$ should be a mixture of all these distinct subensembles. However, $\mathscr{E}$ is by assumption described by the ket (7–1) or, equivalently, by the density matrix

$$\rho = |\psi\rangle\langle\psi| \qquad (7\text{--}14)$$

Then proposition (viii) of Section 6.1 shows that ρ and ρ' are necessarily different (except, of course, in the special case that $|\psi\rangle$ is a product of a vector in the Hilbert space of U and a vector in the Hilbert space of V). Now, if all the Hermitean operators in the Hilbert space of Σ correspond to observables, and if a density matrix describes an ensemble $\mathscr{E}$ of systems, all the matrix elements ρ_{ij} of that density matrix are measurable, as shown in Section 7.1. It is therefore not conceivable that one and the same $\mathscr{E}$ should be describable by more than one density matrix, for obviously, if $\mathscr{E}$ is given, the result of the measurements of any of the elements ρ_{ij} is determined. Therefore the fact that ρ and ρ' should both describe $\mathscr{E}$ constitutes a contradiction. Since it is known that ρ *does* describe the given ensemble $\mathscr{E}$, it follows that ρ' does not. Hence the assumption that the ensembles of the U's and the ensembles of the V's are both separately mixtures of type $\hat{E}$ is necessarily false, since it is from this assumption that we derived the conclusion that $\mathscr{E}$ is describable by ρ'. Q.E.D.

As is apparent, the essence of this proof is that $\mathscr{E}$ is in principle distinguishable from *any* mixture made up of ensembles of the type $\mathscr{E}_{\alpha,m}$. The measurements that allow for such a distinction to be made are, of course, measurements of correlations between observables belonging to the systems U and observables belonging to the systems V.

Since mixtures of the E type and mixtures of the $\hat{E}$ type are in principle operationally different concepts, it is appropriate, at least when fundamental problems are discussed, to differentiate them also in the language [4]. In what follows, the expressions *proper* and *improper* mixtures are used to designate mixtures of the $\hat{E}$ and E types, respectively.*

---

*The names mixtures of *first* and *second* kind are also used.

Remark 1

The ensembles E considered in Eq. (6–1) and in proposition (viii) are made up of a *finite* number of $\hat{E}_\alpha$. However, it is a trivial matter to extend the proof of proposition (viii), and therefore also the present proof, to the case in which the index α varies in a continuous manner (so that the number of E_α's is infinite), as long as the $|\psi_\alpha\rangle$ are normalizable to unity. The case in which the $|\psi_\alpha\rangle$ are normalized to delta functions creates no problem either, because it is an idealized limiting case and should therefore be treated as such.

Remark 2

Both the assumption that every Hermitean operator corresponds to an observable and Assumption Q are necessary for the proof. The necessity for the first one is obvious. The necessity for the second will become apparent upon reflection, particularly in view of the content of Section 7.5. Parenthetically, this can serve as an illustration of the role of mathematical developments in the foundations of physics. Admittedly, this role is prominent, but it is by no means an exclusive one. In these problems nontrivial but at the same time nonmathematical considerations must also be used. It would be an error, therefore, to believe that an ever-increasing formalization of the theory constitutes in itself a universal recipe for answering all the fundamental questions that underlie quantum theory.

7.3   THE HOMOGENEITY OF ENSEMBLES

This section deals with the following question, first investigated systematically by von Neumann [5]: "When is it possible to split an ensemble E into two (or more) different ensembles?"

It is necessary that the concepts involved in the formulation of this question first be made unambiguous. For this purpose, let us make the two statements which follow.

Statement (*a*) bears on the notion of splitting. When we speak of an ensemble E that can be split into two ensembles E_1 and E_2, we do not mean that we can actually perform the operation. What we have in mind is just the abstract possibility of considering ensembles E_1 and E_2, whose total number of elements is the number N of elements in E, and which are such that, for all the observables, the statistical frequencies (i.e., probabilities) of the prospective measurement results are exactly the same on E as they are on the mixture of E_1 and E_2 (by definition the mixture of E_1 and E_2 is just, in this context, the ensemble composed of all the elements of E_1 *plus* all the elements of E_2).

Statement (*b*) is simply a definition of the word "different." Two ensembles are said to be different with respect to some observable A if the statistical frequencies of the prospective measurement results on observable A are different on the two ensembles. Also, two ensembles are said to be different if there exists at least one observable A with respect to which they are different.

From a combination of statements (a) and (b), it follows that an ensemble E can be split into two *different* ensembles, E_1 and E_2, when and only when (i) the prospective statistical frequencies (or, equivalently, the mean values) of all the observables are the same on E and on the mixture of E_1 and E_2, and (ii) these same statistical frequencies or mean values are nevertheless different on E_2 from what they are on E_1 (and also, therefore, from what they are on E).

If it is assumed that every Hermitean operator is the representative of an observable, then, as shown above, the density matrices of the considered ensembles are measurable. If an ensemble E can be split into two ensembles, E_1 and E_2, this then implies that the corresponding density matrices, ρ, ρ_1, and ρ_2, must satisfy

$$N\rho = N_1\rho_1 + N_2\rho_2 \qquad (7\text{-}15)$$

where N, N_1, and N_2 are the numbers of elements in E, E_1, and E_2, respectively, since otherwise the prospective statistical frequencies $\text{Tr}(\rho P_a)$ of some observable A would be different on E from what they are on the mixture of E_1 plus E_2 (P_a is here the projection operator that projects onto the eigenspace relative to the eigenvalue a of A). Under these conditions E can be split into different subensembles if and only if two different density matrices ρ_1 and ρ_2 exist that obey Eq. (7-15). By taking into account the general properties of the density matrices, it is easily verified, however (see Exercises 4 and 5 on page 72), that such matrices can indeed be found if and only if ρ is *not* a one-dimensional projection operator.

Thus, if by definition we call elementary an ensemble E that cannot be split into two different subensembles, we have the following proposition.

Proposition

The elementary ensembles are pure cases (i.e., describable by state vectors), and the pure cases are elementary.

This proposition calls for two remarks.

Remark 1

It should be remembered that the validity of the proposition above hinges on the assumption that all the Hermitean operators correspond to some observable (or, alternatively, on the weaker assumption that the density matrices are measurable quantities). If neither of these assumptions holds, the proposition is false. For example, if in some fixed reference frame only the third components σ_3 of the spin of spin-$\frac{1}{2}$ systems were measurable, then a pure case described, for instance, by an eigenvector of σ_1 would not be elementary. This is so because, according to statement (*a*) above, it could be split into two sub-ensembles, each described by one of the eigenvectors of σ_3.

Remark 2

Conversely, let a density matrix ρ satisfying $\rho^2 < \rho$ be given. The corresponding ensemble E is a mixture, and the mean values of all the observables on this ensemble are of course given by ρ. It is possible to show that all these data can be reproduced by introducing a larger Hilbert space $\mathscr{H}$ than the one in which ρ operates and by representing E by a ket (i.e., by a pure case) in $\mathscr{H}$. This is an application of the well-known Gelfand-Naimark-Segal construction (see, e.g., Ref. [6]). What happens there is, of course, that several Hermitean operators in $\mathscr{H}$ correspond to no observable, so that the pure case in question is not elementary.

Remark 3

When we say that an ensemble E can be split into two different subensembles, E_1 and E_2, this should be understood only in the quite specific, technical sense defined by statement (*a*) above. In particular, we do not mean to imply that the correlations *within* E are then necessarily the same as the correlations within the mixture of E_1 and E_2. Nor do we mean to imply that, if the elements of E have interacted with other systems, the correlations between these systems and the elements of E are correctly reproduced when the mixture of E_1 and E_2 is substituted for E. Indeed, both these statements would be erroneous. In particular, identification of E with a mixture of E_1 and E_2 can usually be made only if the distinction between proper and improper mixtures is overlooked.

7.4   OBSERVABLE HAVING A VALUE ON A SYSTEM*

At a time t let us consider an ensemble E_n of N_n systems S. Let us assume (i) that E_n is *defined* with no reference to the results of observations which have not yet been made on the systems S, and (ii) that nevertheless we *happen somehow* to know for certain that, if an observable A were measured on these S immediately after time t, the result a_n (one of the eigenvalues of A) would be obtained in each element of E_n. It is, at any rate, certainly not obvious that the class of all the E_n satisfying both assumptions should be empty, and we postulate that it is not.

Together with Assumption Q or Q' (see Sections 4.2 and 6.4), assumption (i) guarantees that a density matrix exists which leads to the correct statistical predictions in regard to the results of future measurements of observables per-

---

*The content of Sections 7.4 and 7.5 should be bypassed on the first reading.

taining properly to the systems S. This density matrix, ρ_n, operates in the Hilbert space $\mathscr{H}$ of the systems S; hence, through diagonalization and further proper manipulations, it can always be written as

$$\rho_n = \sum_{k,r} p_{nk} |n, k, r\rangle \langle n, k, r| \qquad (7\text{–}16)$$

where the $p_{n,k}$ and $|n, k, r\rangle$ are its eigenvalues and eigenvectors, respectively. More generally, it can be written in an infinity of ways in the form [see Eq. (6–1)]

$$\rho_n = \sum_{\alpha} q_{n,\alpha} |\varphi_{n,\alpha}\rangle \langle \varphi_{n,\alpha}| \qquad (7\text{–}17)$$

where the $q_{n,\alpha}$ are positive numbers whose sum over α is unity, and where the $|\varphi_{n,\alpha}\rangle$ are normalized vectors of $\mathscr{H}$. At this point let us introduce our assumption (ii). The remark at the end of Section 6.2 then leads to the following conclusion.

Proposition S

Whenever assumptions (i) and (ii) above are true, the vectors $|\varphi_{n,\alpha}\rangle$ appearing in (7–17) [or the vectors $|n, k, r\rangle$ appearing in (7–16)] all necessarily belong to the subspace $\mathscr{H}_n$ of $\mathscr{H}$ that corresponds to eigenvalue a_n of A.

Introducing Condition R

Let us now turn to the main subject of this section. What meaning should be given to this sentence: "A physical quantity A has a definite value in a system S"? Conceptually, one can think of at least three varieties of answer. The first one would be of the mathematical kind: something like, "A has a value in S if S is in an eigenstate of the corresponding operator." Such statements could also be formulated with more elaborate mathematics involving algebras instead of Hilbert spaces and so on. And, of course, we do not mean that any of them would be wrong. On the contrary, they are right by assumption. But what must be stressed is that they are insufficient as definitions and that they must necessarily remain so quite independently of any mathematical refinement. This is so just because they refer to an abstraction: how do we know that the system in question has such and such a state vector or such and such a positive-definite functional on some algebra or . . . attached to it? And what do such propositions mean exactly, that is, in terms of observations? These are the relevant questions [7].

Hence we must consider a second type of answer and regard it as more satisfactory. This type is constituted by the operational definitions. For instance, let us postulate the validity of the inductive inference principle, let us consider an ensemble E of N systems S that have all been prepared in the same well-specified way, and let us consider the case in which measurements of A made on each of $N-1$ among these S have all given the same result a. Then we can infer that a measurement on the Nth system would also give result a, and we can use this to assert that by definition this latter system S has the value a (before the measurement is carried out).

If we were to remain inside the realm of a purely operational philosophy, this definition could be sufficient to cover most, if not all, cases. It so happens, however, that many of us do not consistently remain within that realm. For example, most of us believe that, relatively to other objects, the center of mass of a macroscopic object M can "really" be said to be in a certain region of space and not in others, irrespective of what kind of an ensemble M is a member of. A relevant question then is: Can we formulate such a belief more precisely in order to check whether or not it can be expressed in some self-consistent way? If we can, we will have a third type of definition of what we mean for an observable to have a value a. But the goal is quite ambitious. As a first step (the only one we take in this chapter), it is therefore appropriate to formulate a necessary condition—which we call Condition R—for an observable A defined on a system S to be such that (in quantum mechanics) a definite value can be attributed to it at a time t. Again, as we shall see, this condition can be formulated only if we know something about the preparation of the system S.* But it may also be satisfied in the case in which other systems S similarly prepared did not, upon measurement of A, all lead to the same result a. To formulate it we must make use of the fact that the universality of the quantum rules is postulated and that therefore we already know that a statement such as "A has at time t the value a" on an ensemble of systems has some observable consequences, at least if we interpret it—as is obviously suggested by common usage—as implying that "if A were measured on these systems immediately after t, the result a would be obtained on each of them." These consequences are those that follow from the fact that proposition S above then holds true. They can be expressed as statistical predictions $\{P_B\}$ on the results that measurements of other observables B would give at a time $t' > t$ if they were carried out on these systems.

Keeping these facts in mind, let us introduce an ensemble E of systems S; let us consider the following assertion: "In E, N_n systems (well specified individually)† *have*, at time t, a value a_n for observable A"; and let us make the assumption, which we call *Assumption Z*, that some positive numbers N_n ($\Sigma_n N_n = N$) and some eigenvalues a_n of A exist, such that the corresponding assertion just formulated is valid for each n. Within the limits set by Assumption Z, Proposition S above gives the most general quantum-mechanical

---

*Hence this condition is itself of the conditional type. And if we know nothing about the preparation, we can say that it is always satisfied.

†This simply means what follows: if A were measured individually on each of these N_n systems, the result would be a_n with certainty in every case.

descriptions, at any time $t' > t$, of the subensembles E_n, each of which is composed of the N_n systems that had $A = a_n$ at time t. From these descriptions we can derive some statistical predictions $\{P_B\}_n$ concerning the results that measurements of other observables B would give at t' on the elements of each E_n, and hence also some statistical predictions $\{P_B\}$ concerning the results that the same measurements would give on all the elements of E whenever Assumption Z is valid. And the existence of these predictions is finally what makes it possible to formulate our Condition R.

Condition R

This necessary condition for Assumption Z above to be acceptable can be stated as follows: "In regard to *any* observable B it must be true that the statistical predictions $\{P_B\}$ defined above do not contradict those that are derived directly—for the same time t'—from what is known of the initial preparation of E, without using Assumption Z and by applying the usual rules of quantum mechanics."*

Because of the fact that misunderstandings can quite easily creep in concerning these questions, care was taken above to make every detail explicit. But these developments should in turn not conceal the fact that Condition R is, after all, only a very straightforward and even trivial compatibility condition between Assumption Z and what we can derive from the preparation of the systems.

Comments

1. It should be stressed again that Assumptions Q and Q′ have been used above in deriving the validity of the statistical predictions $\{P_B\}$ from Assumption Z. Hence Condition R is binding only if Assumption Q (or Q′) is made. In other words, it is *not* binding in a hidden variable theory, not even in one that would reproduce exactly the quantum-mechanical prediction.

2. The difficulty with Condition R lies in the words "for any B." The presence of the word "any" is necessary if we adhere at all to common linguistic conventions, for (like any other sentence) the sentence "A has the value a" cannot be true unless all its consequences are true. But, on the other hand, it represents such a strong requirement that the set Σ of the observables "having" definite values might well, under this definition, turn out to be empty in

---

*An example may clarify this. Let the systems S be spin-$\frac{1}{2}$ particles initially prepared in such a way that they are polarized along $O\hat{x}$ and propagate along $O\hat{y}$. A Stern-Gerlach apparatus whose magnetic field is along $O\hat{z}$ separates a beam of N such particles into two beams. Can we assert that at a time t when that separation is already effective there are, roughly, $N/2$ particles in each beam, each particle then having a definite $S_z(+\frac{1}{2}$ or $-\frac{1}{2}$, according to the beam in which it is)? The answer is no, because Condition R is not satisfied for any B, as the hypothetical (but in principle feasible) experiment of beam recombination (see Chapter 15) shows. An example of the opposite case—the one in which Condition R is satisfied—is constructed by considering an initially unpolarized beam.

the majority of cases (since among the B we must reckon the correlations between S and the systems with which S interacted even in the remote past; see Chapters 8 and 12). These problems rank among the major one with which quantum measurement theory tries to cope. They are investigated in detail in Chapter 14 and thereafter.

A Corollary to Condition R

Finally, making use of Eq. (7-17), we can formulate a corollary to Condition R. This states that, in order that Assumption Z can consistently be made, it is necessary that a partition of E into subensembles E_n should exist, such that each E_n is describable as a mixture of pure cases, themselves all describable by kets $|\varphi_{n,a}\rangle$ where each $|\varphi_{n,a}\rangle$ belongs to the subspace $\mathscr{H}_n$ of $\mathscr{H}$ (the Hilbert space of the systems) that corresponds to the eigenvalue a_n of A.

7.5 QUANTUM ENSEMBLES AND SUPPLEMENTARY VARIABLES

The problems concerning the possibility of the existence of hidden variables theories of quantum mechanics (and of the general conditions the latter must obey) are so important that a separate chapter (Chapter 11) has been set aside for them. Here let us merely comment on a special point. Specifically, we introduce the general notion of "supplementary quantities," and we show that this notion reduces to that of hidden variables (to be defined below) in the case in which all the Hermitean operators correspond to observables.

Let us consider an ensemble E of systems S that is a "pure case" in our terminology, or, in other words, that is describable by a ket $|\psi\rangle$ (this means that all the observable consequences that are derived from that $|\psi\rangle$ by means of the general principles hold true). If $|\psi\rangle$ is an eigenvector for a number A, B, . . ., F of commuting observables, corresponding to the eigenvalues a_i, b_j, . . ., f_l, then, in the spirit of Section 7.4, we can say, "A has value a_i, B has value b_j, . . ., F has value f_l on every element of E." In the case in which every Hermitean operator in the Hilbert space $\mathscr{H}$ attached to the systems of the same type as S corresponds to an observable, a set $\{A, B, . . ., F\}$ of compatible observables can be found such that $|\psi\rangle$ is the *only* vector of $\mathscr{H}$ which corresponds to the set $\{a_i, b_j, . . ., f_l\}$ of eigenvalues of these observables. For example, we may choose $A = a_i |\psi\rangle \langle\psi|$, in which case $\{A, B, . . ., F\}$ reduces to just A itself, since $|\psi\rangle$ is the only element of $\mathscr{H}$ that is an eigenvector of A corresponding to the eigenvalue a_i. As we know (see Section 7.3), no partition of E into *quantum* subensembles E_α differing statistically from E exists in such a case. In particular, no partition exists such that some observables G (which we henceforth call "supplementary quantities") would have on each E_α some definite value(s) that would not be the same on all the E_α. [Indeed, if such a partition could be conceived of, the corresponding E_α would be describable by density matrices ρ_α that would not all be equal, and proposition (viii) of Section 6.1 would be violated.] Hence, if we want to assume the existence of supplementary quantities such as G, we must assume also that the subensembles E_α that they label are *not* quantum ensembles; and this means that Assumption Q must be discarded. Correspondingly, the supplementary quantities G can obviously not be associated (in that case) with Hermitean operators in $\mathscr{H}$ in the way that quantum-mechanical observables are. Because of that (and also for historical reason) these G are conventionally called *hidden quantities* or, more commonly, hidden variables. It can be said that the complete set of compatible observables $\{A, B, \ldots, F\}$ determines the state vector up to a phase, and that every other quantity is *hidden*.

Up to this point we have considered exclusively the case in which all the Hermitean operators in $\mathscr{H}$ correspond to observables. In the opposite case, it may well happen that $|\psi\rangle \langle\psi|$ is precisely one of the operators that do not, and that no set $\{A, B, \ldots, F\}$ of commuting observables exists that admits $|\psi\rangle$ as the unique eigenvector common to $A, B, \ldots, F$ and corresponding to a given set of eigenvalues of these observables. In that case it can happen that a partition of E into quantum subensembles E_α can be found such that (i) the set of the observable predictions deduced from $|\psi\rangle$ coincides with the set of the observable predictions deduced from a quantum description of the E_α, and (ii) one or several observables G *have* well-defined values g_α on each of the E_α (in the sense that the density matrices describing any one particular E_α lead to a probability *one* for the value g_α to be found upon measurement of G), the g_α not being all equal. In that case the observables such as G play, with respect to the vector $|\psi\rangle$, the role of *supplementary quantities*, since a knowledge of their values on a system constitutes a supplementary piece of information, not contained in $|\psi\rangle$. On the other hand, the E_α are now quantum ensembles and the G are, correspondingly, ordinary quantum observables, hence there is no reason in this case to attribute the name of "hidden observables" to the G.

Remark 1

If the operator $|\psi\rangle \langle\psi|$ corresponds to an observable, a partition of E into subensembles E_α such as the one considered above cannot exist, even in the case in which not every Hermitean operator corresponds to an observable (see Exercise 8 on page 73).

Remark 2

Even if E can be identified with the union of several E_α at a given time, it is not necessarily true that the possibility of that identification persists at other times. As is almost immediately apparent, a sufficient condition for such a persistence is that, if $\mathscr{U}$ is the time evolution operator, then, for any operator

A that corresponds to an observable, it should be true that $\mathscr{U}^{-1} A \mathscr{U}$ also corresponds to an observable (see Exercise 9).

Remark 3

We have seen in this section that in the case in which not all Hermitean operators correspond to observables there is a difference between the concept of supplementary quantities and the concept of hidden variables. Supplementary quantities that may serve in characterizing a system over and above the state vector can very well not be hidden in that case, so that we must be careful before we say that the introduction of any such quantity is a commitment to a hidden variable theory. On the other hand, the content of Remark 2 shows that the possibility of considering such supplementary quantities that are not "hidden" is probably of no great interest. This is obvious in the case in which such supplementary quantities have only transient existence (see Remark 2). But in the opposite case it is also true, since it is then not clear in what circumstances we should be led to describe the ensemble under study as a "pure case" in the first place, and *then* to split it by means of the supplementary quantities.

For instance, it may happen that the Hilbert space of the type T systems under consideration admits of an orthonormal basis which is such that the matrices describing all the observables of T—the Hamiltonian included—on that basis have their nonzero matrix elements arranged according to the *same* pattern, consisting of nonoverlapping squares lying along the first diagonal. In that case some mixtures admit of an equivalent description as "pure cases" in our terminology; that is, they can be described by means of a unique vector $|\psi\rangle$. Let E be such a mixture. If such a description is chosen for E, the quantities Q whose matrix is diagonal and whose diagonal elements are all equal within any given square play the role of supplementary quantities for E and commute with the Hamiltonian; the ensemble under consideration can be split into several such ensembles, each corresponding to one definite value of Q, and this splitting remains valid at any time. On the other hand, if such a situation holds, no reason can be found that would compel us to describe E by $|\psi\rangle$ in the first place, instead of treating it as a mixture. This fact can also be expressed by saying that a state $|\psi\rangle$ (to be understood as a state presenting, at least at some time, some character of elementarity differenciating it from a mixture) cannot be physically produced. Hence we recognize in such a situation the conditions that prevail when a superselection rule exists: the quantity Q obviously plays the role of the corresponding superselection charge. If the $|\psi\rangle$ *were* used, Q would serve as a supplementary quantity with respect to them. But the common and also the simpler practice consists in not introducing the $|\psi\rangle$ at all.*

---

*This seems especially appropriate in view of the fact that the time evolution operator never leads directly to such $|\psi\rangle$.

These considerations can be illustrated by the well-known example of the electric charge Q. An ensemble consisting of N_p protons (symbol p) and N_n neutrons (symbol n) *could* be described by the state vector

$$|\psi\rangle = N^{-1/2}(N_p{}^{1/2}|p\rangle + N_n{}^{1/2}|n\rangle) \tag{7-18}$$

just as well as by the mixture

$$\rho = N^{-1}(N_p|p\rangle\langle p| + N_n|n\rangle\langle n|) \tag{7-19}$$

in regard to all the observable predictions that can be made on that ensemble. If the description by means of $|\psi\rangle$ were chosen, Q could serve as a supplementary quantity. But such a choice would be highly artificial, since we can only produce such an ensemble by putting together ensembles of protons and of neutrons. It is in that sense (see above) that the statement according to which states such as $|\psi\rangle$ "cannot be produced" can be understood. Q is a superselection charge.

REFERENCES

[1] H. Margenau and J. L. Park, *Int. J. Theor. Phys.* **1**, 211 (1968).
[2] G. Wick, A. Wightman, and E. Wigner, *Phys. Rev.* **88**, 101 (1952).
[3] W. E. Lamb, Jr., *Phys. Today* **22**, (4), 23 (1969)
[4] B. d'Espagnat, *Conceptions de la Physique Contemporaine*, Hermann, Paris, 1965; contribution to *Preludes in Theoretical Physics: In Honour of V. F. Weisskopf*, A. De Shalit, H. Feschbach, and L. Van Hove, Eds., North-Holland, Amsterdam, 1966, p. 185.
[5] J. von Neumann, *Mathematical Foundations of Quantum Mechanics*, Princeton University Press, Princeton, N.J., 1955.
[6] M. Guenin in *Lectures in Theoretical Physics*, Vol. IXA, W. Brittin, A. Barut, and M. Guenin, Eds., Gordon and Breach, New York, 1967.
[7] B. d'Espagnat, *Nuovo Cimento*, **21**B, 233 (1974).

EXERCISES OF PART TWO

EXERCISE 1

Let the two operator-vectors $S^{(1)} = (\tfrac{1}{2})\,\sigma^{(1)}$ and $S^{(2)} = (\tfrac{1}{2})\,\sigma^{(2)}$ describe the spins of two spin-$\tfrac{1}{2}$ particles. Use the identity

$$(S^{(1)} + S^{(2)})^2 = S^{(1)2} + S^{(2)2} + 2(S^{(1)} \cdot S^{(2)})$$

to write the action of the operator $\sigma^{(1)} \cdot \sigma^{(2)}$ on the singlet state $(s = 0)$ and on the triplet state $(s = 1)$ of the two-particles system. Here σ_1, σ_2, σ_3 are the Pauli matrices, the relation $J^2 | J, M \rangle = j(j+1)\,| J, M \rangle$, and the theory of angular momenta are assumed to be known.

EXERCISE 2

Use the result of Exercise 1 in order to write, in terms of $(\sigma^{(1)} \cdot \sigma^{(2)})$, the projection operators into the triplet state and into the singlet state.

Use this result to write the density matrix that describes the singlet state.

EXERCISE 3

Use the procedure of Exercise 2 to show that the density matrix for a particle with spin $\tfrac{1}{2}$ pointing along $\mathbf{n}$ is

$$\tfrac{1}{2}[1 + (\sigma \cdot \mathbf{n})]$$

EXERCISE 4

By means of the general method used in the textbooks to prove the Cauchy-Schwarz inequality, show that, if ρ_1 and ρ_2 are density matrices,

$$\mathrm{Tr}\,(\rho_1 \rho_2) \le 1$$

and the equality holds only for $\rho_1 = \rho_2$.

Use this fact to show that, if ρ describes a pure case and if

$$\rho = a_1 \rho_1 + a_2 \rho_2$$

with $a_1 > 0$ and $a_2 > 0$, ρ_1 and ρ_2 being density matrices, then

$$\rho_1 = \rho_2 = \rho$$

EXERCISE 5

Assume that ρ_1 and ρ_2 operate in a finite-dimensional Hilbert space, and prove the last result of Exercise 4 directly by the method used to prove proposition (viii) of Section 6.1. [*Hint*: Show first that $a_1 + a_2 = 1$; then expand ρ_1 and ρ_2 according to formula (6–1).]

EXERCISE 6

Use the positive-definite character [inequality (6–2)] of ρ to prove the generalized Cauchy-Schwarz inequality $|\rho_{j,i}|^2 \leq \rho_{i,i}\,\rho_{j,j}$
(*Hint*: Introduce $|c\rangle = |e_i\rangle + \lambda|e_j\rangle$.)

EXERCISE 7

Show that the magnitude of the quantity I defined by Eq. (6–49) is always smaller than or equal to one.

EXERCISE 8

Consider the case in which not all the Hermitean operators correspond to observables. Show that, even in this case, if E is an ensemble described by a ket $|\psi\rangle$ and if the operator $|\psi\rangle\langle\psi|$ corresponds to an observable, no partition of E into quantum subensembles E_α such as the ones considered in Section 7.5 can be conceived of. (*Hint*: Use the fact that the observable $|\psi\rangle\langle\psi|$ has probability one of taking the value $+1$.)

EXERCISE 9

Prove the validity of Remark 2 of Section 7.5.

PART THREE

QUANTUM NONSEPARABILITY

The "nonseparability" of quantum systems that have once interacted formally emerges as a very elementary consequence of the general principles of quantum mechanics. It is, however, one that has important conceptual implications, and in this part we therefore investigate it in detail.

Any attempt at a nonsuperficial understanding of this aspect of the theory must necessarily consider the delicate points raised in 1935 by Einstein and his coworkers ([1] of Chapter 8). We describe here the solution to these difficulties that we think is more or less tacitly accepted by the theorists who believe in conventional quantum mechanics. Moreover, since this solution implies a considerable departure from all the familiar ideas about the localization of objects, we are led to consider also a possible alternative, provided by the so-called hidden variables theories. However, quite apart from their inherent intricacy, such theories present conceptual difficulties of their own. More precisely, they introduce considerable nonseparability effects, which are very much akin to the quantum ones and just as unpalatable.

CHAPTER 8

THE EINSTEIN-PODOLSKY-ROSEN PROBLEM

This problem, or "paradox" as it is sometimes called, has both formal and conceptual aspects. There is some advantage in considering the two types separately, as much as is possible, from each other.

8.1 THE FORMAL ASPECTS

Let us consider two types of quantum systems U and V, and the Hilbert spaces $\mathscr{H}^{(U)}$ and $\mathscr{H}^{(V)}$ attached to them. We are interested in investigating the case in which a system of type U interacts during a noninfinite time interval Δt with a system of type V. For example, these two systems collide and the collision time is Δt.

Let us consider an ensemble of such collision events. Let us moreover specify that, initially, the ensemble of the U systems is so prepared that it can be described before the collision by a state vector $|f_i\rangle \in \mathscr{H}^{(U)}$. Let us similarly specify that the ensemble of the V systems can be described before the collision by a state vector $|\phi_0\rangle \in \mathscr{H}^{(V)}$. The interaction between U and V can then be such that after the collision the ensemble of the U systems can still be described by the vector $|f_i\rangle$, while the ensemble of the V systems is describable by a vector $|\varphi_i\rangle$ of $\mathscr{H}^{(V)}$. We then write symbolically

$$|f_i\rangle \phi_0\rangle \rightarrow |f_i\rangle |\phi_i\rangle \tag{8-1}$$

More generally, the interaction between U and V can be such that

$$|f_i\rangle |\phi_0\rangle \rightarrow |f_i'\rangle |\phi_i\rangle \tag{8-1a}$$

with $|f_i'\rangle \neq |f_i\rangle$. In (8–1) and (8–1a) the arrow represents the effect of the time evolution operator on the state vector of the composite system, which is in $\mathscr{H}^{(U)} \otimes \mathscr{H}^{(V)}$. Before the interaction this state vector is of course a product of state vectors. In the particular case here considered, it is a product of state vectors also after the interaction. This has the consequence that if, for instance, the vectors $|f_i\rangle$ and $|f_i'|$ are eigenstates of some observable A of U, and if the vectors $|\phi_j\rangle$ are eigenstates of some observable B of V:

$$A|f_i\rangle = a_i|f_i\rangle \tag{8-2}$$

$$A|f_i'\rangle = a_i'|f_i'\rangle \tag{8-2a}$$

$$B|\phi_j\rangle = b_j|\phi_j\rangle \tag{8-3}$$

then the collision (8-1a) can be said to "change the values of A and B." A changes from a_i to a_i', and B changes from b_0 to b_i. The process of measurement of a physical quantity A which pertains to U by means of an instrument V whose pointer coordinate is B, when sufficiently schematized, provides an example of such a case (a_i' can be either different from or identical to a_i, according to whether or not the measurement process induces a perturbation on the value of A on U). It may be noted parenthetically at this stage that in the case under consideration the description of systems U and V as each having *definite* values of A and B, respectively, is valid without restriction, that is, irrespective of what kinds of experiments are planned for the future on these systems.

The first (elementary) point we want to make is that, among the phenomena of collision (or of temporary interaction between systems), those that can be described by means of (8-1) and (8-1a) are only very special cases. This is a simple and unavoidable consequence of the most fundamental principle of quantum mechanics, namely, the linearity of the laws of evolution. As a matter of fact, the linearity property implies that, as a consequence of (8-1),

$$(|f_i\rangle + |f_k\rangle)|\phi_0\rangle \rightarrow |f_i\rangle|\phi_i\rangle + |f_k\rangle|\phi_k\rangle \tag{8-4}$$

where the arrow again describes the evolution in time from before to after the collision event. Now the left-hand side of (8-4) is still a product of a state vector of $\mathcal{H}^{(U)}$ and a state vector of $\mathcal{H}^{(V)}$, but the right-hand side is not.

It is very easy to imagine examples in which the left-hand side of (8-4) can be physically realized. This is the case when

$$|f_i\rangle + |f_k\rangle \tag{8-5}$$

is itself an eigenvector of some observable C attached to the systems of U type; for instance, if A is the z component of a spin $\frac{1}{2}$, C is the corresponding x component. In such cases the left-hand side of (8-4) can be prepared just as easily as the left-hand side of (8-1), simply by making preliminary measurements on U which include a measurement of C as a substitute for the measurement of A. The physical interpretation of the left-hand side of (8–4) is then similar to the one above: before the collision, the observables B of V and C of U both have definite values. Does the right-hand side of relation (8–4) similarly possess a simple, intuitive physical interpretation? Since it is *not* expressed as a product of a vector in $\mathcal{H}^{(U)}$ and a vector in $\mathcal{H}^{(V)}$, we may guess that the answer is "no".

Thus these considerations lead to the suspicion that, in the general case, when two systems have interacted in the past, it is not possible to attribute to each of them (or to any ensemble of these if the experiment is repeated) any definite state vector. Correspondingly, we also suspect that it is not possible to attribute to each system definite (though possibly unknown) values for a complete system of compatible observables that would pertain to that system alone. However, as long as we keep to the simple arguments above, such conclusions cannot yet be considered as final, because of the fact that the state vector for the composite $U + V$ system is essentially only an element of a mathematical formalism. In particular, it has not been proved that this state vector stands in a one-to-one correspondence to the "physical reality" of the systems (whatever meaning this expression may be given). As is often the case in physics (let us remember classical probability distributions), it might be, for instance, that several state vectors would correspond to the *same* physical conditions, and that the difference between them would simply reflect differences in the knowledges of the users.

Thus, in order to check the validity of the indications gathered above, we must investigate whether or not alternative descriptions of the final state of the systems may be invented that are compatible with the attribution of a definite state vector and/or of a definite set of physical properties both to the ensemble of systems U and to the ensemble of systems V separately. To study this question we can use the methods an the results of Chapter 6. Since these are based on the use of ensembles, let us consider the ensemble $\mathcal{E}_0$ (pure case) of composite $U + V$ systems, which is described by

$$|\psi_0\rangle \equiv (|f_1\rangle + |f_2\rangle)|\phi_0\rangle \qquad (8\text{–}6)$$

After the collision has taken place, this ensemble $\mathcal{E}_0$ has changed into the ensemble $\mathcal{E}$, described by

$$|\psi\rangle \equiv |f_1\rangle|\phi_1\rangle + |f_2\rangle|\phi_2\rangle \qquad (8\text{–}7)$$

We then have to investigate whether $\mathcal{E}$ can be described as composed of sub-ensembles E_n, each of which would be characterized by the attribution of a definite state vector to the ensemble of its systems U and of a definite state vector to the ensemble of its systems V (these state vectors could of course be different from one another for different E_n). Clearly, this question is a special case of the one already investigated in Section 7.2, and we can therefore use here the results obtained in that section. In the case that all the Hermitean operators of the Hilbert space $\mathscr{H}^{(U)} \otimes \mathscr{H}^{(V)}$ correspond to observables, these results can be stated as follows. The general assumption that lies at the basis of conventional (i.e., without hidden variables) quantum mechanics is Assumption Q (Section 4.2), and this assumption implies that the answer to the question just posed is negative. It is *not* possible to consider $\mathscr{E}$ as being the union of several subensembles, in each of which the systems U would themselves constitute an ensemble describable by a state vector (and similarly with respect to the systems V). In other words, it is not possible to consider each constituent system U (or V) as having a complete set of definite properties.

To conclude, we may state that, in the realm of conventional quantum mechanics (Assumption Q), when two quantum systems, each of which could originally be considered as having a complete set of definite properties, have once interacted, it is generally no longer possible to think of either of them as *having* a complete set of definite properties of its own. Here the words "definite properties" precisely mean specific values of observables pertaining only to that system and independent of future experiments that will done on the system. The important point about this conclusion is that it holds true even for systems that have interacted but no longer interact at the time when they are considered. This is the essence of what is sometimes called the "quantum nonseparability" of systems.

Let it be stressed once again that the argument of Section 7.2, showing that each $\mathscr{E}_{\alpha,m}$ must be describable by a density matrix $\rho_{\alpha,m}$ [Eq. (7–13)], depends on Assumption Q, and that such an assumption does not hold in the case of the hidden variables theories—not even those that "exactly reproduce" all the predictions of quantum mechanics, for even then such exact reproduction is achieved only when the hidden variables are distributed completely at random over the ensemble. Even if this is the case for $\mathscr{E}$, we have no guarantee that it is necessarily the case for all the $\mathscr{E}_{\alpha,m}$. For this reason the result obtained above is binding only in the realm of a theory in which hidden variables are assumed not to exist.

8.2   CONCEPTUAL ASPECTS

The difference between the conceptual framework of classical physics and that of quantum physics which is due to the nonseparability property of the latter theory is better understood in specific examples. Let us therefore consider the three following ones.

Example A (Classical Physics)
A car equipped with two axles is sent into free space. Each axle has pre-

viously been given a spinning motion, and the angular momenta of the axles are equal and opposite. The system as a whole has no angular momentum. At some time t_0 an explosion takes place in its middle and tears the car into two pieces, front and rear. Let us label these U and V; U and V have equal and opposite spinning motions, which we describe by means of angular momentum or *spin* vectors $\sigma^{(U)}$ and $\sigma^{(V)}$. Observers can subsequently make measurements of the various components of $\sigma^{(U)}$ and $\sigma^{(V)}$ defined in a given reference system. They may also compare their results in order to investigate the correlations.

Example B   (Quantum Physics)

A spin-zero particle with positive parity decays at time t_0 into two electrically neutral spin-$\frac{1}{2}$ particles U and V, which have the same intrinsic parity [5]. The interaction that is responsible for the decay conserves parity. A system of shutters makes it possible to select without affecting the spins the U particles that travel in some definite direction and the V particles that travel in the opposite direction. We call $\sigma^{(U)}$ $(\sigma^{(V)})$ the vector that is twice the spin vector of U (V).

Example C   (Quantum Physics)

Antiprotons are brought to rest in liquid hydrogen. The cases in which the antiproton-proton systems then formed decay into two kaons are selected.

In Example B, just as in Example A, observers may measure the various components of $\sigma^{(U)}$ and $\sigma^{(V)}$, compare their results, and establish correlations. In Example C similar measurements may be made on the hypercharges and PC quantum numbers of the kaons.

Our purpose is to use these specific examples to test the validity, both in classical and in quantum mechanics, of two *conceptual principles*. By this term we simply mean ideas regarding Nature that are more or less implicit in everybody's mind but that could, nevertheless, conceivably be rejected. These conceptual principles are labeled (*a*) and (*b*) below.

(*a*)   *Reality Principle* (Einstein, Podolsky, Rosen)

The very concept of a *reality* may be criticized by philosophically minded empiricists (positivists). A widespread opinion, however, is that somehow a physical reality does exist, which should not just amount to the talks *we* give about it. As a matter of fact, most people think that *elements of reality* exist that are not influenced by our knowledge. If these very general ideas hold true, a meaningful question is this: "What, if any, are the *elements of reality* that man can *know* by experience?"

In the days of classical physics this question was already a formidable one, but for physicists its degree of significance increased dramatically as soon as it became clear that most measurements of microscopic quantities are *bound* to alter somehow the system on which they are made. Men have always (and wisely) considered that what makes it easiest to discriminate between real and imaginary phenomena is that the former exhibit regularities that render them predictable. Hence, if we can, by means of suitable observations, foresee the result of the measurement of a given quantity, we are led to believe that an element of reality underlies it. However, if it is true that many measurement processes act on the quantity to be measured, an element of doubt creeps in: Is it not possible that the measurement itself may *create* the measured value? This is more or less what Procrustes did in ancient Greece. The bed he made use of had a certain length L. Hence he could predict with absolute certainty that anybody whom he let go had exactly a length equal to L. Could he logically deduce from this that the concept of a man's *height* correctly reflects some natural regularities and that therefore it corresponds to something real? Certainly not.

In their search for a criterion that would make it possible to recognize elements of reality, Einstein, Podolsky, and Rosen [1] had to take such considerations into account. This led them to propose the following formulation.

If, without in any way disturbing a system, we can predict with certainty (i.e., with probability equal to unity) the value of a physical quantity, then there exists an element of physical reality corresponding to that quantity.

In what follows we call that element a *property* of the system.

(b) Principle of the Separability of Mechanically Isolated Systems

From the content of the foregoing subsection, we can easily guess that this second principle, which remains somewhat implicit in the work of Einstein et al., may well be the source of insuperable difficulties. It can, however, be formulated in such ways as to appear almost self evident and, consequently, very difficult to reject. We choose the following formulation.

If a physical system remains, during a certain time, mechanically (including electromagnetically, etc.) isolated from other systems, then the evolution of its properties during this whole time interval cannot be influenced by operations carried out on other systems.

Here, as above, *a property* of a system is, of course, an attribute that can be attached to this system without any restriction bearing on the future. In particular, it cannot depend at a time t on what experiments are planned for later times. The existence of a system as such (i.e., independently of other systems) is to be considered as one of its properties and should, therefore, also obey principle (b).

Let us now test the two principles, (a) and (b), against the laws of classical and quantum physics by means of the chosen examples.

In regard to classical physics and Example A, it is easily seen that no difficulty occurs. At any time subsequent to the time t_0 of the explosion, each spin $\sigma^{(U)}$ and $\sigma^{(V)}$ has a perfectly well-defined direction in space, although this direction (i) may vary from one composite $U + V$ system to another one, and (ii) initially at least is not known. When, at a time $t_1 > t_0$, the observer makes his first measurement, on $\sigma_3^{(U)}$, for instance, he does not thereby influence in any way the values of the components of $\sigma^{(V)}$, even if his measurement changes the direction of $\sigma^{(U)}$. Principle (*b*) is therefore not violated.

Furthermore, we may, for example, define two quantities, $Q^{(U)}$ and $Q^{(V)}$, which are equal to ± 1 according to whether $\sigma_3^{(U)}$ or $\sigma_3^{(V)}$, respectively, is positive or negative. Then, because of the fact that the total spin is zero, a measurement of $\sigma_3^{(U)}$ provides a knowledge of $Q^{(V)}$ [$Q^{(V)} = -1$ $(+1)$ if $\sigma_3^{(U)} > 0$ (< 0)] without in any way perturbing V. Under these conditions principle (*a*) asserts that there exists an element of reality corresponding to $Q^{(V)}$. This, again, is correct, since this element of reality is just the objective fact that $\sigma_3^{(V)}$ *is* negative or positive. It may also be noted that this element of reality corresponding to $Q^{(V)}$ already existed at times t prior to t_1, since it was in fact produced at the very time of the explosion. In this we see again a perfect—even if completely trivial—agreement with the content of principle (*b*). Finally, in regard to the correlations between the possible results of measurements done on $\sigma^{(U)}$ and on $\sigma^{(V)}$, they of course result in particular from the fact that, for all times $t > t_0$, $\sigma_i^{(U)}$ and $\sigma_i^{(V)}$ *have* precise values (and also, of course, from the fact that these values are correlated).

Let us now consider Example B. Let $u_\pm$ and $v_\pm$ be the eigenvectors of $\sigma_3^{(U)}$ and $\sigma_3^{(V)}$, respectively. Here the separate conservation laws of total angular momentum and of parity force the total spin of the system $U + V$ to be zero, as in Example A. This means that the spin state vector ψ describing the ensemble $\mathscr{E}$ of composite systems $U + V$ at any time subsequent to t_0 and prior to the time t_1 of the first spin measurement is

$$\psi = 2^{-1/2}(u_+v_- - u_-v_+) \qquad (8\text{--}8)$$

When the first measurement, that of $\sigma_3^{(U)}$, for example, is made, this wave function is reduced accordingly. After that measurement the N systems V (N is the total number of systems in $\mathscr{E}$) are split into two subensembles, E_+ and E_-, each involving approximately $N/2$ systems. One of these, E_+, contains all the systems V associated with the systems U for which the result of the first measurement was positive. Subensemble E_+ is described by a ket

$$v_- \qquad (8\text{--}9)$$

as can be seen by applying the principle of wave packet reduction through an incomplete measurement (Rule 10a of Chapter 3). Alternatively (that is, without applying Rule 10a), the same result may be obtained as follows. According to the general statistical rule of quantum mechanics, Eq. (8–8) implies that the joint probability for observing $\sigma_3{}^{(U)} = +1$ *and* $\sigma_3{}^{(V)} = +1$ is zero. Thus an ensemble of $U + V$ systems for which a measurement of $\sigma_3{}^{(U)}$ gives the value $+1$ must be describable by a density matrix which gives a statistical frequency of zero for the observation $\sigma_3{}^{(V)} = +1$. The subdensity matrix which is obtained by "partial tracing" on the U variables and which refers to the corresponding ensemble of subsystems V must evidently have the same property. It is, however, a 2×2 matrix, and the only such matrix with that property is

$$\begin{pmatrix} 0 & 0 \\ 0 & 1 \end{pmatrix} \tag{8-10}$$

corresponding to a pure case described by Eq. (8–9).

Similarly, of course, the other subensemble E_- of subsystems V which results from the measurement of $\sigma_3{}^{(U)}$ is describable by the ket

$$v_+ \tag{8-11}$$

Let us now focus our attention on one particular system V, and let us consider it at some time $t_1' > t_1$ when it has not yet interacted with any measuring system. It is then a member of one of the two subensembles E_+ and E_-—E_+, for example. The measurement that has been made on $\sigma_3{}^{(U)}$ therefore makes it possible to predict with certainty the value that will be observed for $\sigma_3{}^{(V)}$ when that quantity is measured.* On the other hand, this possibility has been gained (at time t_1) without in any way perturbing system V, which was isolated from U at time t_1. Under these conditions principle (a) (the *reality principle*) informs us that an element of reality exists, which is associated with $\sigma_3{}^{(V)}$. In other words, the particular system V under consideration really *has* (quite independently of our knowledge and of our experimental program) a definite physical property which is adequately described by the fact that $\sigma_3{}^{(V)} = -1$ for this system.

Let us now consider the same system V, not only at time t_1' but during the whole time interval (t_1'', t_1'), characterized by the inequalities

$$t_0 < t_1'' < t_1 \tag{8-12}$$

and (as above)

$$t_1 < t_1' < t_2 \tag{8-13}$$

where t_2 is the time at which V interacts with the instrument that measures

*Quantum mechanics provides some information on an individual system only when this system is a member of an ensemble described by a ket that is an eigenvector of some observable quantity. This is precisely the case here.

$\sigma_3{}^{(V)}$. During that whole time interval, V remains mechanically isolated. Principle (*b*) then informs us that the evolution in time of the properties of V is exactly the same as if U did not exist, or, in other words, as if V were alone. Now, if V is isolated and alone, and if its $\sigma_3{}^{(V)}$ has a well-defined value at some time, the quantum laws of evolution tell us that it has this same well-defined value at any, earlier or later, time (the equations of motion are time reversible in this case). Thus, if principle (*b*) is true, we cannot escape the conclusion that even at time $t = t_1''$ the particular V system we are considering already had the physical property (element of physical reality) that is described by the statement "$\sigma_3{}^{(V)}$ *has* the value -1." Let it be stressed once more that, if principle (*b*) is correct, this statement should be considered as being "objectively true" in the sense that it must necessarily hold irrespective of what measurements we may decide to perform at any later time.

If we now consider again the ensemble E_+, made of such systems V, we must conclude from the above that its spin properties were at time $t = t_1''$ identical to what they are at time $t = t_1'$. In fact, the only difference between the two situations is that at time $t = t_1'$ somebody *knows* the value of $\sigma_3{}^{(V)}$ for all the members of this ensemble, whereas at time $t = t_1''$ nobody had such knowledge. A change in the knowledge of somebody who has no interaction with given systems does not, however, affect the objective physical properties of these systems. Thus, also at time $t = t_1''$, the ensemble E_+ is, in regard to its spin properties, one that can be described by the matrix (8–10) or, alternatively, by the ket (8–9). Therefore it is a pure case. Similarly, at time $t = t_1''$ the ensemble E_- is describable by the ket v_+.

Let us now consider at time t_1'' the ensemble $\mathscr{E}_+$ of the $U + V$ systems whose V systems compose E_+. From the initial condition that the total spin is zero, it is quite generally known that, for any subensemble of the total ensemble of $U + V$ systems, the measurements of the spin components of U and V along any axis must add up to zero. Indeed, this holds true in any formalism. The density matrix associated with $\mathscr{E}_+$ should therefore predict $\sigma_3{}^{(U)} = +1$ with dispersion zero, since, from the above, it must predict $\sigma_3{}^{(V)} = -1$ also with dispersion zero. As $\sigma_3{}^{(U)}$ and $\sigma_3{}^{(V)}$ together constitute a complete system of compatible observables on the $U + V$ spin systems, it follows that $\mathscr{E}_+$ is a pure case, describable by

$$u_+v_- \tag{8–14}$$

(here we have applied the theorem which states that no mixture is equivalent to a pure case). Similarly, the ensemble $\mathscr{E}_-$ of all the other $U + V$ systems is also a pure case, describable by

$$u_-v_+ \tag{8–15}$$

The total ensemble $\mathscr{E}$ of the $U + V$ systems must therefore be a (proper) mixture of these two pure cases. This conclusion, however, is most certainly false, since in this particular example it introduces a privileged direction not present in the initial conditions. Moreover, we know that $\mathscr{E}$ is itself a pure case [which can be described by (8–8)]. Thus, quite generally, we can again apply the theorem which states that no mixture exists that gives the same statistical predictions as a pure case. (The reasons why Bohr [2] can in this connection legitimately ignore that theorem are shown in Chapter 9.) In the present instance, the statistical frequency with which simultaneous measurements of $\sigma_1^{(U)}$ and of $\sigma_1^{(V)}$ are both expected to give $+1$, for example, is zero according to (8–8), whereas it would be $\frac{1}{4}$ according to the present mixture.

The conclusion is that a systematic application of principles (a) and (b) leads to a contradiction with the predictions of quantum mechanics. Let it be noted that in the argumentation above, although the concept of an element of reality *is* applied in several places to single systems, the laws of quantum mechanics are applied either to *ensembles* of systems or to individual systems, but in the latter case only when the consequence that is drawn is a certitude according to these laws.

Finally, Example C is a straightforward transposition of Example B. As is well known (see, e.g., Refs. [3] and [4]), the antiprotons $\bar{p}$ at rest in hydrogen associate with protons p and constitute electrically bound p$\bar{p}$ systems that eventually reach the S states. These systems subsequently annihilate via strong interactions. Some of them decay into two kaons. Because of the conservation of C, P, and hypercharge in the strong interactions, this two-kaon system is easily found to be describable by the state vector

$$\psi = \frac{K_1 K_2 - K_2 K_1}{\sqrt{2}} \equiv \frac{K\bar{K} - \bar{K}K}{\sqrt{2}} \tag{8–16}$$

where

$$K_1 = \frac{K + \bar{K}}{\sqrt{2}} \tag{8–17}$$

and

$$K_2 = \frac{K - \bar{K}}{\sqrt{2}} \tag{8–18}$$

are one-kaon states with definite values of C (and therefore also of PC). In the approximation in which PC is conserved, K_1 and K_2 are the states that decay into two-pion and three-pion systems, respectively. Equation (8–16) shows that, if one of the kaons has been observed to decay into two pions, the other kaon is known to have $PC = -1$, that is, to be a K_2. Let us again assume that principles (*a*) and (*b*) are valid. In the case where an ensemble of such pairs is considered and the kaons are selected by means of two opposite slits, the same argumentation as above then leads to a similar and similarly unavoidable conclusion: that before any decay or interaction has taken place, the ensemble is already a proper mixture of $K_1 K_2$ systems and of $K_2 K_1$ systems [the first (second) symbol in the products corresponds to the particle going through the "left" ("right") slit]. This conclusion, however, is in contradiction to the predictions from (8–16). For example, if strangeness measurements are carried out on the two kaons by means of the strong interactions, our conclusion would make us predict that the probabilities of simultaneously observing two K or two $\bar{K}$ are both $\frac{1}{4}$, whereas (8–16) predicts that they are both zero.

The specific interest of Example C is that here the experiment considered is not simply a *gedanken experiment*. It is feasible (on the other hand, it could almost be considered as having already been done, since we know quite generally that no strong interaction violates hypercharge conservation). Another example considered by Bohm and Aharonov [5], which bears on the two-photon decay of the neutral pion, leads to a similar conclusion. Therefore the provisional conclusion reached at this stage is that, in spite of its character of apparent self-evidence, principle (*b*) on page 81 must really be abandoned. In other words, before time t_1, U and V can in no way be considered as being *two* systems. In spite of the possible separation of the corresponding wave packets (moving shutters may be introduced), they are—at least as regards spin—to be considered as forming only *one* system. This is the phenomenon that we already refered to as *nonseparability*.

8.3   RELATIVISTIC ASPECTS

The interaction of quantum systems with instruments, and the corresponding reduction of the state vector, raise interesting problems when viewed in conjunction with relativity. These problems are most easily investigated by using special examples in which the nonseparability of distant components of a system play the most prominent part.

Anticipating a little the content of Chapter 14, let us describe the initial states of the instruments A and B with which U and V interact by the state vectors ϕ_0 and χ_0, respectively. Then an interaction of U with A taking place at time t_1 can be symbolized by

$$\phi_0 u_\pm \rightarrow \phi_\pm u_\pm \tag{8–19}$$

where the $\phi_\pm$ describe instrument A after the interaction has taken place. Let us note parenthetically that (8–19) leads to a kind of phenomenological substitute for the reduction of the wave packet. To see this, let us first generalize (8-19) by considering another instrument A', identical to A in its structure and function, with which U interacts at time t_2. Let A' be described by ϕ'. The two successive interactions are then described together by

$$\phi'_0\phi_0 u_\pm \underset{t=t_1}{\rightarrow} \phi'_0\phi_\pm u_\pm \underset{t-t_2}{\rightarrow} \phi'_\pm \phi_\pm u_\pm \tag{8–20}$$

a system of relations which, by superposition, give (α_+, α_- being complex numbers).

$$\phi'_0\phi_0(\alpha_+ u_+ + \alpha_- u_-)$$

$$\underset{t=t_1}{\rightarrow} \phi'_0(\alpha_+\phi_+ u_+ + \alpha_-\phi_- u_-) \tag{8–21}$$

$$\underset{t=t_2}{\rightarrow} \alpha_+\phi'_+\phi_+ u_+ + \alpha_-\phi'_-\phi_- u_- \tag{8–21a}$$

Relation (8–21a) shows that, if the measurement on A gives the result $+$, the measurement on A' can give only the result $+$. Since this is also what a reduction of the wave packet taking place at time t_1 would predict, we see that, in a way, the simple description of the system-instrument interaction here considered [Eq. (8–21)] correctly reproduces the phenomena predicted by such a wave packet reduction. To be sure, this description has important limitations (they are discussed at length in Part Four). However, it is a useful one in the present context.

Closing the parenthesis, let us now introduce also system V. As above, let us first consider the following simple nonrelativistic problem. An ensemble of systems made of two parts U and V, both of which have spin $\frac{1}{2}$, is initially described, as far as the spins are concerned, by the state vector

$$u_+ v_- \tag{8–22}$$

System U interacts with instrument A at time t_1, and system V interacts with instrument B at time $t_2 > t_1$. How is the time evolution to be described? The obvious answer is the time sequence

$$\phi_0\chi_0 u_+ v_- \underset{t=t_1}{\rightarrow} \phi_+\chi_0 u_+ v_- \underset{t=t_2}{\rightarrow} \phi_+\chi_- u_+ v_- \tag{8–23}$$

If, instead, the initial state vector is

$$u_- v_+ \tag{8–24}$$

the corresponding time sequence is of course

$$\phi_0\chi_0 u_- v_+ \underset{t=t_1}{\rightarrow} \phi_-\chi_0 u_- v_+ \underset{t=t_2}{\rightarrow} \phi_-\chi_+ u_- v_+ \tag{8–25}$$

The superposition principle then implies that the solution of the same problem bearing on the initial state vector

$$2^{-1/2}(u_+v_- - u_-v_+) \tag{8-26}$$

is the sequence

$$2^{-1/2} \, \phi_0\chi_0(u_+v_- - u_-v_+)$$

$$\underset{t=t_1}{\rightarrow} \; 2^{-1/2} \, \chi_0(\phi_+u_+v_- - \phi_-u_-v_+) \tag{8-27}$$

$$\underset{t=t_2}{\rightarrow} \; 2^{-1/2}(\phi_+\chi_-u_+v_- - \phi_-\chi_+u_-v_+) \tag{8-27a}$$

Relation (8–27a) shows that if the measurement on A gives the result $+$ the measurement on B can only give the result $-$, as a wave packet reduction would predict. Thus, in analogy with (8–21), (8–27) suggests that the transition from pure case to mixture (or, in other words, the wave packet reduction) takes place at time t_1 also in regard to system V.

In other words, (8–27) suggests that system V emerges from the nonseparable whole $(U + V)$ precisely at time t_1. This seems surprising at first sight. We can, for instance, adjust automatic shutters on both sides of the source in such a way that the two wave packets which fly in opposite directions are completely separated and distant at time t_1, when the interaction of U with its instrument A takes place. Our knowledge of relativity then makes it difficult for us to believe that the wave packet reduction takes place instantaneously also at the distant place where V is. This is all the more true since we know from the foregoing discussion that the wave packet reduction should not be naively considered as a purely subjective event, describing only a decrease in our ignorance. This would be the case only if we could consistently think of V as already having a definite though unknown σ_3 value—independent of B—before $t = t_1$, and we know that we cannot. Although the relativistic quantum theory falls very much outside our general program, it is appropriate that we make a short incursion into its field in order to clear up this point.

For this purpose let us first agree to call an interaction between one of the systems U or V and the corresponding apparatus (A or B) simply an *event*. Let us then consider the case in which the two events that constitute the experiment under consideration are separated by a spacelike interval. Under these circumstances, it is possible to find a Lorentz transformation along AB (which we take as our $O\hat{y}$ axis) such that the time order of the two events is reversed. Then, to an observer O' at rest in the new frame, the experiment appears as if V reached B before U reached A. If the usual assertion is made that the transition from pure case to mixture is induced by the *first* instrument-system interaction, a question obviously arises as to which of the two interactions should here be considered as being the "first" one. If that transition (the wave packet reduction) is considered as being associated in some way with the intrinsic properties of the composite system (or of the ensemble of such systems), the temptation is great to consider that the choice of the event that induces it is determined once and for all by the intrinsic properties of that system and is therefore independent of the reference frame of the observer. By arguing along these lines, we would conclude that the wave packet reduction is produced by the instrument (or local observer) that interacts first *in the reference frame in which both instruments are at rest* and that this statement, having a general validity, also holds for O' or for any other observer. However, it is easy to show that this is not the case, and that, on the contrary, the wave packet reduction should be considered by a moving observer as being induced *either* by A or by B, according to whether the A–U interaction or the B–V interaction happens to take place first *as viewed by that observer.*

To show this, we could perform the appropriate Lorentz transformation on the state vectors appearing in Eq. (8–27) Alternatively, we can use a simple argument based on the correspondence principle and on the linearity of the law of evolution.

The latter procedure is grounded on the observation that the two successive processes described quantum-mechanically by Eq. (8–23) can be given a classical description in terms of observables of U, V, A, and B, having at each time definite values. In that description, the observable associated with A (and described by ϕ) changes first, and the observable associated with B (and described by χ) does not change until later. This, however, is true in the frame where both A and B are at rest but not in other reference frames. In other words, for O', the classical description of the same overall phenomenon implies that the variable associated with B is the one which is modified first. Because of the correspondence principle,* the quantum description that is valid for O' must reproduce this feature of the classical description. For O' the correct quantum description must therefore be one that can be symbolized, at least qualitatively, by

$$\phi_0\chi_0 u_+ v_- \xrightarrow[t'=t'_2]{} \phi_0\chi_- u_+ v_- \xrightarrow[t'=t'_1]{} \phi_+\chi_- u_+ v_- \tag{8-28}$$

A similar remark holds, of course, for Eq. (8–25). Applying these results and the linearity of the laws of evolution, we then immediately deduce the time evolution of any linear superposition of (8–23) and (8–25). For O' the sequence of Eq. (8–27) should thus be replaced by

---

*As a matter of fact, we use here a "natural" extension of the correspondence principle that can be formulated as follows: "If a classical description of a quantum phenomenon can be found in a given reference frame, then the transformed description must be correct in the transformed reference frame."

$$2^{-1/2}\,\phi_0\chi_0(u_+v_- - u_-v_+)$$

$$\xrightarrow[t'=t'_2]{}\ 2^{-1/2}\,\phi_0(\chi_-u_+v_- - \chi_+u_-v_+) \qquad\qquad (8\text{--}29)$$

$$\xrightarrow[t'=t'_1]{}\ 2^{-1/2}\,(\phi_+\chi_-u_+v_- - \phi_-\chi_+u_-v_+) \qquad\qquad (8\text{--}29a)$$

where t'_1 and t'_2 are the Lorentz transformations of t_1 and t_2, respectively. In the wave packet reduction language (8–29a) shows that, to the extent that we could replace (8–27) by a mixture, we can similarly replace (8–29) by a mixture. However, it is then apparent that for O' the reduction of the wave packet is induced at time t'_2 and therefore by instrument B.

The considerations of Section 8.2 showed that, when the usual assumption governing quantum mechanics, namely, Assumption Q, is made, the wave packet reduction cannot be interpreted as a purely subjective phenomenon, that is, as a mere increase in our knowledge. *The conclusion we have just reached shows that, on the other hand, we should not go too far in the other direction, that is, in interpreting the reduction process as a physical one in a narrow and naive sense, for if, in a spatially extended system (such as $U + V$) a physical process originates from one region of that system, this is an objective fact that cannot depend on the frame of reference that *we use* in considering the system (physical signals follow timelike trajectories; hence their direction of propagation cannot be reversed by a change in the reference frame). As we shall see, this is a powerful argument against an interpretation of the theory that is sometimes called the "macro-objectivist" interpretation. The latter consists in asserting that a microsystem which is initially in a superposition of, for example, two states goes into one or the other through a particular physical process governed by mere chance, when it interacts with some special macrosystems such as instruments. If that interpretation were true, then in a given experimental arrangement the physical process in question would in general be associated with *one* well-defined instrument (i.e., the "first" one with which the microsystem interacts). Again, this "objective" association could in no way depend on the reference frame from which *we* look at the whole system. But, as we have seen, precisely the reverse is true.

More generally, the puzzle with which we have to struggle is constituted by the fact that, since the wave function (nonrelativistic or relativistic) is a nonlocal entity, its collapse is a nonlocal phenomenon. According to the formalism (both nonrelativistic *and* relativistic), this phenomenon propagates instantaneously. In that sense we may say that the wave packet reduction is a noncovariant process. Again, this would create no difficulty if, like the reduction of probabilities in classical phenomena, this collapse were of a purely

---

*The role of Assumption Q (Section 4.2) in these considerations is made explicit in Section 11.2.

subjective nature. But we have seen quite strong arguments in favor of the thesis that it is not.

One possibility for escaping such difficulties would seem at first sight to remain open. This is to assume that, when relativistic effects become important, the physical process which, in nonrelativistic quantum mechanics, is associated with wave packet reduction must be slightly dissociated from it. Although, considered as a mathematical artifice, the wave packet reduction is instantaneous, the corresponding physical process would propagate with light velocity from the place where U interacts with A to the place where V is. Unfortunately, the experimental fact that correlations are observed between the two gamma rays that are produced in π^0 decay disproves such an assumption.

Problems connected with relativity will not be considered in subsequent chapters, and the question raised by quantum measurement theories will be investigated solely in the light of the conditions imposed by the rules of ordinary quantum mechanics and by macroscopic experience. However, we should always keep in mind that important limitations on the possible description schemes are imposed quite independently by relativity requirements.

Note added to the special edition: The foregoing discussion rests on a (partial) equivalence between the predictions from eqs. such as (8–21) and (8–27) and those from a wave packet reduction. For a more thorough examination (leading to essentially equivalent conclusions) see Y. Aharonov and D.Z. Albert, Phys. Rev. *D21* 3316 (1980) and *D24* 359 (1981).

REFERENCES

[1] A. Einstein, B. Podolsky, and N. Rosen, *Phys. Rev.* **47**, 777 (1935).
[2] N. Bohr, *Phys. Rev.* **48**, 696 (1935).
[3] M. Schwartz, *Phys. Rev. Lett.* **6**, 556 (1961); B. d'Espagnat, *Nuovo Cimento* **20**, 1217 (1961).
[4] R. Armenteros et al., in *Proceedings of the 1962 International Conference on High Energy Physics at CERN,* J. Prentki, Ed., CERN, Geneva, 1962.
[5] D. Bohm and Y. Aharonov, *Phys. Rev.* **108**, 1070 (1957).

Further Reading

P. A. Schilpp, Ed., *Albert Einstein: Philosopher, Scientist,* Library of Living Philosophers, Evanston, Ill., 1949.

Y. Aharonov and D. Bohm, *Nuovo Cimento* **17**, 904 (1960).

W. H. Furry, *Phys. Rev.* **49**, 393, 475 (1936).

E. Schrödinger, Die Naturwissenschaften **23**, 807, 823, 844 (1935).

CHAPTER 9

POSSIBLE VIEWS ON NONSEPARABILITY

Let us first examine a suggestion that is somewhat too simple. It is asserted from time to time that the solution to the Einstein-Podolsky-Rosen problem is really entirely trivial. The defenders of this view assert that the apparent paradox stems entirely from the fact that the state vector is unduly regarded as an objective property of the system. They argue that if the state vector *is* objective its abrupt change, at the place where V is and at the precise moment when a measurement is made on some other, distant system U, is paradoxical indeed. However, they claim that nothing compels us to regard the state vector as being objective. They then point out that if it is *not* (i.e., if its nature is simply that of a piece of *information* about the system) a measurement made somewhere can quite conceivably change it anywhere else: there is no paradox in this (any more, indeed, than there is in the abrupt changes of classical probabilities under similar circumstances). These advocates conclude that the Einstein-Podolsky-Rosen problem is therefore a false one, and that it could arise only from the unjustified idea of some people that state vectors are objective properties of the systems they represent.

The remark may seem pertinent, but the conclusion does not follow. This becomes obvious if it is merely observed that, as a matter of fact, no serious description of the Einstein-Podolsky-Rosen problem uses the incriminated *postulate* that the state vector is objective. Indeed these descriptions are, as we saw, essentially based on Assumption Q. As soon as this assumption is made, the nonseparability effects follow, as we have seen, not from any a priori assumption bearing on a possible correspondence of the state vector with "Reality" but—more convincingly—from a straightforward application of the *rules of calculation* of quantum mechanics to some specific problems. Such effects cannot therefore be eliminated from the theory unless these rules are themselves changed.

This leads us to consider an altogether different, though related question: "Does the hypothesis that hidden variables do not exist *imply* in some way or other that the state vector is objective?"

This new question is a delicate one. It has a semantic aspect, of course, since its very formulation implies that the word "objective" has been defined. But since the definition of that word is necessarily formulated differently by the supporters of various epistemological doctrines, what this semantic aspect really reflects is a deep and, indeed, as yet unsolved issue in all our natural philosophy. Very roughly speaking, this issue bears on realism versus positivism. And we should probably not expect to gain any clear-cut answer to the question just formulated before we have made our choice between these two conflicting viewpoints, for the choice partially determines the answer.

The most straightforward case is that of *realism* (described in some detail in Chapter 19). More particularly, let us here focus our attention on a special version of realism. This version incorporates a causality principle, which is what essentially matters in this discussion; it is formulated as follows. "At least *some* physical systems exist that have intrinsic properties. Here 'intrinsic properties' means properties that are fully independent of what an experimentalist will do with the system at later times. In particular, an element S of an ensemble that, for some reason or other, can be described by a state vector is such a system."

In the realm of that philosophy the intrinsic properties that characterize in an absolute way (i.e., independently of what measurements are to be made at later times) the system S are (i) the complete set of observables A, B, . . ., F that determine the state vector up to a phase, and (ii) possibly, some supplementary quantities. Moreover, as shown in Section 7.5, in the case in which all the Hermitean operators of the system correspond to observables the latter quantities can only be hidden variables. In the case just mentioned, if hidden variables are assumed not to exist, it can properly be said that all the physical properties of the elements of an ensemble which can be described by a state vector are identical from one element to the next. It is thus apparent that the state vector has here a function which is altogether different from the functions of the probability densities or of the statistical frequencies of classical physics. In classical physics the statistical frequencies refer to ensembles of systems some properties of which (e.g., velocities) differ from one system to the next one. Similarly, the probability distributions refer, at least in part, to our *knowledge* of the systems. In the present formulation of quantum mechanics the state vector refers, on the contrary, to an ensemble of systems whose properties are all the same by assumption. Moreover the corresponding *ray* (the state vector up to a phase) cannot be changed without inducing a change in the physical quantities.

If we now call *objective* the intrinsic physical quantities, then we must say that the physical quantities A, B, . . ., F are objective; and, since the ray is in a one-to-one correspondence with a set of values of A, B, . . ., F, we must also say that the ray is objective.

Realism, however, is not universally accepted. Moreover, more elaborate forms of realism could conceivably be invented that would not embody the particular causality principle stated above. Then the absence of hidden variables does *not* imply that the state vector (or, more precisely, the ray) is an objective quantity. Let us show this for Bohr's epistemology, where the situation is very simple [1]. Bohr's views are discussed in greater details in Chapter 21, but here we need only the central point in Bohr's answer to Einstein et al. In substance, this is that strictly speaking a microsystem has *no* properties that are intrinsic in any "normal" sense of the word. According to this author, this statement is true even when the system is an element of an ensemble that can be described by a state vector. Only the composite system composed of the microscopic system *S plus* the instrument *A* that is prepared so as to serve for a measurement on *S* possesses definite properties of its own. (See also Ref. [2].) Under these conditions it is clear that the ray describing the ensemble to which *S* belongs cannot be uniquely associated with any objective reality, since it refers exclusively to observables of *S* alone.

necessarily very different when Bohr's viewpoint is chosen from what it is when the views of some even moderate realism are adhered to. In the latter case, nonseparability means, as already shown, that when two systems (two microsystems, for instance) have interacted in the past they should really be considered as composing but *one* system. This holds as long as no observation is made and is a consequence of the fact that no complete set of physical attributes can be thought of as being *possessed* by any one of the two systems without some quantum predictions being violated. As soon as an observation is made, however (we do not discuss here the mechanisms thereof but consider it as a whole), the two systems can again be regarded as two separate entities. This occurs instantaneously in the reference frame of any observer. If, in the time interval between two measurements, several systems interact with one another, the nonseparability of course extends to all of them.

Among these strange aspects of the theory, it is comforting, however, to see two familiar milestones still standing erect. One is the possibility of keeping a nonrelativistic *causality principle* in the technical sense of the term: no property at a time *t* is determined—or even affected—by the events that may occur thereafter. The other one is the freedom to imagine a reality that would underlie the phenomena. By reality we mean here, not something that is necessarily fully separable (as in the world of classical objects), but yet something that we may think would still exist, with its own laws and perhaps internal configurations, even if no human beings were present in the world.

In Bohr's philosophy, nonseparability is also present, but it has a very different meaning. Indeed the whole argumentation that was developed in Chapter 8 is of no value from Bohr's standpoint, simply because it is based on the explicit assumption that the properties of the systems involved—be they *U*, *V* or *U plus V*—are at a given time *t* fully independent of the measuring instruments with which all these systems will later interact. As already stressed, this is an assumption that—in effect at least—Bohr rejects. Consequently, in Example B of Chapter 8 for instance, Bohr would presumably assert that both $\sigma_3^{(U)}$ and $\sigma_3^{(V)}$ already have* perfectly definite values *before* any instrument interacts with the considered systems. In other words, he would not consider that the two systems in the quantum Example B are in any way less separable from each other than are the two systems in the classical Example A. For him, such a standpoint does not imply the contradictions with the quantum rules that have been analyzed above because, if before the first system-instrument inter-action we change our mind as to what observables we shall measure, we must also change our instruments. By so doing, however, we modify the list of the physical quantities that should be considered as having quite sharp, though unknown, values. For example, if we decide to measure $\sigma_1^{(U)}$ and $\sigma_1^{(V)}$, we must now say that these quantities, not the σ_3, have definite unknown values at a time when neither U nor V has as yet interacted with any instrument. The nonseparability of the quantum world is thus transferred, in this philosophy, from the system of the microparticles to the systems that each of these micro-objects constitutes with its instrument.

In regard to Lorentz invariance, the transition from the pure case to the mixture—the reduction of the wave packet—raises less acute problems in Bohr's philosophy than it does in other approaches; it is of course no more an objective, real fact than are the wave packets themselves, and it is in a way immediately effective as soon as the instruments have been chosen (see, how-ever, Chapter 21). On the other hand, causality is lost; the list of all the prop-erties which can legitimately be considered as having definite values and as being attached to a system U at a time t depends on the future, since it can be changed by changing the corresponding instruments at some time t' such that $t < t' < t_1$ (t_1 is the time when U interacts with its instrument). Moreover, Bohr's philosophy obviously implies some restrictions on our freedom to imagine an *independent* external reality. The reason for this is that this concep-tion endows the instruments with virtues that clearly transcend their status as mere physical systems (they define in advance properties on other systems)—indeed, it ultimately defines instruments with reference to *our desires*.

REFERENCES

[1] N. Bohr, Quantum Physics and Philosophy, in *Philosophy in Mid-Century*, R. Kliban-sky, Ed., Florence, 1958.
[2] L. Rosenfeld in *Louis de Broglie, Physicien et Penseur*, André George, Ed., Albin Michel, Paris, 1953.

---

*In the restricted sense in which we may speak *by convention* of observables *having* values on a microsystem; see Chapter 21.

Further Reading

P. K. Feyerabend, *Z. Phys.* **148**, 551 (1955).

A. Shimony, *Am. J. Phys.* **31**, 755 (1963).

N. R. Hanson, *The Concept of the Positron,* Cambridge University Press, London, 1963.

N. Bohr and L. Rosenfeld, *Kgl. Dan. Vidensk. Selsk. Mat. Phys. Medd.* **12** (8) (1933); *Phys. Rev.* **78**, 194 (1950).

B. Ferretti, *Nuovo Cimento* **12**, 558 (1954).

CHAPTER 10

INDIVIDUALS, ENSEMBLES, STATES

Is it possible to formulate quantum mechanics in terms of individual systems rather than in the language of ensembles? Can density matrices that are not projection operators also be said to describe "states" and, if so, in what sense? Do such reformulations throw new light on the nonseparability problem? These are the questions that are considered in this chapter.

10.1 QUANTUM MECHANICS OF INDIVIDUAL SYSTEMS

Most of the predictions of quantum mechanics are of a statistical nature and therefore make sense only for ensembles. To this general rule, however, an obvious exception exists. This occurs when it is known that a given system S is an element of a pure case described by a state vector:

$$|a, b, \ldots, f\rangle \qquad (10\text{--}1)$$

where the notation means that the considered state vector is an eigenvector of the complete system of compatible observables $A, B, \ldots, F$. It is then known with certainty that, if the observables $A, B, \ldots, F$ are measured on the *individual* system S, the values $a, b, \ldots, f$ will be found.

In the opinion of many physicists (see also Condition R in Section 7.4) these are just, by definition, the conditions that make it legitimate to assert that the individual system *has* the properties in question. Let us follow this line; that is, let us consider that the system S *has* the value a for property A, etc. We can then associate a state vector $|a, b, \ldots, f\rangle$ with this individual system. As for the observables $L, M, \ldots$ that do not commute with $A, B, \ldots, F$, we simply consider them as indeterminate on S. (This noncommittal statement enables us to avoid using a disputed concept, that of the probability of an isolated event.) In order to reconstruct the full axiomatics of quantum mechanics on that basis, we still need a supplementary proposition that should lead to the usual frequency rule. Finkelstein [1] and Hartle [2] have independently pointed out that this proposition can also be formulated in a nonstatistical way.

The idea is to consider a system Σ composed of a large number N of noninteracting systems S that all correspond with one and the same normalized state vector $|\psi\rangle$, and to introduce new observables L', M', ... defined on Σ. These new observables are also properties of an *individual* system, namely, Σ, but in the limit of large N they are at the same time, by definition, identical to the mean values of L, M, ... on Σ when Σ is considered as an ensemble. Of course, Σ is described by

$$|\Psi\rangle = |\psi, 1\rangle \, |\psi, 2\rangle \ldots |\psi, N\rangle \qquad (10\text{--}2)$$

where $|\psi, n\rangle$ is a replica of $|\psi\rangle$ in the Hilbert space of the nth system S. As for the assumption that here replaces the usual statistical postulates, it requires that the observables L', M', ... correspond to the operators

$$L' = N^{-1} \sum_{n=1}^{N} L_n; \qquad M' = N^{-1} \sum_{n=1}^{N} M_n \ldots \qquad (10\text{--}3)$$

where L_n, M_n, ... are nothing more than the operators L, M, ... when operating on the ket vectors of the Hilbert space of the nth system S. Let us then define

$$\Delta_L = \|(L' - l')|\Psi\rangle\| \qquad (10\text{--}4)$$

where l' is a number, and *let us assume that $\Delta_L \to 0$ when $N \to \infty$.* Since

$$(\Delta_L)^2 = \frac{1}{N} \langle \psi|(L - l')^2|\psi\rangle$$

$$+ \frac{N-1}{N} [\langle \psi|L|\psi\rangle - l']^2 \qquad (10\text{--}5)$$

this requires that

$$l' = \langle \psi|L|\psi\rangle = \bar{L} \qquad (10\text{--}6)$$

because of the presence of the second bracket in Eq. (10–5). Conversely, if (10–6) holds, Eq. (10–5) shows that $\Delta_L \to 0$ as $\sigma_L/N^{1/2}$, where σ_L is the root mean square of $L - l'$ in the conventional formulation. Equation (10–4) then shows that in the limit $N \to \infty$ the observable L' will, upon measurement, be found with certitude to have the value (10–6) on the composite system Σ. If Σ is considered as a statistical ensemble, this is just the mean value rule, from which, as we know, the rule for the statistical frequencies of the results of

prospective measurements is easily derived. This shows that a consistent formulation of quantum mechanics can be given in which state vectors are associated with *individual* systems.

We do *not* make use of this formulation in the critical discussions of the
measurement theories that appear in the following chapters. We shall, however, find it useful to refer to it occasionally when describing new proposals.

What has been shown above is the following. If a physical system S is such
that a complete system of compatible observables can be said to have definite
values on S (so that a ket $|\psi\rangle$ can be associated with S), and if, moreover, this
system is associated with many other systems that are identical to it in this
respect, then the association of each of these systems with $|\psi\rangle$ lead to the usual
statistical formulas, provided only that, with definition (10–4), $\Delta_L \to 0$ when
$N \to \infty$.

As we know, it is not the case that *any* isolated system can be associated with a
state vector. Indeed, if such an association were separately made in regard to each
of the systems U and V of Example B of Chapter 8, the predictions that would
follow would be those which (in the conventional formulation) result from
describing the ensemble of the systems $\Sigma = U + V$ as a mixture, not as a pure
case. In other words, in the nonstatistical formulation we must say that there are
systems—such as *a* system U or *a* system V—*which cannot be associated with any
state vector.* By generalizing the procedure described above, it is however possible
to associate a density matrix with each such individual system [2]. This matrix
is, of course, the one which is obtained by partial tracing from the density
matrix that describes Σ. In the considered example, it is thus proportional to the
unit 2×2 matrix.

The formalism presently considered has some similarity with classical
physics in that it is able to describe *single* systems by means of suitable mathematical entities. In connection with it, however, two questions should be
considered. The first one is: "To what extent is such a description complete?";
and the second one (which in a way is a restatement of the first) is: "Is that
description objective?"

To the first question, we can immediately answer that for systems which
have no state vector a description by means of a density matrix is quite obviously not complete. For example, let us assume that in Example B of Chapter
8 we know only the exact density matrices of system U and of system V. Since,
as a matter of fact, both these matrices are proportional to the unit matrix, in
this case we obviously cannot derive from that knowledge any correlation
between prospective measurements of $\sigma_3{}^{(U)}$ and $\sigma_3{}^{(V)}$. Upon measurement,
however, such correlations are found.

In regard to the second question, it is appropriate to observe that, if we
work inside the framework of an even moderate realism, that is, if we define
objective as "corresponding to properties truly possessed by the system,"
then—as pointed out in Chapter 9—either hidden variables exist or a descrip-

tion of a system by means of a ray *is* objective. In both cases, nonseparability effects are unavoidably present. In the case of the hidden variables theories, this is shown in Chapter 11. In the other case these effects result from the fact that in Example B of Chapter 8, for instance, an operation of measurement made on U gives to V (which is far away) a state vector that it did not previously have. If we tried to deny nonseparability by arguing [2] that the state vectors are not objective and that the system V has a definite though unknown value for $\sigma_3{}^{(V)}$ even before any measurement is made on U, then either we would *ipso facto* introduce hidden variables (and thus incur the difficulties analyzed in Chapter 11) or we would have to accept a nonseparability of the kind considered by Bohr (i.e., a nonseparability between object and instrument). Thus, inside the framework of the present formalism, just as in the or a standpoint that is closer to Bohr's doctrine. In the first case, we must either say that the state vectors are objective entities (even those associated with composite, extended systems such as the above $\Sigma = U + V$) or believe in the existence of hidden variables. In the second case, the state vector is definitely not objective. In either case, however, surprising instantaneous but distant effects are unavoidably present. They are of the types discussed at length in Chapters 8, 9, and 11. In other words, the formulation of quantum mechanics in terms of individual systems and its formulation in terms of ensembles are entirely equivalent in that respect.

Hope was expressed [3] that the axiomatics based on the quantum calculus of propositions might open the way toward a general definition of the quantum state of an individual system. However, it was soon recognized [4] that this leads to separability and thereby to violation of some of the quantum-mechanical predictions.

10.2   THE QUANTUM STATES

In contemporary literature, the concept of *state* is not wholly unambiguous. Some authors associate it with the notion of an individual system and say, for instance, that a given system is in a given state. Others associate it exclusively with ensembles. The situation is further complicated by the fact that for some theorists the concept of state refers only to ensembles (or individuals) that can be described by some state vector, whereas for other authors it refers just as well to ensembles (or individuals) that cannot be associated with any state vector and for which only a density matrix description is therefore acceptable. For the latter authors a density matrix thus describes a state, while for the former ones it describes at best a *mixture* of states. Any definition is of course acceptable. Nevertheless, because of the possible misinterpretations that could originate from preconceived ideas the use of the word "state" was avoided up to this point. In what follows it is used only occasionally, in

circumstances where its meaning follows from the context. A few supplementary comments on the possible definitions of this notion may, however,
be of some use.

First of all, as has been repeatedly pointed out, if a density matrix p is not a
projection operator, not *one* but an infinity of proper mixtures that have p as
their density matrix can be thought of. Are these proper mixtures equivalent
to each other? Quite obviously the definition of a state of a system is largely
dependent on the answer to this question. Just as obviously, however, such an
answer depends a great deal on exactly what we mean when we assert that two
ensembles are equivalent.

To investigate this point further, let us consider three methods that can
be used for preparing an unpolarized beam of spin-$\frac{1}{2}$ particles. Method I
consists in mixing, by means of suitable magnets, two completely polarized
beams, one along the $O\hat{z}$ direction, the other one along the opposite direction.
Method II is identical to Method I except that the $\pm O\hat{x}$ directions replace
the $\pm O\hat{z}$ directions. Finally, Method III consists in constructing a beam with
the V particles of Example B of Chapter 8 (i.e., the V particles are decay
products of spin-zero particles that decay into two spin-$\frac{1}{2}$ particles through
a spin-conserving interaction). Considered as ensembles, these three beams
have the same density matrix:

$$p = \tfrac{1}{2}\begin{pmatrix} 1 & 0 \\ 0 & 1 \end{pmatrix}$$

Consequently, a mathematician can of course decide that, for this reason,
these three beams should be considered as *equivalent by definition*. These
beams are also physical systems, however. Ensembles of them can, in turn, be
prepared, and they can, therefore, be subjected to statistical measurements.
Now the list of all the experiments that *can* be made on given physical systems
does not depend on the free will of the theorist. In the present case, for instance, the fluctuations of the values of

$$\Sigma_z = \sum_{n=1}^{N} (\sigma_z)_n$$

can be experimentally measured on the beams prepared by any of the three
methods [$(\sigma_z)_n$ and Σ_z are twice the components along $O\hat{z}$ of the spin of the
n^{th} particle and of the total spin, respectively]. When the number N of particles
in each beam becomes very large, these fluctuations are characterized by the
standard deviations

$$\sigma_I = 0, \qquad \sigma_{II} = \sigma_{III} = \tfrac{1}{2}\sqrt{N}$$

so that the beams prepared by Method I can easily be distinguished from the other ones by suitable experiments. Of course, interchanging the roles of $O\hat{z}$ and $O\hat{x}$ then provides a method for distinguishing Case II from the other two. Similarly, it should be recalled here that correlation measurements between the U and V systems also offer in principle unambiguous tests for Case III.

Under these conditions, there can be no doubt that, quite independently of what we call "ensembles" (this can be a matter of convention) at any rate, the beams themselves prepared by Methods I to III are objectively different from each other. This statement proceeds from considering these beams as physical systems rather than as abstract entities, but, after all, this is the point of view that is physically significant.

Returning now to the problem of defining the concept of a quantum state, we see that it is largely, although not wholly, a matter of free convention. If, for instance, we decide that a density matrix ρ is a representative of a state, then, in the case in which $\rho^2 \neq \rho$, we must at the same time remember that, according to this definition, two physical systems S and T of the same type can be in the same "state" even when a set of identical replicas of S and a set of identical replicas of T differ with respect to some observables. As a matter of fact, the association of the word "state" with a given density matrix does not even seem to be markedly appropriate in the case of *proper* mixtures.

REFERENCES

[1] D. Finkelstein, *Trans. N. Y. Acal. Sci.* **25,** 621 (1962–63).
[2] J. B. Hartle, *Am. J. Phys.* **36,** 704 (1968).
[3] J. M. Jauch and C. Piron, *Helv. Phys. Acta* **42,** 842 (1969).
[4] J. M. Jauch in *Foundations of Quantum Mechanics, Proceedings of the Enrico Fermi International Summer School,* Course IL, Academic Press, New York, 1971.

CHAPTER 11

THE "HIDDEN VARIABLES" APPROACH

The results described in Chapter 8 easily give the impression that any theory whatsoever which, in all circumstances, attributes definite properties to non-interacting physical objects necessarily violates *some* of the predictions of quantum mechanics. The very existence of the hidden variables theories should, however, prompt us to have a second look at the problem. As already mentioned, Bohm [1], for example, has constructed a deterministic theory of this type, in which any physical object always *has* definite properties of its own. This theory, moreover, exactly reproduces the quantum-mechanical predictions when the values of the hidden variables are averaged over. It therefore constitutes a counterexample to the statement formulated above.

We must first of all understand why the existence of such a theory is possible at all. Then we must also inquire whether or not some "price has to be paid" for such an achievement. It is shown below that the price is indeed a rather steep one. The theory necessarily implies the existence of a violent nonlocality effect which, as a matter of fact, may reasonably be considered as similar to—and even more surprising than—the nonseparability of the orthodox theory.

11.1 THE EXISTENCE PROBLEM

This problem must first be examined in a very general context, since originally it was formulated quite independently of the difficulty raised by Einstein, Podolsky, and Rosen. In classical statistical mechanics the statistical character of the theory arises from the fact that the macroscopic states are averages over better-defined states, for which the results of all possible measurements are assumed to be determined exactly. The question was whether the statistical character of quantum mechanics could be thought of as proceeding from a similar cause. In other words, it was asked whether the quantum states (ensembles) could be considered as averages over states for which the results of all possible measurements would be entirely determined. These hypothetical *dispersion-free* states would be specified, not only by the state vector but also by additional variables, called "hidden" since we cannot actually *prepare* ensembles corresponding to such states (otherwise quantum mechanics would be violated).

For a long time it was thought that the answer to the question raised was already known and was negative. This opinion was based on a theorem due to von Neumann [2], which shows that the very idea that dispersion-free ensembles could exist cannot be reconciled with the general principles of quantum mechanics. Von Neumann's proof made use only of general assumptions such as the following one.

Linearity assumption: Any real linear combination

$$C = \alpha A + \beta B$$

of any two Hermitean operators A and B represents an observable, and the expectation value $\bar{C}$ of C is the linear combination

$$\alpha \bar{A} + \beta \bar{B}$$

of the expectation values of A and B.

If we consider types of systems that involve no superselection rules (see Section 7.1), the statement about C expressed in the *linearity assumption* above is in fact necessarily correct for any ensemble described by a state vector or by a density matrix. However, as Bell [3] pointed out, nothing compels us to extend the validity of the linearity assumption to the hypothetical dispersion-free ensembles, which cannot be produced physically.* Von Neumann's proof, however, requires this extension. Hence we must conclude that this proof is not sufficient to rule out the possibility of a hidden variables description of quantum mechanics.

To illustrate this point let us outline an unrealistic but elementary hidden variables model [4, 5] that, in its limited field of application, *does* indeed reproduce *all* the quantum predictions when the hidden variables are suitably averaged over. The model bears on the spin of a spin-$\frac{1}{2}$ particle. It describes this spin by means of two unit vectors, **q** and **λ**. It states that, if an individual particle is made to interact with a Stern-Gerlach measurement device whose field gradient is oriented along the unit vector **a**, then:

(*a*) the result of the measurement is equal to

$$\tfrac{1}{2} \operatorname{sign}(\boldsymbol{\lambda} \cdot \mathbf{a}')$$

where "sign x" means the sign of the quantity x, and the unit vector **a'** is

---

*For these ensembles the expectation values coincide, of course, with *the* observed value.

obtained by rotating **a** toward **q** in the plane (**q**, **a**) and in such a way that its angle θ' with **q** is equal to

$$\theta' = \pi \sin^2 \frac{\theta}{2}$$

(in the latter expression θ is the angle between **a** and **q**); and

(*b*) both λ and **q** are changed. For instance, let us consider the case of an apparatus that lets pass through, only particles for which the measurement we are considering gives the result $+\frac{1}{2}$. Then the new vector **q** coincides with **a**. As for the new vector λ, it is obtained from the initial one by a rotation about the **q** × **a** axis, keeping the polar angle fixed, to a new azimuthal angle given by

$$\varphi' = \frac{\varphi}{1 - \theta'/\pi} + \theta - \frac{\theta'}{2(1 - \theta'/\pi)} \qquad (11\text{–}1)$$

Let then, a measurement of the spin component along **a** be made on an ensemble of N systems whose vectors **q** (identifiable with the polarization direction) all coincide and whose vectors λ are distributed uniformly on the hemisphere ($\lambda \cdot$**q**) ≥ 0. As is easily verified, the apparatus then lets a number of particles equal to

$$N \cos^2 \frac{\theta}{2} \qquad (11\text{–}2)$$

pass through, and the resulting ensemble is composed of particles whose vectors **q** coincide with **a** and whose vectors λ are distributed uniformly on the hemisphere ($\lambda \cdot$**a**) ≥ 0. These results guarantee that, if the initial distribution of the vectors **q** and λ is as described, the quantum-mechanical predictions for any succession of measurements of the spin components along arbitrary directions are necessarily satisfied in spite of the fact that the model involves hidden variables.

For completeness, the following fact must nevertheless be mentioned. In the case in which the number of dimensions of the Hilbert space associated (in quantum mechanics) with a system S is larger than two, a restriction holds in regard to the classes of the hidden variables theories which can describe S without contradicting the verifiable predictions of quantum mechanics. As shown, for example, in Ref. [3], it is a consequence of a theorem due to Gleason that, in such a case, the result of a measurement of an observable A must be allowed to depend, not only on the values of the hidden variables but also on which other variables compatible with A are measured along with A—if any. For this reason, it is sometimes said that such theories are "contextualistic." [6] That property of the acceptable hidden variables models already makes them less similar to the classical descriptions than we might like them to be. But—as shown in Section 11.4—they have in that respect also other features (related to contextuality but more specific) that are even more surprising. For that reason the contextualistic aspects of the hidden variables theories are not studied here. The interested reader can find an extensive analysis of them in Belinfante [6], for instance, where further references can be found.

11.2 HIDDEN VARIABLES AND THE SEPARABILITY PROBLEM

In Chapter 8 a proof of nonseparability was given in terms of a specific example (Example B). The quantum-mechanical predictions for the spin components (including their correlations) along two orthogonal directions are sufficient ingredients in that proof. In other words, to obtain the desired result it is not necessary to consider the predictions along any other direction. Since our present purpose is to understand why *this* proof does not apply to a theory with hidden variables, it is therefore sufficient to consider a simple hidden variables theory of spin $\frac{1}{2}$, which exactly reproduces the quantum-mechanical predictions *along two orthogonal axes*. A theory of this kind is obtained if such a spin is described by means of one unit vector, λ. A measurement of a spin component, $\sigma_3/2$, along a given axis, which we take as our third axis, is then assumed to give either the result $+\frac{1}{2}$ or the result $-\frac{1}{2}$, according to whether λ_3 is positive or negative. A measurement of σ_3 on an ensemble E of systems whose spin vectors λ are initially isotropically distributed separates E into two subensembles, E_+ and E_-. In E_+, for example, λ is uniformly distributed on the upper hemisphere (see Figure 11–1) so that the model predicts a probability $+1$ for obtaining the result $\sigma_3 = +1$ if σ_3 is measured on E_+,

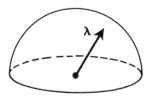

Figure 11–1.   An element of E_+.

and a probability 1/2 for obtaining the result $\sigma_1 = +1$ if it is σ_1 that is measured. Thus, in regard to the measurements along axes 1 and 3, the predictions from E_+ coincide exactly with those from the quantum ensemble described by the state vector u_+.

When this theory is used, the two-particle system studied in Chapter 8 coincides with the classical Example A of that Chapter. The reality and separability principles are then trivially satisfied, as already shown. Moreover, it is clear that, in that model, *even before any measurement is made*, the ensemble $\mathscr{E}$ of all the systems $U + V$ can be thought of as constituted of two ensembles, $\mathscr{E}_+$ and $\mathscr{E}_-$, where $\mathscr{E}_+$ ($\mathscr{E}_-$) is the ensemble of all the systems $U + V$ whose constituents "U systems" have a spin vector $\boldsymbol{\lambda}^{(U)}$ with a positive (negative) third component. Correlatively (since $\boldsymbol{\lambda}^{(U)} = -\boldsymbol{\lambda}^{(V)}$) the constituent "$V$ systems" of the elements of $\mathscr{E}_+$ have $\lambda_3^{(V)} < 0$ (see Figure 11–2). If E_+ is the ensemble of the systems U that are present in $\mathscr{E}_+$, then the $\boldsymbol{\lambda}^{(U)}$ in E_+ are isotropically distributed in the upper hemisphere. In regard to the physical quantities σ_1 and σ_3, in which we are interested, the predictions from E_+ therefore coincide with those from the quantum ensemble u_+. The same holds true, of course, for the corresponding ensemble $\hat{E}_-$ of V particles, whose predictions coincide with those of v_-.

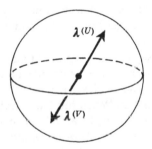

Figure 11–2.   An element of $\mathscr{E}_+$.

Now the essential point is that, in spite of the fact that E_+ thus "coincides" with u_+ and $\hat{E}_-$ with v_- (in the restricted sense considered here), $\mathscr{E}_+$ does *not* coincide, even in this sense (i.e., in regard to the possible measurements to be made on components 1 and 3 of $\sigma^{(U)}$ and $\sigma^{(V)}$), with u_+v_-. The reason for this fact is, in the present model, elementary and obvious (it holds also in more difficult cases). It proceeds from the fact that, although $\boldsymbol{\lambda}^{(U)}$ is distributed at random in E_+, as well as $\boldsymbol{\lambda}^{(V)}$ in $\hat{E}_-$, the variables $\boldsymbol{\lambda}^{(U)}$, $\boldsymbol{\lambda}^{(V)}$ are *not* distributed completely at random in $\mathscr{E}_+$, since the correlation $\boldsymbol{\lambda}^{(U)} = -\boldsymbol{\lambda}^{(V)}$ is always present. The predictions of u_+v_- which bear on the quantities that interest us

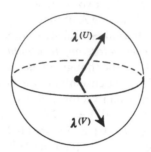

Figure 11–3.   An element of $\mathscr{E}'_+$.

are reproduced exactly in this hidden variables theory, not by $\mathscr{E}_+$, but by another ensemble $\mathscr{E}'_+$, which is similar to $\mathscr{E}_+$ except for the decisive fact that $\lambda^{(U)}$ and $\lambda^{(V)}$ are completely uncorrelated within each constituent (Figure 11–3). This is the reason why the ensemble $\mathscr{E}$, although it *is* a proper mixture of $\mathscr{E}_+$ and $\mathscr{E}_-$, still gives, in regard to those quantities, the same predictions as the quantum pure state $(u_+v_- - u_-v_+)/\sqrt{2}$, instead of giving those that would emerge from a quantum mixture of u_+v_- and u_-v_+, as one would naively expect.* A straightforward extension of the model described in Section 11.1 leads to similar conclusions.

Summarizing, we may say that the argument against separability developed in Chapter 8 is based not only on the assumption that a few particular statistical predictions of quantum mechanics are correct but also, as already stressed, on Assumption Q (Section 4.2). When hidden variables are supposed to exist, such an assumption cannot be maintained, since $\mathscr{E}_+$, for instance, is *not* a quantum ensemble. Hence the question of whether separability can hold in hidden variables theories must be made the subject of separate investigations. It is shown in the next section that nonseparability holds (because of different reasons) also in the case in which the opposite assumption—that of the existence of hidden variables—is made.

11.3   NONSEPARABILITY IN HIDDEN VARIABLES THEORIES

The simple description of a spin $\frac{1}{2}$ by means of a unit vector used above does not reproduce the predictions of orthodox quantum mechanics for an ensemble with $\sigma_3 = +1$ except in the direction of the third axis and in the

---

*In a language more intuitive than rigorous, we have here a check on a fact already pointed out, namely, that because of their internal correlations the hidden variables in $\mathscr{E}_+$ are not distributed sufficiently at random for the equivalence with the quantum ensemble u_+v_- to hold.

directions orthogonal to that axis. At the price of a slight complication, however, a model can be given which does reproduce exactly all these predictions (see Section 11.1). Similarly, models that reproduce exactly all the quantum-mechanical predictions when the hidden variables are averaged over can be constructed also for more elaborate systems. Bell [4], however, has shown that all of the hidden variable theories (such as Bohm's) which, statistically, reproduce exactly the predictions of quantum mechanics—in the sense defined in the foregoing section—have a general property in common. This property is that for *some* experiments it necessarily happens that the readings on a measurement apparatus M_1, located at a certain place, depend on the settings of another apparatus M_2, located at another, possibly very distant, place and not connected in any apparent way with the first one. The readings on M_1 can accordingly be modified—for given values of the hidden variables— simply by changing the positions of some pieces of equipment in M_2.

It is sufficient to show that this holds true on a particular example. A convenient one is again Example B of Chapter 8, that is, the example of a zero-spin particle that decays with spin conservation into two spin-$\frac{1}{2}$ particles, U and V. These particles fly far apart from each other, and what is shown is that, if an ensemble of such systems is considered, and if M_1 is a magnet that measures $(\mathbf{a} \cdot \boldsymbol{\sigma}^{(U)})$ ($\mathbf{a}$ is a unit vector that characterizes the direction of the magnet M_1), then the results of a measurement of $(\mathbf{b} \cdot \boldsymbol{\sigma}^{(V)})$ at some other, possibly quite distant place depend on the direction of $\mathbf{a}$. As Bell states, the conclusion is that "in a theory in which parameters are added to quantum mechanics to determine the results of individual measurements without changing the statistical predictions there must be a mechanism whereby the setting of one measuring device can influence the reading of another instrument, however remote." Moreover, the signal involved must propagate instantaneously, so that such a theory could not be Lorentz invariant.

We now summarize the proof of this property. Let λ be a set of hidden variables that provide a description of the system $U + V$. If U and V are assumed to propagate in a subquantic medium of some sort, the hidden variables associated with this medium should be incorporated into this set, as well as (more generally) any other hidden variable that does not depend on the orientations ($\mathbf{a}$, $\mathbf{b}$, . . .) of the instruments. Let $\rho(\lambda)$ be the probability density distribution of these λ on an ensemble that corresponds to the quantum-mechanical initial state under consideration. As above, let $\mathbf{a}(\mathbf{b})$ specify the orientation of the Stern-Gerlach magnet $M_1(M_2)$ used for measuring the corresponding spin components of $U(V)$. It is not necessary to restrict the present considerations to hidden variables theories that are fully deterministic. Moreover, we can also assume that the result of a particular measurement depends partly on the values of some hidden variables which are associated with the corresponding apparatus (and whose probability distribution depends exclusively on the local conditions). We can then argue *a contrario;* that is, we

assume that the result $A_a(B_b)$ (expressed in units of $\hbar/2$) of a measurement of
the spin component of $U(V)$ along $\mathbf{a}(\mathbf{b})$ does *not* depend on $\mathbf{b}(\mathbf{a})$, and we show
that this leads to a contradiction of the predictions of quantum mechanics.

To that end [7], let us call $P(\mathbf{a}, \mathbf{b})$ the mean value of the product A_aB_b. Under
the assumptions stated above,

$$P(\mathbf{a}, \mathbf{b}) = \int d\lambda\, \rho(\lambda)\, \bar{A}(\mathbf{a}, \lambda)\, \bar{B}(\mathbf{b}, \lambda) \tag{11–3}$$

In Eq. (11–3), $\bar{A}(\mathbf{a}, \lambda)$ $[\bar{B}(\mathbf{b}, \lambda)]$ is the expectation value of the result that
would emerge from a measurement of A_a $[B_b]$ on a system $U(V)$ characterized
by definite values of *those* hidden variables λ that do *not* depend on $\mathbf{a}$ and $\mathbf{b}$
[those that *do* depend on $\mathbf{a}$ and $\mathbf{b}$ (i.e., on the apparatuses) are already averaged
over].

With the units chosen, $A_a = \pm 1$ and $B_b = \pm 1$. Hence

$$|\bar{A}| \le 1; \qquad |\bar{B}| \le 1 \tag{11–4}$$

Let $\mathbf{a}'$, $\mathbf{b}'$ be alternative directions of the instruments. Then

$$P(\mathbf{a}, \mathbf{b}) - P(\mathbf{a}, \mathbf{b}') = \int d\lambda\, \rho(\lambda)\, [\bar{A}(\mathbf{a}, \lambda)\, \bar{B}(\mathbf{b}, \lambda) - \bar{A}(\mathbf{a}, \lambda)\, \bar{B}(\mathbf{b}'\, \lambda)]$$

$$= \int d\lambda\, \rho(\lambda)\, \bar{A}(\mathbf{a}, \lambda)\, \bar{B}(\mathbf{b}, \lambda)\, [1 \pm \bar{A}(\mathbf{a}', \lambda)\, \bar{B}(\mathbf{b}', \lambda)]$$

$$- \int d\lambda\, \rho(\lambda)\, \bar{A}(\mathbf{a}, \lambda)\, \bar{B}(\mathbf{b}', \lambda)\, [1 \pm \bar{A}(\mathbf{a}', \lambda)\, \bar{B}(\mathbf{b}, \lambda)]$$

Then, because of inequalities (11–4),

$$|P(\mathbf{a}, \mathbf{b}) - P(\mathbf{a}, \mathbf{b}')| \le \int d\lambda\, \rho(\lambda)\, [1 \pm \bar{A}(\mathbf{a}', \lambda)\, \bar{B}(\mathbf{b}', \lambda)]$$

$$+ \int d\lambda\, \rho(\lambda)\, [1 \pm \bar{A}(\mathbf{a}', \lambda)\, \bar{B}(\mathbf{b}, \lambda)]$$

or

$$|P(\mathbf{a}, \mathbf{b}) - P(\mathbf{a}, \mathbf{b}')| \le 2 \pm [P(\mathbf{a}', \mathbf{b}') + P(\mathbf{a}', \mathbf{b})] \tag{11–5}$$

It is to be noticed that inequalities (11–5) are derived here without assuming
that the total spin of the system $U + V$ is zero. Hence they are valid independently of this assumption. However, if we consider now the case in which
the total spin of $U + V$ *is* zero, then

$$P(\mathbf{a}', \mathbf{a}') = -1 \tag{11–6}$$

(this is predicted by quantum mechanics and/or by any theory that defines the angular momenta of composite systems) so that with the special choice $\mathbf{b}' = \mathbf{a}'$ inequalities (11–5) reduce to,

$$|P(\mathbf{a}, \mathbf{b}) - P(\mathbf{a}, \mathbf{b}')| \leq 1 + P(\mathbf{b}, \mathbf{b}') \qquad (11\text{–}7)$$

Inequalities (11–7) constitute the original form [4] of the Bell inequalities. The generalized form (11–5) was first derived by Clauser, Horne, Shimony, and Holt [8].

Under the assumptions already stated (essentially, that A_a should not depend on $\mathbf{b}$, and conversely), inequalities (11–7) should hold for any unit vectors $\mathbf{a}$, $\mathbf{b}$, $\mathbf{b}'$. Let us investigate whether or not quantum mechanics satisfies this condition. For this purpose, let us observe that in that theory $P(\mathbf{a}, \mathbf{b})$ is given by the formula

$$P(\mathbf{a}, \mathbf{b}) = T_r\,[M(\mathbf{a} \cdot \boldsymbol{\sigma}^{(U)})\,(\mathbf{b} \cdot \boldsymbol{\sigma}^{(V)})] \qquad (11\text{–}8)$$

where M is the density matrix of the $S = 0$ state of $U + V$. Since the eigenvalue $S = 0$ is not degenerate, M is the projection operator into this state:

$$M = 4^{-1}\,(1 - \boldsymbol{\sigma}^{(U)} \cdot \boldsymbol{\sigma}^{(V)}) \qquad (11\text{–}9)$$

Expression (11–8) is then easily calculated. It is found that

$$P(\mathbf{a}, \mathbf{b}) = -(\mathbf{a} \cdot \mathbf{b}) \qquad (11\text{–}10)$$

so that inequalities (11–7) now read as

$$|(\mathbf{a} \cdot \mathbf{b}) - (\mathbf{a} \cdot \mathbf{b}')| \leq 1 - (\mathbf{b} \cdot \mathbf{b}') \qquad (11\text{–}7a)$$

To study whether these inequalities hold for any unit vectors $\mathbf{a}$, $\mathbf{b}$, $\mathbf{b}'$, let us choose as the z axis the normal to the $(\mathbf{b}, \mathbf{b}')$ plane. Then, if θ_α, φ_α denote the spherical coordinates of a unit vector $\boldsymbol{\alpha}(\alpha = \mathbf{a}, \mathbf{b}, \text{ or } \mathbf{b}')$, (11–7a) can be rewritten as

$$|\sin\theta_a|\,|\sin(\phi_a - \frac{\phi_c + \phi_b}{2})| \leq |\sin\frac{\phi_c - \phi_b}{2}| \qquad (11\text{–}11)$$

for any values of the coordinates except $\phi_b = \phi_c$. It is clear, however, that, if $\phi_c - \phi_b$ is chosen to be small, θ_a and ϕ_a can always be chosen so that the inequality is not satisfied. The conclusion is that under the assumption that A depends only on $\mathbf{a}$ and λ and that B depends only on $\mathbf{b}$ and λ the kind of theories under consideration—whatever choice is made for the distribution $\rho(\lambda)$ of the hidden variables—cannot reproduce exactly all the predictions of conventional quantum mechanics. In fact, as can readily be inferred from the proof given above and as is shown in more detail by Bell [4], such theories cannot even reproduce these predictions with an arbitrarily small error.

This completes the proof of the statement made at the beginning of this section. In order for the deterministic* hidden variables theories to reproduce exactly the predictions of quantum mechanics for all possible experimental arrangements, they must, in particular, be such that in the arrangement here considered the result A of the measurement made on U depends not only on λ —a set of parameters describing the microscopic system—and on **a**—a parameter describing the position of the magnet used for obtaining A—but also on the other parameters in the theory, namely, **b**—a set of three parameters that describes the position of the magnet used for making measurements on V. This dependence exists in spite of the fact that these two magnets may be far apart, and it does not decrease with the distance between the magnets, since, indeed, it does not depend at all on this distance and apparently propagates instantaneously from one place to the other (see Section 11.6 for further discussion). It therefore violently contradicts a very primeval and common-sense idea which was clearly expressed by Einstein and which we may call the principle of separability.

Principle of Separability

Let $[U]$ and $[V]$ be two systems that have once interacted but are no longer interacting. Then "the real factual situation of the system $[V]$ is independent of what is done with the system $[U]$, which is spatially separated from the former" [9].

It is not our purpose to investigate in detail the hidden parameters theories, or to review the many interesting and important contributions that have been made to this problem. It may conceivably be that a modification of quantum mechanics will ultimately be found that preserves hidden parameters. Up to now, however, this is nothing but a hope, and in this book we are mainly interested in the nonmodified quantum physics, that is, in formalisms that exactly reproduce the observable predictions of conventional quantum mechanics. In this respect the proofs described above are of fundamental importance, since they show that all such formalisms that claim to reintroduce determinism by means of hidden variables have necessarily a very strange and artificial large-scale nonlocality built into them. It is true that attempts at interpreting conventional quantum mechanics in terms of an "independent reality" (to be reviewed below) also lead to very surprising consequences. What has been shown in this section is, however, that the hidden variables picture necessarily gives rise to consequences at least as artificial and unpleasant. Since, moreover, it introduces a great deal of complication into the theory, particularly when indistinguishability and particle creation are taken into

---

*Or locally stochastic. See above, before and after Eq. (11–3).

account, there seem to be good arguments, at least at the present time, in favor of the standpoint of the majority of physicists, who regard these attempts at reintroducing determinism into physics as—to say the least—unpromising.

The nonseparability of the hidden variables theory of Bohm was pointed out by Bell [3] and more recently was stressed by Bohm himself [16].

11.4 ANOTHER PROOF

Several authors have proposed other proofs of Bell's result. It is appropriate to describe here the one which is probably the simplest. It is due to Wigner [10].

Let us consider again the phenomenon studied in Section 11.3. Let the vectors $\mathbf{a}_i$ ($i = 1, 2, 3$) designate three directions in space, and let $S_i^{(U)}(S_i^{(V)})$ be the component of $S^{(U)}$ ($S^{(V)}$) along $\mathbf{a}_i$. Let us again consider an ensemble of $(U + V)$ systems whose total spin is zero. The hidden variables λ determine the future history of any single $(U + V)$ system under well-specified conditions. Hence in a deterministic theory it is meaningful to define the *number*

$$(\sigma_1, \sigma_2, \sigma_3; \tau_1, \tau_2, \tau_3) \tag{11–12}$$

as the number of systems (in the ensemble) on which a measurement of any one of the three observables $2S_i^{(U)}$ *would* lead to the result σ_i if it were carried out and on which, similarly, a measurement of any $2S_i^{(V)}$ *would* lead to the result τ_i if it were carried out *instead*. ($\sigma_i = \pm 1$, $\tau_i = \pm 1$, in normal units $\hbar = 1$). Then the number of systems on which measurements of, for instance, $2S_1^{(U)}$, $2S_2^{(U)}$, and $2S_1^{(V)}$, would give the results σ_1, σ_2, and τ_1 is obviously

$$(\sigma_1, \sigma_2, \ldots; \tau_1, \ldots) = \sum_{\sigma_3, \tau_2, \tau_3} (\sigma_1, \sigma_2, \sigma_3; \tau_1, \tau_2, \tau_3) \tag{11–13}$$

for the events corresponding to $\sigma_i = \pm 1$ (or to $\tau_i = \pm 1$) are mutually exclusive. From now on we replace by dots the σ's or the τ's that are left unspecified, and we use expressions similar to Eq. (11–13) for the corresponding symbols.

Our deterministic theory should reproduce all the verifiable predictions of quantum mechanics. One of the latter is that

$$(+1, \ldots; +1, \ldots) = 0$$

Expanding the first member of this equation as done in Eq. (11–13), we find that

$$\sum_{\sigma_2, \sigma_3, \tau_2, \tau_3} (+1, \sigma_2, \sigma_3; +1, \tau_2, \tau_3) = 0 \tag{11–14}$$

But all the terms in the sum are positive or zero, so that Eq. (11–14) implies that they are all equal to zero. By repeating this argument for other values of i, we find that all the numbers $(\sigma_1, \sigma_2, \sigma_3; \tau_1, \tau_2, \tau_3)$ for which one σ_i is equal to the corresponding τ_i are equal to zero.

Other predictions of quantum mechanics are as follows:

$$(+1 \ . \ . \ ; \ . +1 \ .) = \tfrac{1}{2} N \sin^2 \frac{\theta_{12}}{2} \tag{11–15a}$$

$$(+1 \ . \ . \ ; \ . . +1) = \tfrac{1}{2} N \sin^2 \frac{\theta_{13}}{2} \tag{11–15b}$$

$$(. \ . +1; \ . +1 \ .) = \tfrac{1}{2} N \sin^2 \frac{\theta_{32}}{2} \tag{11–15c}$$

where θ_{ij} is the angle between $\mathbf{a}_i$ and $\mathbf{a}_j$.

Now, because of our previous result,

$$\begin{aligned}(+1 \ . \ . \ ; \ . +1 \ .) &= (+1, -1, . \ ; -1, +1, .) \\ &= (+1, -1, +1; -1, +1, -1) \\ &+ (+1, -1, -1; -1, +1, +1)\end{aligned} \tag{11–16}$$

But the first term in this sum also intervenes in the expansion of $(. \ . +1; \ . +1 \ .)$ and is therefore smaller than this number or, at most, equal to it, for the same reasons as above. Similarly, the second term also intervenes in the expansion of $(+1 \ . \ . ; \ . . \ +1)$ and is therefore smaller or equal to this number. In other words,

$$(+1 . . ; . +1 .) \leq (+1 . . ; . . +1) + (. . +1; . +1 .)$$

If our theory is to reproduce the observable predictions of quantum mechanics, this implies that

$$\tfrac{1}{2} \sin^2 \frac{\theta_{12}}{2} \leq \tfrac{1}{2} \sin^2 \frac{\theta_{13}}{2} + \tfrac{1}{2} \sin^2 \frac{\theta_{32}}{2} \tag{11–17}$$

However, this inequality is not valid for *any* choice of the directions $\mathbf{a}_1$, $\mathbf{a}_2$, and $\mathbf{a}_3$. Hence no hidden variables theory of the general type considered here can reproduce all the verifiable predictions of quantum mechanics.

With respect to Wigner's proof, it may finally be asked whether it applies exclusively to the hidden variables theories that obey separability or whether it is more general. The answer is that, for the proof to go through, the number

(11-12) must be definable. This implies that the number of systems $U + V$ on which a measurement of $2S_1^{(U)}$ would give the result σ_1 does not depend on whether it is $2S_2^{(V)}$ or $2S_3^{(V)}$ that is measured on V, at the distant place where V is. Hence this proof has the same range of validity as Bell's proof reported above. It does not exclude the nonseparable hidden variables theories.

11.5   SEARCH FOR POSSIBLE VIOLATIONS OF QUANTUM MECHANICS

Several times in the past, physical principles once considered as being fundamental—and therefore exact—have turned out to be mere approximations. It is far from inconceivable that the same might happen to the fundamental principles of quantum mechanics, in which case we should not expect all their experimentally verifiable consequences to be correct. On the other hand, many of these consequences have already been verified. But it is a rather remarkable fact that up to quite recent years none of these verifications really provided a decisive clue as to the question raised by Einstein about separability. Indeed, the distant correlation experiments required for such tests were always done under such conditions (e.g., angles 0 or 90 degrees) that some separable hidden variables theory could always be imagined that would give the same predictions as ordinary quantum mechanics [8]. Under these conditions it is not unreasonable to try to discover whether it is Einstein's separability principle or the fundamental set of quantum-mechanical principles that is correct. For this purpose it is necessary to test Bell's inequalities (11-7) or their generalization (11-5).

It is not easy to experiment with spin-$\frac{1}{2}$ particles; however, although photons have a spin 1, their polarization effects are formally quantized in the same way as the components of a spin-$\frac{1}{2}$ vector (with the difference that to an angle α in the one case corresponds an angle 2α in the other case). This opens the possibility of testing inequality (11-5) or (11-7) on suitable pairs of photons [8].

Any attempt at discussing systematically the conditions under which such experiments are conclusive, or what kind of information each of them conveys exactly, would fall outside the subject of this book. The main points of such discussions can be found in Refs. [7] and [8]. Here let us merely note that experiments have already been carried out on photons [11, 12, 13] and on protons [14], and others are in progress. For several reasons the results obtained should probably still be considered as being preliminary ones. On the whole, they seem to favor quantum mechanics and nonseparability, although some discrepancies appear to remain. It is to be expected that the next few years will bring decisive progress in this field.*

Note added to the special edition: Considerable advances have now been made on this question. See in particular A. Aspect *et al.* Phys. Rev. Letters *49*, 91 and 1804 (1982).

11.6 FURTHER DISCUSSION AND COMMENTS

In all the preceding chapters a distinction was carefully made between the verifiable predictions of quantum mechanics (which are unambiguous), on the one hand, and both its formalizations and its conceptual interpretations (which are varied and/or controversial), on the other hand. This distinction is essential and should be retained here. Among the verifiable predictions let us consider especially the *elementary, verifiable quantum predictions,* which we define as the predictions of quantum mechanics that bear on systems composed of a small number of stable particles, and let us make the following "Assumption Y."

Assumption Y
The elementary, verifiable quantum predictions are all correct.

Then the content of this chapter can be summarized by the assertion that, in regard to the hidden variables theories, Assumption Y is not compatible with *the principle of separability,* formulated here toward the end of section 11.3.

The importance of such a conclusion makes it worthwhile to scrutinize its content. For this purpose we must analyse carefully the exact meaning that can possibly be given to the words and concepts entering into the formulation of Assumption Y and of the principle of separability.

In regard to Assumption Y, it should be observed that the predictions it mentions bear on the results of measurements and that, up to now, any instrument that can be used for such purposes must be set up a rather long time in advance in the position it will occupy when it actually interacts with the measured quantum system. There is, therefore, a meaningful content to the following question, which we call "Question W": "Do the quantum predictions referred to in Assumption Y also include those bearing on hypothetical measurements made with 'versatile' instruments?" By "versatile instrument" we mean one which could be set up in a time comparable to the time that microsystems take for propagating through macroscopic distances (room size). For the time being, there seems to be no clear-cut answer to Question W, since no versatile instrument is really in use. The answer "yes" will be favored by theorists who are most impressed by the generality and scope of the principles of quantum mechanics.*

Note added to the special edition: In view of the experimental developments mentioned in the foregoing footnote the answer "no" must now be considered as unrealistic. For more thorough and up-to-date reviews of the whole subject see J.F. Clauser and A. Shimony *Rep. Prog. Phys.* *41*, 1881 (1978) and B. D'Espagnat *Phys. Rep. 110*, 201 (1984).

If we answer "yes" to Question W, a consequence of the investigations reported in Section 11.3 is that Assumption Y is definitely *not* compatible with the principle of separability, for we could imagine that the measurements giving $A(\lambda)$ and $B(\lambda)$ are made with versatile instruments. Under such conditions, if the principle of separability were valid, the probability distribution $\rho(\lambda)$ considered in Section 11.3 could not possibly depend on the direction, **a**, **b**, or **c**, characterizing the instrumental setup. And, similarly, the probability distribution associated with the instruments would depend exclusively on the local conditions. Hence the conditions that entail the validity of Eq. (11–3) would be satisfied, and Bell's inequalities would follow, contradicting Assumption Y.

As already pointed out in Section 8.3, the simultaneous existence of non-separability effects and of the principle that the velocity of propagation of interactions is finite raises nontrivial conceptual problems. One possible way to study these problems is to make use of the fact that hidden variables theories exist which reproduce correctly the verifiable predictions of quantum mechanics (even in regard to "versatile" measurement processes). As we know now, these theories *must* be of the contextualistic, nonseparable variety (in the "restricted" sense). Let us imagine that we have one such theory, Theory T, at our disposal and let us use it as a model.

Does theory T predict that "effects are propagated faster than light"? A priori, we are tempted to answer "yes" without any mental restriction, since the theory *is* nonseparable in the restricted sense. But before we make such a judgment we should investigate the exact meaning that we give to the word "effect." Do we mean "observable effect"? And, if this is the case, what do we understand exactly by an "observable effect"?

If we start making such inquiries, one possibility is that we shall arrive to the following conception. When relativity asserts that no signal—or no effect—propagates faster than light, what it really means is that no *instruction* can be sent to a distant place faster than light. If *this* is our point of view, then we must inquire whether of not a Theory T exists which is based on hidden variables, is contextualistic and nonseparable, and does not allow for such (faster-than-light) propagations of *orders*.

Now, in what is reported in this book, nothing can be found that indicates that such a theory does not exist. In particular, since we have only quantum ensembles at our disposal, the experiments studied above—with correlated particles U and V—do not make it possible to send any instruction from the place where U is to the place where V is. Within the conception studied here, we can say, therefore, that there is no contradiction between relativity and nonseparability. However, it remains true, of course, that Einstein's "separability" principle (let it be recalled that the wording, not the name, is his) is violated (even in the restricted sense) by any acceptable Theory T. This means that the events occurring at the place where $S_a^{(U)}$ is measured would really be modified if, *keeping the hidden variables λ as they are,* we were to change **b**.

But clearly such an operation is impossible. And it is this impossibility (due to the fact that the λ's are uncontrollable) that finally makes all the foregoing considerations consistent, and permits the principle of finite propagation of instructions to hold. Hence we arrive to the conclusion that the "principle of finite velocity of interactions," as it is usually called, is presumably of a mere phenomenological nature. It does not bear on the "reality itself" (here represented by the hidden variables, together with $\mathbf{a}$, $\mathbf{b}$, A_a, and B_b), but it holds in regard to our possibilities of action with respect to such a reality. Of course, it is then but a question of semantics whether we choose to keep the word "reality" for qualifying the set of the hidden variables and so on, or whether we prefer to restrict it to the phenomena on which we, as human beings, can exert some action (a choice that the similarities between the words *Wirklichkeit* and *Wirken* might, for instance, induce a German physicist to make quite spontaneously). But this treacherous path had led us right into philosophy, and at the present stage we do not yet want to explore these quicksands!

REFERENCES

[1] D. Bohm, *Phys. Rev.* **85**, 166, 180 (1952).
[2] J. von Neumann, *Mathematical Foundations of Quantum Mechanics,* Princeton University Press, Princeton, N. J., 1955.
[3] J. S. Bell, *Rev. Mod. Phys.* **38**, 447 (1966).
[4] J. S. Bell, *Physics* **1**, 195 (1964).
[5] J. F. Clauser, *Am. J. Phys.* **39**, 1095 (1971).
[6] A. Shimony in *Foundations of Quantum Mechanics, Proceedings of the Enrico Fermi International Summer School,* Course IL, Academic Press, New York (1971), F. J. Belinfante, *A Survey of Hidden Variables Theories,* Pergamon Press, (1973). See also W. Ochs, *Z. Naturforsch.* **26a**, 1740 (1971).
[7] J. S. Bell in *Foundations of Quantum Mechanics, Proceedings of the Enrico Fermi International Summer School,* Course IL, Academic Press, New York, (1971).
[8] J. F. Clauser, M. A. Horne, A. Shimony, and R. A. Holt, *Phys. Rev. Lett.* **23**, 880 (1969).
[9] A. Einstein in *Albert Einstein: Philosopher, Scientist,* P. A. Schilpp, Ed., Library of Living Philosophers, Evanston, Ill.
[10] E. P. Wigner, *Am. J. Phys.* **38**, 1005 (1970).
[11] L. Kasday in Ref. [7].
[12] S. J. Freedman and J. F. Clauser, *Phys. Rev. Lett.* **28**, 938 (1972).
[13] R. A. Holt, Unpublished Harvard thesis (1973); R. A. Holt and F. M. Pipkin, Harvard Preprint (1974), to be published; G. Faraci, S. Notarigo, A. Penisi, and D. Gutkowski, *Lett. Nuovo Cimento* **9** (15), 607 (1974).
[14] M. Lamehi and W. Mittig, Report from Department de Physique Nucléaire, Saclay, 1974, and Proceedings of the Colloquium "Demi-siècle de la Mécanique Quantique," Strassburg, 1974, to be published.

[15] K. R. Popper, in *Perspective in Quantum Theory,* W. Yourgau and Avander Merwe, Eds., The MIT Press, Cambridge, Mass.

[16] D. Bohm and B. J. Hiley, Birkbeck College (University of London) Preprint (1974).

Further Reading

L. de Broglie, *J. Phys. Rad.* **20,** 963 (1959).

D. Bohm and J. Bubb, *Rev. Mod. Phys.* **38,** 453 (1966).

S. Kochen and E. Specker, *J. Math. Mech.* **17,** 59 (1967).

J. F. Clauser and M. A. Horne, *Phys. Rev. D***10,** 526 (1974).

CHAPTER 12

DIRECT STUDY OF SOME PRINCIPLES

Considered as rules for making verifiable predictions, the great principles of quantum mechanics have proved to be extremely fruitful and have never been contradicted by experiment. However, no system can claim both exactitude and eternal validity; and it is likely that these principles will sooner or later meet the fate that befell those of Newtonian mechanics. Indeed, such considerations have often been used as an argument for not worrying too much about the conceptual problems raised by the quantum principles. If these principles are eventually to be superseded by some more refined and more exact set of axioms, then—it is said—we can reasonably hope that the conceptual difficulties that accompany them will correlatively disappear, or at any rate become less acute.

This chapter is intended to show that the problems originally raised by the quantum principles cannot all be dismissed by such a sweeping sort of analysis, because some of them turn out to be, in fact, independent of these principles. Indeed, the material described here can be understood by a reader who knows nothing about quantum mechanics, provided only that he accepts as empirical evidence the quantization of spin components and the existence of composite systems having a total spin equal to zero.

12.1 SOME ASPECTS OF THE CALCULUS OF PROPOSITIONS

Strictly speaking, a knowledge of the classical and nonclassical theories of logical propositions is not required for an understanding of the analysis of the nonseparability of systems that constitutes the main subject of this chapter (Section 12.2 onward). On the other hand, some very elementary notions about that matter may help to place the general problem in appropriate perspective. For this reason such notions are briefly reviewed in this section. Let it be stressed that the present summary is intended, not as a systematic survey, but only as an introduction to the sections that follow.

Classically, physical propositions bear upon physical systems. A proposition *a* bearing on a system *S* is a qualitative statement about *S*. It can only be either true or false. The fact that *a* is true for a given *S* is a property of that system. As for the notion of "physical systems" itself, it is such a primitive one that usually a few examples are considered sufficient to define it: a stone, an electron, a voltmeter—all these are "physical systems" or, for short, "systems."

Let *a* and *b* be two propositions bearing on a system *S*. It can happen that, whenever *a* is true, *b* is true also. In this case it is said that *a implies b,* and this is noted as

$$a \subseteq b \qquad (12\text{--}1)$$

Whenever $a \subseteq b$ and $b \subseteq a$, it is said that *a* and *b* are equivalent. This is noted as

$$a = b$$

Obviously if $a \subseteq b$ and $b \subseteq c$, then $a \subseteq c$: the relationship $\subseteq$ is *transitive*. Note also that $a \subseteq a$ and that not all of the propositions are related to one another by a relationship such as (12–1). In other words, the set $\mathscr{L}$ having *a*, *b* as elements is only *partially ordered*.

When $a \subseteq b$, the statement described by *a* is obviously equally or less general (less "often true") than the statement described by *b*. This remark leads to the observation that there exists a proposition so general that any proposition implies it. This proposition, which asserts really nothing at all, is labeled I. Similarly, one can think of a proposition, labeled $\varnothing$, that is so restrictive that it is never true. It is then not inconsistent (and it is a useful common practice in logic) to say that, if $\varnothing$ were true, it would imply *a*, where *a* is any proposition of the set $\mathscr{L}$ under consideration. The foregoing statements can be summarized as follows:

$$\varnothing \subseteq a \subseteq I \qquad \forall\, a \in \mathscr{L} \qquad (12\text{--}2)$$

Moreover, with any *a* and *b* of $\mathscr{L}$ it is possible to associate two other propositions, denoted as

$$x = a \cap b = b \cap a \qquad (12\text{--}3)$$

and

$$y = a \cup b = b \cup a \qquad (12\text{--}4)$$

and called, respectively, the "meet" or "lower bound" and the "join" or "upper bound" of *a* and *b* such that

(a) $x \subseteq a$ (12–5a)

                                            $x \subseteq b$ (12–5b)

and if

                              $z \subseteq a$ and $z \subseteq b$

then

                                       $z \subseteq a \cap b$ (12–5c)

(b) $a \subseteq a \cup b$ (12–6a)

                                       $b \subseteq a \cup b$ (12–6b)

and if

                              $a \subseteq w$ and $b \subseteq w$

then

                                       $a \cup b \subseteq w$ (12–6c)

Indeed the propositions x and y are, respectively, just the propositions "a and b" (which is true if and only if both a and b are true) and "a or b" (which is true if either a or b or both are true). Any partially ordered set satisfying (12–2) and conditions (a) and (b) above is by definition a *lattice*. Hence the considered set of propositions, $\mathscr{L}$, is a lattice. Moreover, to any proposition $a \in \mathscr{L}$ we can always associate another proposition $a' \in \mathscr{L}$, so that a' is false whenever a is true, and conversely. Proposition a' is called the *orthogonal complement* to a. It satisfies

                                       $(a')' = a$ (12–7)

                       $a \cap a' = \varnothing ; \quad a \cup a' = I$ (12–8)

                       $a \subseteq b$ implies $b' \subseteq a'$ (12–9)

We describe this situation by asserting that $\mathscr{L}$ is an orthocomplemented lattice.

Up to this point we have essentially used very general formal properties of the entities called propositions, namely, the partial ordering that the implication relationship introduces between them and the possibilities of defining as propositions the "meet" and "join" of two propositions. Let us now take into account the fact that, as stated above, classical propositions are supposed to describe objective properties of the systems and to be either true or false. Then we can construct the well-known tables of "truth values," of which a specimen is given here.

a	b	$a \cap b$	$a \cup b$	
yes	yes	yes	yes	
yes	no	no	yes	
no	yes	no	yes	(12–10)
no	no	no	no	

With the aid of such tables it is easy to check that the equalities

$$a \cup (b \cap c) = (a \cup b) \cap (a \cup c) \tag{12–11}$$

$$a \cap (b \cup c) = (a \cap b) \cup (a \cap c) \tag{12–12}$$

hold separately for all the possible cases and are therefore identically true. Equations (12–11) and (12–12) are called the *distributive identities*.

In addition to the classical propositions on a system, another example of an orthocomplemented lattice that satisfies the distributive identities is the set of the parts of a set. Moreover, it can easily be shown from (12–5) and (12–6) [e.g., with the help of (12–11) and (12–12)] that

$$a \cap (b \cap c) = (a \cap b) \cap c \tag{12–13a}$$

$$a \cup (b \cup c) = (a \cup b) \cup c \tag{12–13b}$$

$$a \cap a = a; \quad a \cap b = b \cap a \tag{12–13c}$$

$$a \cup a = a; \quad a \cup b = b \cup a \tag{12–13d}$$

$$a \cup (a \cap b) = a \cap (a \cup b) = a \tag{12–13e}$$

Now, by definition any set in which two operations $\cup$ and $\cap$ can be defined so that they obey relationships (12–3), (12–4), (12–11), (12–12) and (12–13) is, as is well known, a Boolean algebra. For this reason, a lattice $\mathscr{L}$ that satisfies the distributive identities is called a "Boolean lattice." The lattice of the classical propositions on a system is thus an "orthocomplemented Boolean lattice" (and so is the set of the parts of a set).

If we now turn from classical to quantum physics, we immediately observe that we should expect considerable changes in the formulation described above. At any rate, such changes seem unavoidable if we want to preserve the *experimental* significance of propositions, for we know that in quantum mechanics many observables are not compatible with one another. Obviously the safest method here is to associate experimental propositions with ideal *measurements* of two-valued observables (observables having but two distinct eigenvalues), that is, to *projectors in subspaces of the Hilbert space $\mathscr{H}$ of the system* [1].

Let, then $\mathcal{H}_a$ be the subspace of $\mathcal{H}$ associated with a given proposition a. The apparatus A that serves to measure a can be identified with an ideal filter. If the system S under consideration is describable by a ket contained in $\mathcal{H}_a$, we know with certainty beforehand that it will pass the filter. Under such circumstances we shall say that a is *true* on S.

Let us now consider another filter A' that acts in the opposite way: any system that will pass A will be stopped by A', and conversely [A and A' could, e.g., be constituted by the same beam separator and by screens whose apertures would be complementary to one another; if one of them is open on A (A'), it is shut on $A'(A)$]. Obviously, we know with certainty beforehand that the system S considered above cannot pass A'. Let us then also consider the orthogonal complement $\mathcal{H}_{a'}$ of $\mathcal{H}_a$ in $\mathcal{H}$. It is natural to associate the corresponding proposition a' with A' and to call it the orthogonal complement of a: a' satisfies (12–7), and it will be verified below that it satisfies (12–8) and (12–9) also. On the other hand, it must be observed that systems S can be found on which neither a nor a' is true. They are described by kets that belong neither entirely to $\mathcal{H}_a$ nor entirely to $\mathcal{H}_{a'}$. This situation contrasts with the one that prevails in the classical case, where it is (implicitly or explicitly) postulated that for any system S (and for any a defined on systems of "type S"), either a or a' is true. The difference is obviously related to the fact that classically systems are most accurately described by points in phase space, while quantum mechanically they are most accurately described by vectors in a Hilbert space: whereas a point necessarily lies either inside a given domain in space *or* inside its complement, a vector does not necessarily lie either inside a given subspace *or* inside its orthogonal complement. If it is agreed to define falsehood by stating that "a is false if a' is true," it can happen that a quantum proposition is neither true nor false on a given system.* This makes it impossible to use the usual truth tables in dealing with quantum propositions.

Just as in the case of classical propositions, the set of quantum propositions on a system S can be given the structure of orthocomplemented lattices. For this purpose it must first be observed that the association of the propositions with subspaces of $\mathcal{H}$ makes it possible to introduce quite naturally a partial order in the set, defined by

$$a \subseteq b; \quad \text{if } \mathcal{H}_a \subseteq \mathcal{H}_b$$

(the second use of the symbol $\subseteq$ means that $\mathcal{H}_a$ is embedded in $\mathcal{H}_b$), a relation that indeed is consistent with the significance of the verb "to imply" and with the rules of quantum mechanics. It is then easy to introduce propositions I and $\varnothing$ as was done in the classical case. Finally, "meet" and "join" can also

---

*This is the reason why the expression "many-valued logic" is sometimes used to describe the quantum calculus of propositions.

be defined quite naturally by assuming that there exist propositions associated with the intersection $\mathscr{H}_a \cap \mathscr{H}_b$ and with the linear sum $\mathscr{H}_a \cup \mathscr{H}_b$ of any two subspaces $\mathscr{H}_a$ and $\mathscr{H}_b$ and by defining these propositions to be $x = a \cap b$ and $y = a \cup b$, respectively. It is then easily verified that relations (12–5), (12–6), (12–8), and (12–9) are satisfied. Hence the set of propositions considered here constitutes indeed an orthocomplemented lattice.

On the other hand, since the truth tables cannot be used here, such lattices have no a priori reason to be Boolean. In fact, they are not, for they do not satisfy the distributive identities. This can easily be verified on the lattice of propositions associated with two-dimensional Hilbert space (spin $\frac{1}{2}$). Let $\mathscr{H}_a$, $\mathscr{H}_b$, and $\mathscr{H}_c$ be three distinct one-dimensional subspaces of $\mathscr{H}$ that are not mutually orthogonal. Then $b \cap c = \varnothing$, $a \cup b = \mathrm{I}$, $a \cup c = \mathrm{I}$, so that the left-hand side of (12–11) is a, whereas the right-hand side is I. Similarly, the left-hand side of (12–12) is a, whereas the right-hand side is $\varnothing$.

Although the mathematical aspects of this formalism are highly satisfactory, its operational significance is not quite as simple as one might hope. This is due to the fact that operational definitions of "meet" and "join" are not easily given. However, these two notions can be related by

$$a \cup b = (a' \cap b')' \tag{12–14}$$

so that we can consider only the meet. For this, a possible operational definition (see Ref. [2]) is through an infinite alternate sequence of "*a* filters" and "*b* filters." If for some reason we know with certainty that a system will pass such an infinite alternate sequence, we say by definition that $a \cap b$ is true on it. The objection that is sometimes made to this definition is that it is not *as* operational as one would like it to be, since infinite sequences of instruments cannot be built. But otherwise it satisfies all the conditions that it should. On the whole, therefore, it seems to be quite a useful definition.

As already mentioned, the quantum calculus of propositions has been successfully used as an alternative and more general basis for the foundations of quantum mechanics, that basis being in fact somewhat more general than the usual Hilbert space axiomatics used elsewhere in this book [3, 4]. Now the usual axiomatics do not allow us, as we have seen, to consider systematically all the observables as being just physical quantities *possessed* by microsystems. The question may be asked whether, by any chance, the new axiomatics based on the quantum calculus of proposition would be more flexible in this respect. Indeed, the fact that it is based on statements such as "*a is* true on such and such a microsystem *S*" is encouraging in that regard. It seems at first sight to open the prospect that perhaps—at the price of introducing quantum propositions constituting non-Boolean lattices—a way has at last been found to speak of the *microsystems themselves* and not merely of what we shall observe on them.

One of the results that will emerge from the content of the following sections is that such hopes are largely unfounded and that indeed the axiomatics based on the quantum calculus of propositions does not fare appreciably better in this respect than the conventional Hilbert space axiomatics.

12.2 A SET OF "NATURAL" ASSUMPTIONS CENTERED ON SEPARABILITY

It is a truism that the advent of the modern physical theories—relativity, quantum mechanics, quantum electrodynamics, S matrix theory, and so on—has induced us to abandon many familiar intuitive concepts. When we are asked why, our standard (and quite appropriate) answer is that one may well be skeptical about the possibility and usefulness of developing some alternative theoretical framework that (i) would incorporate and use these old concepts and (ii) would be as successful as each of the present-day theories in all their respective domains.

On the other hand, a motivated skepticism is far from being equivalent to a disproof. All the successful theories mentioned above are built on elaborate sets of axioms that are justified only a posteriori, that is, by agreement between some of their consequences and observed facts (and by the absence of discrepancies). But it should be remembered that the appearance of two or more theories using very different basic concepts and yet accounting equally well for a given set of experimental data is not quite a rare event in physics. Hence the mere *existence* of the successful theories referred to does not establish that such and such a concept (or general view, or the like) which they reject is indeed to be discarded once and for all, as being definitely inadequate. For this reason it is quite often asserted that in such a domain we cannot make any absolute statement. Quite frequently, it is even stated as an obvious truth that all the judgments we can form on these matters are dependent not only on the facts *but also on the general axioms of the existing theories.*

Still, if not for our *practice* of physical research, at least for our *understanding* of the whole subject, we would like to know for certain as many items as we can concerning the adequacy or inadequacy of given concepts of general ideas. In particular, we would be satisfied, if we could, about some given concept or idea, establish not only that it is *useless* at present (i.e., within the framework of the present-day theories) but also that it is *false* in that it leads unavoidably to a contradiction with the data.

For that purpose, we stress again that a mere reference to the existing theories is not enough. How then should we proceed? Obviously by trying, as much as possible, to short-circuit these theories, by attempting to compare directly—or as directly as we can—the concept or idea with the experimental facts.

Now, if our purpose is really to study but *one* concept or *one* particular idea—in isolation, so to speak—the above program is probably overambitious and cannot in fact be fulfilled. But at least it can be applied, as we show below, to a given *set* of concepts and ideas (assumptions). The result of course is weaker, since when we have shown that this set of concepts and assumptions directly contradicts known facts, we can only conclude that one or more of these concepts or assumptions must be rejected, without being able to specify which one. Still, if this set contains only notions and ideas that are *all* deeply ingrained in our minds, even this weaker result is interesting.

In this section a set of concepts and assumptions is introduced [5] which, admittedly, is not consistent with the interpretations of quantum mechanics hitherto studied (Assumption Q and the nonseparable hidden variables assumption). The question in which we are interested is this: Are we *sure* that this set is *absolutely* inacceptable, that is, that it will remain so in *any* future theory or interpretation thereof? It is shown in Sections 12.3, 12.4, and 12.5 that we can answer this question positively provided only that we accept a reasonable idealization of a few recent experimental results.

The set of concepts and assumptions that we want to falsify directly, without reference to quantum mechanics, is the following one.

In regard to the concepts, we merely assume that we can use the words "system" and "proposition" in the usual way. As in the foregoing Section, propositions are defined operationally. A proposition *a* pertaining to a type *T* of systems *S* is defined by specifying the class of measuring apparatus that corresponds to it. As for the concept of systems, it includes as above microscopic systems (see Chapter 1 and Section 12.1). In particular, it includes of course *stable* systems, that is, systems that do not change spontaneously. In the class of stable systems we shall in fact have to consider only the subclass of stable particles, which is a well-specified one. Therefore we shall need no especially elaborate and extensive definition of what a stable system is in complete generality. Any subsystem of a system is itself a system. This of course remain true, just as in classical physics, in the case in which correlations exist between the subsystems constituting a system. Such correlations can be due either to *interactions* actually taking place between the subsystems or to the fact that several systems *U*, *V*, . . . have interacted in the past. In what follows, the latter case is the one in which we shall be especially interested. Fortunately, it is also the one in which the decision of calling *U* and *V* "subsystems" of the "system" *U* + *V* is most obviously identifiable as a mere *convention*, which means nothing special about these systems.

Many systems that we are led to consider are somehow localized within some domain of space-time, where the probability of finding them is not vanishingly small. Two such systems are said to be *isolated* from one another if the domain associated with any one of them is external to the light cone associated with the other system (the notion extends also to other cases, but we need not consider them). In the nonrelativistic approximation we consider as isolated any system that lies arbitrarily far apart from any other. Because of its previous interactions, a system that is isolated at some time can of course possess correlations with other systems. Examples of such cases can easily be found in classical macroscopic physics (see, e.g., Example A in Chapter 8). If, following Einstein, we believe in the principle of separability (see Section 11.3), we must believe that, whenever a system is isolated (in the above sense), it cannot be acted upon by manipulating other systems.

Here, instead of postulating separability alone, let us incorporate it—as already suggested—into a more detailed set of assumptions which we proceed to formulate.

One of the ideas concerning physical systems that is most deeply ingrained in our general conceptions about Nature is that, in some cases at least, some propositions are true about these systems, and that, when this is the case, it is so even if nobody is actually going to try to become conscious of the fact. Let us formulate precisely this idea in the following way.

Assumption 1

It is meaningful to associate to any proposition a defined on types T of systems a family $F(a)$ of systems S of types T, $F(a)$ being defined by two conditions: (i) the systems S that belong to $F(a)$ are those and only those such that, if a were measured on S by any method, the result "yes" would necessarily be obtained; and (ii) the fact that a given S belongs to $F(a)$ is an intrinsic property of S (i.e., it does not depend on whether or not S will interact with some instrument devised so as to measure a).

Remark 1

Assumption 1 apparently conflicts with some at least of the conventional interpretations of quantum mechanics. In particular, it seems difficult to reconcile it with some of the views of the Copenhagen school concerning the role of the instruments and the inseparable wholeness they are supposed to constitute with the object. On the other hand, this particular aspect of the Copenhagen interpretation has always remained somewhat controversial, even in the opinion of some physicists who consider themselves as being substantially in agreement with the conception of that school. Indeed, some of the latter physicists seem to have hoped to be able to restore the validity of our Assumption 1 by using a non-Boolean logic or a non-Boolean calculus of propositions of the type described in Section 12.1. One of the points we expect to make in this chapter is, as suggested above, that such hopes cannot be maintained, and that this is true quite independently of any theory.

Remark 2

The possibility that systems of types T might exist that belong neither to $F(a)$ nor to $F(a')$ is clearly not excluded by Assumption 1, nor is even the possibility that some systems might belong to no family of that sort at all.

In particular, we do not assume that if a is not true it is false. Indeed, we do not even define a meaning for the latter epithet as applied to a proposition bearing on a system.

Remark 3
No determinism—either manifest or hidden—is postulated.

Definition 1
If S belongs to $F(a)$, a is said to be true on S.

Definition 2
Let a system S be isolated between times t_a and t_b. Then a is said to be *persistent* on S between t_a and t_b if the condition that a is true at time t_1 entails that it is true also at time t_2, for any t_1 and t_2 satisfying

$$t_a < t_1 < t_2 < t_b$$

Assumption 2
Let $t_a < t_1 < t_2 < t_b$, let S be a stable system isolated between t_a and t_b, and let a be persistent between t_a and t_b. Then, if a is true at t_2, it is also true at t_1.

Remark 4
Assumption 2 is again one of those whose consistency with the general principles of quantum mechanics can be questioned quite seriously (see Chapter 13). However, as stressed above, we are not concerned at present with subtle points in the axiomatics of quantum mechanics. Anyhow, Assumption 2 seems quite natural in view of Definition 2 and of our general opinion that in such matters some kind of time-reversal principle should hold. This is particularly true if we believe separability to hold good, for then no "isolated" system can be acted upon, as already mentioned.

Assumption 3
If it can be shown that, at a time t, a proposition a is true on a system S under specified conditions C, then a is also true at time t under any other conditions C' obtained from C by modifying at time t or later the experimental devices with which S (or any other system) *will* interact at times subsequent to t.

Assumption 4
If a is true on S, then it is also true on any system $S + S'$ of which S is a part. Conversely if a is a proposition defined on systems of the type of S, if it bears on S and if it is true on $S + S'$, it is true on S.

12.3 CONSEQUENCES

Let us consider the experiment discussed in Chapters 8 and 11. A spin-zero particle decays at time t_a into two particles, U and V, of equal spin S by means of a spin-conserving interaction. Let $\{e_i\}$ be unit vectors defining directions in space. Let v_i be the proposition "$S^{(V)}(e_i) = \mu$" and let u_i be the proposition "$S^{(U)}(e_i) = -\mu$," ($\mu = -S, -S + 1, \ldots, S$) where $S^{(W)}(e_i)$ is the projection along e_i of the spin of particle $W(W = U$ or V). Propositions u_i and v_i can be defined by means of suitably oriented Stern-Gerlach devices. It is then apparent that v_i' is the proposition "$S^{(V)}(e_i) \neq \mu$," and similarly for u_i'. On the other hand, if we were to measure u_i, v_i, in any order we would always get either two answers "yes" or two answers "no." This can be considered as a definition of the statement that the composite system $U + V$ has total spin zero (all measurements are assumed here to be "ideal"), and we can consider it as an experimental fact that systems $U + V$ prepared as stated above *do* have spin zero. Combined with Assumption 1, the fact that upon measurement of u_i and v_i we would certainly get either two yeses or two nos implies that, if the composite system $U + V$ belongs to $F(u_i)$, it also belongs to $F(v_i)$, and conversely.

Let us consider the case in which, at a time $t_2 > t_a$, u_i is measured on U by means of some instrument A. Let us assume first that the result "yes" is obtained. Then, for the reason already mentioned, it can be stated with certainty that a measurement of v_i on the corresponding system V would also give the result "yes." According to Assumption 1, V therefore belongs, after time t_2, to family $F(v_i)$. Since v_i is a persistent proposition on V from $t = t_a$ to $t = \infty$, Assumption 2 has the consequence that that particular V belongs to $F(v_i)$ also at any time t_1 satisfying $t_a < t_1 < t_2$. Assumption 4 then shows that the composite system $U + V$, of which the considered V is a part, also belongs to $F(v_i)$. Because of the strict spin correlation established at time t_a, it thus also belongs to $F(u_i)$.

Let us now assume that the result of the measurement made on U at time t_2 is "no." Exactly the same argumentation then leads us unavoidably to the conclusion that in this case the composite $U + V$ system belongs at time t_1 to families $F(u_i')$ and $F(v_i')$.

Instead of considering only one composite system $U + V$, let us now consider N such systems, all identically prepared and all subjected to a measurement of u_i at t_2. For each of them the result of that measurement is necessarily "yes" or "no", so that each of these systems necessarily falls into one of the cases considered above. The previous argument shows, therefore, that at time t_1, under the conditions of the experiment and if Assumptions 1, 2, and 4 are correct, the composite systems $U + V$ all belong *either* to $F(u_i)$ and $F(v_i)$ *or* to $F(u_i')$ and $F(v_i')$. If now we also take Assumption 3 into account, we must conclude that this situation *would also hold* if the measurement hitherto assumed to be made on U at time t_2 were *not* made at all, or were replaced by some other one. But then the same argumentation can be repeated over again with reference to a new pair, u_j, v_j, of propositions. Hence the conclusion is that in the special case of the decay considered here we have to deal with a situation in which it so happens that any composite $U + V$ system (i) must belong either to $F(v_i)$ or to $F(v'_i)$; (ii) must belong also to $F(u_i)$ in the first case and to $F(u'_i)$ in the second one; and (iii) belongs as a matter of fact to an infinity of such families at the same time, since e_i can be chosen in an infinity of ways. Although we may question these conclusions, the point is that we may not do this without giving up one or several of Assumption 1, 2, 3, and 4.

Remark 1

This argumentation closely parallels the one developed by Einstein, Podolsky, and Rosen [6] to show that quantum mechanics is incomplete. But it is used here with somewhat different assumptions and for a different purpose, since our objective is *not* to test any assumption (e.g., completeness) concerning the axioms of quantum mechanics. As a consequence—in contradistinction with what was the case in the article cited above—the results obtained in this section do not yet constitute a difficulty in regard to the assumptions we want to test, since no contradiction exists between them and the experimental facts that are used here as reference. In particular, they are fully compatible with the experimental facts usually described under the heading "the spin components along different directions are not simultaneously measurable." Admittedly, the results in question imply, for instance, that *if* u_i were measured at t_1 on some system U the answer "yes" *would* be obtained, and that *if* u_j were measured *instead* on the same U the answer "yes" *would* also be obtained. But it asserts nothing about any *actual* sequence of such measurements (concerning which the problem of the perturbation created by the first instrument would have to be taken into account); and, what is even more significant, it does not give us any operational means for effectively sorting out from the statistical ensemble a system U possessing these features.

Indeed, under these circumstances it would even seem at first sight that the special character imparted to the considered composite systems $U + V$ by our assumptions has no observable implication whatsoever. If this conclusion were correct, it would reinforce the view that sets of assumptions of this sort are "legitimate but metaphysical." But, as we show below, a complete elucidation of the bearing of Bell type inequalities [7, 8] must lead us—on the contrary—to relinquish this view, since such inequalities (i) *can* be falsified and (ii) *are* consequences of the results derived directly in the present section from the considered set of assumptions.

Remark 2

Some formulations (see, e.g., [2]) introduce the notion of atomic propositions. When a is atomic, then, if x is a proposition

$$\emptyset \subseteq x \subseteq a \Rightarrow x = \emptyset \qquad \text{or } x = a$$

It might seem that the results of this section preclude the possibilities of u_i or v_i being atomic on U or V, respectively, since the assertion $x =$ "u_i and u_j" (which was shown to hold on some U's) entails u_i, while being different from $\emptyset$. But the conclusion does not follow, since (as pointed out in Remark 1) assertion x is not operational and therefore is not a proposition.

On the other hand, this makes clear a point that could be important for the development of the theories collected under the name of "quantum logic" or "quantum calculus of propositions." This point is that any such theory that makes use, implicitly or not, of our set of assumptions implicitly contains "built in" significant assertions, such as x above, that are different from propositions.

12.4 INEQUALITIES

The semipositive definite character of the probabilities (that they cannot be negative) has many consequences, some of which have perhaps not yet been completely exploited, particularly in conjunction with strict correlation phenomena. Here we derive Bell's inequalities and some generalizations thereof as simple—nay, almost trivial—consequences of that semipositive definiteness (these inequalities consequently apply for a wide range of physical theories and phenomena, including macroscopic, classical ones).

Through the use of the concepts of measure, conditional probabilities and so on (and of the corresponding shorthand notations), the following derivations could easily be formulated in concise, abstract terms. However, to do this would conceal, rather than reveal, their intrinsic simplicity and (what is more important) their corresponding generality. Therefore let us use instead the very simple notion of number of systems in an ensemble. The number of elements in a statistical ensemble is an inherently nonnegative quantity, and the number of elements in the union of two disjoint ensembles is the sum of the numbers of elements of the two constituents. These two trivial but indisputable statements are essentially all we need, and by formulating them in such a concrete manner we hope to show in a convincing way that the basis of the following deduction is extremely difficult to reject.

Let us consider an ensemble E of system V of a given type T. Let $\{v_1 \ldots v_i \ldots v_n\}$ ($n \geq 3$) be a set of propositions—$\{v_1' \ldots v_i' \ldots v_n'\}$ being the set of their orthogonal complements—defined on systems of type T and such that every element of E belongs, for any value of the index i, either to $F(v_i)$ or to $F(v_i')$, $F(v_i)$ and $F(v_i')$ being the families of systems defined in Assumption 1. In classical physics, ensembles E satisfying such conditions can be constructed in an extremely wide variety of cases, as already mentioned. But even when propositions of a type more general than the classical ones are considered, it may happen (in particular cases) that such ensembles can be considered also. An example is provided by the ensembles $E = E_v$ of the systems V considered in Section 12.3. This, as shown in that section, is a consequence of the set of assumptions introduced in Section 12.2. Hence the following considerations apply also to E_v as soon as Assumptions 1 to 4 of Section 12.2 are made, which we assume to be the case.

Let us choose the approach originally used by Wigner (see Section 11.4) in order to deal with the hidden variables problem: with each element V of E let us associate a sequence $\sigma_1 \ldots \sigma_i \ldots \sigma_n$ of dichotomic quantities σ_i which have the values $+1$ (denoted as $+$) if V belongs to $F(v_i)$ and -1 (denoted as $-$) if V belongs to $F(v_i')$. Let us first consider three v_i only, and let then

$$n(\sigma_1, \sigma_2, \sigma_3)$$

be the number of systems V in E that have the specified values of $\sigma_1, \sigma_2, \sigma_3$. Although we cannot know n, it has a well-defined value according to our assumptions (supplemented with the considerations of Section 12.3) in all the cases we consider. Moreover, in all these cases

$$\sum_{\sigma_1, \sigma_2, \sigma_3} n(\sigma_1, \sigma_2, \sigma_3) = N \tag{12–15}$$

where N is the total number of elements of E ($N \to \infty$).

Let $M(i, j)$ ($i, j = 1, 2, 3$) be the mean value on E of the product $\sigma_i \sigma_j$, so that, of course,

$$-1 \leq M(i, j) \leq 1$$

and

$$M(i, j) = N^{-1} \sum_{\sigma_1, \sigma_2, \sigma_3} \sigma_i \sigma_j n(\sigma_1, \sigma_2, \sigma_3) \tag{12–15a}$$

Proposition 1

$$|M(i, j) - M(j, k)| \leq 1 - M(k, i) \qquad \text{for } i \neq j \neq k \tag{12–16}$$

Proof. The quantity

$$M(i, j) - M(j, k) = N^{-1} \sum_{\sigma_1, \sigma_2, \sigma_3} \sigma_j(\sigma_i - \sigma_k) n(\sigma_1, \sigma_2, \sigma_3) \tag{12–17}$$

contains no term with $\sigma_i = \sigma_k$ —hence only terms with $\sigma_k = -\sigma_i$, and

$$M(i, j) - M(j, k) = 2N^{-1} \sum_{\sigma_1, \sigma_2, \sigma_3}' \sigma_j \sigma_i n(\sigma_1, \sigma_2, \sigma_3) \tag{12–18}$$

the symbol Σ' meaning that all the terms in which $\sigma_k = \sigma_i$ must be excluded from the summation, and only these. Similarly,

$$1 - M(i, k) = N^{-1} \sum_{\sigma_1, \sigma_2, \sigma_3} (1 - \sigma_i \sigma_k) \, n(\sigma_1, \sigma_2, \sigma_3) \qquad (12\text{–}19)$$

also contains no term with $\sigma_k = \sigma_i$ and can be rewritten (with the same convention) as

$$1 - M(i, k) = 2N^{-1} \sum{}' n(\sigma_1, \sigma_2, \sigma_3) \qquad (12\text{–}20)$$

In Eqs. (12–18) and (12–20) the summations bear on the same terms; however, in (12–20) all these terms are positive, whereas in (12–18) some of them can be negative. Hence (12–16) follows. Q.E.D.

Since any composite system $U + V$ that belongs to $F(v_i)$ also belongs to $F(u_i)$, as we have shown, $M(i, j)$ can be known experimentally. Indeed, when U and V are spin $\frac{1}{2}$ particles,

$$M(i, j) = -P(i, j) \qquad (12\text{–}21)$$

where $P(i, j)$ is the mean value of the (observable) product of $S_{ei}^{(U)}$ and $S_{ej}^{(V)}$. Equation (12–16) therefore gives rise to Bell's inequalities [7]:

$$|P(i, j) - P(j, k)| \leq 1 + P(k, i) \qquad (12\text{–}22)$$

Proposition 2

$$M(12) + M(23) + M(31) \geq -1 \qquad (12\text{–}23)$$

Proof. The left-hand side, which we designate by A, can be written as

$$A = N^{-1} \sum_{\sigma_1, \sigma_2, \sigma_3} (\sigma_1 \sigma_2 + \sigma_2 \sigma_3 + \sigma_3 \sigma_1) n(\sigma_1, \sigma_2, \sigma_3) \qquad (12\text{–}24)$$

$$A = (2N)^{-1} \sum_{\sigma_1, \sigma_2, \sigma_3} [(\sigma_1 + \sigma_2 + \sigma_3)^2 - 3] n(\sigma_1, \sigma_2, \sigma_3) \qquad (12\text{–}25)$$

Since $\sigma_i = \pm 1$, the quantity inside the square brackets can take only the values $+6$ (for $\sigma_1 = \sigma_2 = \sigma_3$) and -2 (otherwise). Hence

$$A \geq - N^{-1} \sum_{\sigma_1, \sigma_2, \sigma_3}{}'' n(\sigma_1, \sigma_2, \sigma_3) \qquad (12\text{–}26)$$

where Σ'' is a summation extended to all the terms for which not all three σ's are equal. Obviously $\Sigma'' \ldots \leq N$, and (12–23) follows. For the observable quantities $P(i, j)$, (12–23) gives the new inequality

$$P(12) + P(23) + P(31) \leq 1 \qquad (12\text{–}27)$$

Remark 1
In the special but important case (used above as an example) in which U

and V are spin $\frac{1}{2}$ particles in a state of zero total spin (and in similar experiments using photons), it can be shown that (12–27) combined with (12–22) is equivalent to an inequality derived by Gutkowski and Masotto [9] and relating with one another not the P's but the corresponding numbers of systems (probabilities). On the other hand, the G-M inequality is based on the fact that the probabilities of the results $\sigma_i = \pm 1$ are equal. Experiments could probably be imagined in which such an equality would not hold, but for which (12–23) (or (12–27)) would still be valid.

Remark 2

Note that $P(i, j') = -P(i, j)$ if $\mathbf{e}_j = -\mathbf{e}_{j'}$. When such relations are taken into account, all the inequalities (12–22) and (12–27) can be deduced from just one of them; for example, from the symmetrical one (12–27). Similarly, inequalities (12–30) below are then not independent. See Section 12.5.

Remark 3

The effect of the strict spin correlation between U and V is twofold: (i) together with Assumptions 1 to 4, it integrates any ensemble of systems V into the class of those any element of which belongs either to $F(v_i)$ or to $F(v_i')$, and (ii) it has the effect that the quantities $M(i, j)$ become observable, by means of the $P(i, j)$.

Proposition 3

Let

$$
\begin{aligned}
Nx &= n(+, +, +) + n(-, -, -) \le N \\
Ny &= n(+, +, -) + n(-, -, +) \le N \\
Nz &= n(+, -, -) + n(-, +, +) \le N \\
Nt &= n(+, -, +) + n(-, +, -) \le N
\end{aligned}
\tag{12–28}
$$

Then

$$
A = \frac{1 - M(12)}{2} = z + t
$$

$$
B = \frac{1 - M(13)}{2} = y + z
\tag{12–29}
$$

$$
C = \frac{1 - M(23)}{2} = y + t
$$

or

$$
y = \frac{B + C - A}{2}; \qquad z = \frac{A + B - C}{2}; \qquad t = \frac{A + C - B}{2}
$$

The only independent inequalities (or equalities) satisfied by x, y, z, and t on these grounds are

$$x \geq 0, \qquad y \geq 0, \qquad z \geq 0, \qquad t \geq 0 \tag{12–30}$$

$$x + y + z + t = 1 \tag{12–31}$$

Hence the only inequalities that the additive and the semipositive definite nature of the entities "numbers of systems" can generate for linear combinations of A, B, C are those derived from (12–30) and (12–31) by substitution. The three last inequalities (12–30) give inequalities (12–16) (Bell's inequalities); (12–31) and the first inequality (12–39) together give inequality (12–23); the first inequality (12–20) gives no information. It follows that inequalities (12–16) and (12–23) exhaust the list of inequalities satisfied by linear combinations of the $M(i, j)$ as a consequence of additivity and semipositive definiteness.

Proposition 4

Let us consider a fourth unit vector $\mathbf{e}_4$ and the corresponding proposition v_4. Then

$$-2 \leq M(12) + M(13) + M(24) - M(34) \leq 2 \tag{12–32}$$

Proof. Let the symbols Σ' and Σ'' denote summations over the possible values of the σ's from which the terms having, respectively, $\sigma_2 = \sigma_3$ and $\sigma_2 = -\sigma_3$ are excluded. Let the middle member of (12–32) be denoted by B. With obvious notations, B can be written as

$$B = N^{-1} \sum_{\sigma_1, \sigma_2, \sigma_3, \sigma_4} [\sigma_1(\sigma_2 + \sigma_3) + \sigma_4(\sigma_2 - \sigma_3)] \, n(\sigma_1, \sigma_2, \sigma_3, \sigma_4)$$

$$B = N^{-1}[2\Sigma''\sigma_1\sigma_2 \, n(\sigma_1, \sigma_2, \sigma_3, \sigma_4) + 2\Sigma'\sigma_4\sigma_2 n(\sigma_1, \sigma_2, \sigma_3, \sigma_4)]$$

Hence

$$-2N^{-1}\,[\Sigma''n(\ .\ .\ .\ ) + \Sigma'n(\ .\ .\ .\ )] \leq B \leq 2N^{-1}[\Sigma''n(\ .\ .\ .\ ) + \Sigma' \, n(\ .\ .\ .\ )]$$

and therefore (since the ensembles Σ' and Σ'' are disjoint)

$$-2 \leq B \leq 2 \qquad \text{Q.E.D.}$$

Inequalities (12–32) and the inequalities derived by permuting the symbols give rise to the so-called generalized Bell's inequalities between the $P(i, j)$. These inequalities were first derived within the hidden variables conception by Clauser, Horne, Shimony, and Holt [7]. Within that conception they hold true even if a strict correlation does not hold between U and V in the sense in which this concept was introduced in Section 12.3. On the contrary, if we only assume the validity of the set of assumptions listed in Section 12.2, these inequalities are valid only in cases in which strict correlations hold (e.g., in the case of a total spin zero in our example).

More generally, let us now consider the case in which m distinct propositions u_i are taken into account. We first consider the case in which m is odd. Then we have the following proposition.

Proposition 5

$$\sum_{i<j} M(i,j) \geq \frac{1-m}{2}, \qquad m \text{ odd}, \qquad i,j = 1, \ldots, m \qquad (12\text{--}33)$$

and, correspondingly,

$$\sum_{i<j} P(i,j) \leq \frac{m-1}{2}, \qquad m \text{ odd}, \qquad i,j = 1, \ldots, m \qquad (12\text{--}34)$$

Proof. The left-hand side of (12–33) is

$$(2N)^{-1} \sum_{\sigma_1,\ldots,\sigma_m} [(\sigma_1 + \cdots + \sigma_m)^2 - m]\, n(\sigma_1, \ldots, \sigma_m) \qquad (12\text{--}35)$$

The smallest term among those written inside square brackets in (12–35) has the value $1 - m$. Inequality (12–33) follows.

When the parity of m is not specified, the inequality obtained by this method is less stringent:

$$\sum_{i<j} M(i,j) \geq -\frac{m}{2} \qquad (12\text{--}36)$$

The corresponding inequality for the $P(i,j)$ is unlikely to be falsified by experiments made on the $U + V$ systems such as those considered in Section 12.3, since—contrary to (12–34)—it is always satisfied by spin-$\frac{1}{2}$ systems obeying quantum mechanics. This follows from the fact that $P(i,j)$ can then be written as

$$P_{q.m.}(i,j) = -(\mathbf{e}_i \cdot \mathbf{e}_j)$$

so that

$$\sum_{i<j} P_{q.m.}(i,j) = -2^{-1}[(\mathbf{e}_1 + \cdots + \mathbf{e}_m)^2 - m] \leq \frac{m}{2} \qquad (12\text{--}37)$$

the equality being realized for $\mathbf{e}_1 + \cdots + \mathbf{e}_m = 0$.

On the other hand, the derivation of inequality (12–36) can be applied to the more general case in which $M(i, j)$ is the mean value of the product $X_i X_j$ of two random variables. Denoting by σ_i the values taken by the X_i, we have

$$\sum_{i<j} M(i, j) = N^{-1} \sum_{\sigma_1,\ldots,\sigma_m} \sum_{i<j} \sigma_i \sigma_j \, n(\sigma_1, \ldots, \sigma_m)$$

$$= (2N)^{-1} \sum_{\sigma_1,\ldots,\sigma_m} [(\sigma_1 + \cdots + \sigma_m)^2 - (\sigma_1^2 + \cdots + \sigma_m^2)]$$

$$\times \, n(\sigma_1, \ldots, \sigma_m)$$

and hence

$$\sum_{i<j} M(i, j) \geq - \frac{\sum_i M(i, i)}{2}, \qquad i, j = 1, \ldots, m \qquad (12\text{--}38)$$

In the case in which the X_i are centered and have equal root mean squares, (12–38) reduces to the inequality

$$\sum_{i<j} r(i, j) \geq - \frac{m}{2}, \qquad i, j = 1, \ldots, m \qquad (12\text{--}39)$$

between the correlation coefficients $r_{ij} = M(i, j)[M(ii) \cdot M(jj)]^{-1/2}$. Inequalities (12–38) and (12–39) belong essentially to ordinary probability theory, and they should be used as such. Inequalities (12–16), (12–23), (12–32), and (12–33) can also be used within the same framework. When the X_i can be considered as constituting together a stationary stochastic function of the index i, inequality (12.39) reflects the well-known fact that any correlation function of such a stochastic function is positive-definite.[*]

---

*To avoid possible misunderstandings, it may be relevant to describe at length the well-known general notion that is at the basis of the argumentation of this section, to make explicit its operational nature, and to explain in detail how it is utilized.

(a) *The Notion of a Representative Sample*

Let us consider a statistical ensemble E_N of N systems S, all prepared in the same way; let us consider also an observable A; and let $n_\alpha^{(A)}$ be the number of results $A = a_\alpha$ that would be obtained if A were measured on each S by means of some suitable device. On the other hand, let us consider M of these N system ($M < N$), and let $m_\alpha^{(A)}$ be the number of those that would give the result a_i under the measurements just mentioned. Then it is assumed that, for N and M both sufficiently large, $m_\alpha^{(A)}/M \approx n_\alpha^{(A)}/N$.

Let us now consider another observable, B, defined on the same class of systems. Let again $n_\beta^{(B)}$ be the number of results $B = b_\beta$ that would be obtained if B were measured on the N systems S introduced above. How can we possibly obtain a knowledge of $n_\beta^{(B)}/N$? Certainly not by using the M systems already used for measuring $n_\alpha^{(A)}/N$, for these systems have interacted with the devices that served to measure A and hence their preparation is no longer the one that defines E_N. Nevertheless, we can get a knowledge of $n_\beta^{(B)}/N$ (even if we have already obtained one of $n_\alpha^{(A)}/N$), simply by using a representative sample made of M systems selected among those in E_N which did *not* serve in measuring A. In such a way we can obtain simultaneous experimental knowledges of the mean values of *any* pair—or

12.5 DISCUSSION: EXPERIMENTAL

As stressed in preceding chapters, a careful distinction must be made between, on the one hand, the verifiable predictions of quantum mechanics (which are unambiguous in every case) and, on the other hand, both its formalization ant its conceptual interpretations (which are varied and/or con-

---

multiple—of observables A, B, . . . defined on systems the preparation of which is sufficiently specified.

(b) *The Role of the Concepts and Assumptions Listed in Section 12.2*
In the physical cases considered in this section, the systems S are the composite systems $U + V$ (made of two spin-$\frac{1}{2}$ particles U and V in a state of zero total spin), and the observables A, B, . . . just referred to are the products $S^{(U)}$, $S^{(V)}$ of the spin components of particles U and V along specified directions. The $P(i,j)$ (Bell's original notation) are the *mean values* of these observables on any very large statistical ensemble E_N of systems $U + V$ with zero total spin. From paragraph (a) it follows that experimentally the $P(i,j)$ relative to several pairs of directions i, j can be simultaneously known, from experiments made with sufficiently large samples of systems $U + V$.

On the other hand, it should be clear that the Bell-type inequalities for these $P(i,j)$ cannot be derived by the method used in this section without making use of assumptions (such as those of Section 12.2, for instance). Admittedly, we can *always* write

$$P(i,j) = N^{-1} \sum_{\sigma_i, \sigma_j} \sigma_i \sigma_j s(\sigma_i, \sigma_j)$$

where $s(\sigma_i, \sigma_j)$ is the number of systems which are such that a measurement of $S^{(U)}$ along i would give $\sigma_i/2$ *and* a measurement of $S^{(V)}$ along j would give $\sigma_j/2$. But without introducing some assumptions we simply cannot write in any sensible way a formula like

$$s(\sigma_i, \sigma_j) = \sum_{\sigma_k} s(\sigma_i, \sigma_j, \sigma_k)$$

since we have no way of giving a meaning to $s(\sigma_i, \sigma_j, \sigma_k)$—not, at any rate, if we want to make such a number independent of the time ordering of the measurements and of the fact that three interactions (with instruments) are viewed instead of two. The point is, of course, that we are allowed to consider only measurements composed of *one* measurement on U only and of *one* measurement on V only; otherwise, as stressed above in the general case, the first measuring apparatus may cause pertubations of the second result on the same particle. Hence it is not a "logical scandal" that the Bell type inequalities are indeed violated by experiment (as it seems).

The role of the "concepts and assumptions" of Section 12.2 is precisely to make it possible to relate the $P(i,j)$ to quantities $M(i,j)$ that are defined *on V alone*. As explained in the text, they give a meaning to $n(\sigma_1, \sigma_2, \sigma_3)$ and to $M(i, j)$. In conjunction with conventional logic applied to the families F of systems V, they lead to Eqs. (12–15) and (12–15a). The $n(\sigma_1, \sigma_2, \sigma_3)$ and also the $n(\sigma_1, \sigma_2, \cdot) \equiv \sum_{\sigma_3} n(\sigma_1, \sigma_2, \sigma_3)$ and so on are defined on E_v alone. They do not refer to measurements that are actually made. Hence, the former cannot be known. The latter can, however, since $n(\sigma_1, \sigma_2, \cdot) = s(\sigma_1, -\sigma_2)$ *because of the strict $U - V$ correlation.*

Seen from another viewpoint, these concepts and assumptions have the consequence that the $U + V$ systems under consideration (not necessarily *other* systems) possess local hidden variables labelling the families F. The inequalities follow. To verify that there is no a priori contradiction between the existence of a classical hidden variables description and the non-Boolean structure of the lattice of propositions solve Exercise 5 (p. 157).

troversial). In Section 11.6 we defined what we mean by *elementary* verifiable predictions and by *versatile* instruments. If we believe that all the elementary verifiable predictions of quantum mechanics are correct, even when they bear on results supposedly obtained using versatile instruments, then the foregoing sections of this chapter force upon us (with no commitment to a particular formalism or to a particular interpretation) the conclusion that the set of the assumptions listed in Section 12.2 cannot be kept, since such inequalities as (12–22), (12–27), and—more generally— (12–34), which follow from these assumptions, are violated in some cases by the said verifiable predictions. Such cases include those in which the systems U and V considered in Section 12.3 are spin-$\frac{1}{2}$ particles and in which the unit vectors e_i are chosen in some special ways. This was shown by Bell [7] in regard to inequalities (12–22). In respect to inequalities (12–27) and (12–34), it follows, for instance, from the fact that the "balanced" configurations $\Sigma_i\, e_i = 0$ corresponds to the equality sign in relation (12–37) and hence to a violation of (12–34).

If we do *not* take it for granted that all the elementary verifiable predictions of quantum mechanics are true, we must rely on direct experiment. Fortunately, in connection with the hidden variables problem, experiments have been made (see Section 11.5) which can be used (in principle) for such purposes, and others are in progress. However, at the time of writing there seems to be some experimental discrepancies between the results. This is a supplementary motivation for varying the tests, and such variation can be achieved by using inequalities (12–34) also—for instance, in their simplest version, which is (12–27). But independently of this, it should be pointed out that the set of the experiments that are suitable for testing the hidden variables hypothesis does not coincide exactly with the set of experiments that are suitable for testing the set of assumptions under discussion in this chapter. For example, the hypothesis that local hidden variables exist can be tested (by using the generalized Bell's inequalities already mentioned), even in the case in which the correlation between the spins of U and V is not strict in the sense in which this concept is used in Section 12.3. On the contrary, for testing the validity of the set of assumptions listed in Section 12.2 the strict correlation effect is essential. This can also be done with photons, as in Refs. [10], [11], [12], and [13], and if the results of Ref. [11] are taken at face value they certainly seem to contradict the consequences that we have derived from the said set.

It may be noted here that the condition that the spins of U and V should be $\frac{1}{2}$, or that U and V should be photons, is by no means necessary for the validity of the considered tests. For instance, let U and V be two particles of spin S. Let u_i, v_i, u_i', v_i' be defined as in Section 12.3, and let $n(\sigma_1, \sigma_2, \sigma_3)$ and $n(\sigma_1, \sigma_2, \sigma_3, \sigma_4)$ be defined as in Section 12.4. Then, under the assumptions listed in Section 12.2, the mean values $M(i, j)$ of the products $\sigma_i\, \sigma_j$ should satisfy inequalities (12–16), (12–23), (12–32), and (12–33), as shown above. Moreover, because of the strict correlation between the spins of U and V, these quantities can be measured by counting the relative number of cases in which

$$(a)\ S_{ei}^{(V)} = \mu \quad and \quad S_{ej}^{(U)} = -\mu$$

$$(b)\ S_{ei}^{(V)} \neq \mu \quad and \quad S_{ej}^{(U)} \neq -\mu, \quad etc.$$

Of course the $M(i, j)$ are no longer the opposites of the mean values of $S_{ei}^{(U)} \cdot S_{ej}^{(V)}$. Still, the values of $M(i, j)$ as predicted by quantum mechanics can be calculated by means of the usual rules. As in the spin-$\frac{1}{2}$ case, a variety of configurations can then be found, in which these predictions violate the inequalities mentioned above. In the special case in which photons are used, the dichotomic quantities σ_i are conveniently chosen to be $+1$ if the photon passes a polarizer oriented along e_i and -1 if it fails to pass such a polarizer.

In regard to the relationships between the various inequalities noted in Remark 2 of Section 12.4, they should not be interpreted as asserting that inequalities (12–22) and (12–27) are experimentally equivalent. For example, in the symmetrical case $e_1 + e_2 + e_3 = 0$, (12–27) is violated by the predictions of quantum mechanics, whereas (12–22) is not. On the other hand, these relationships imply some symmetries that can be experimentally useful. In the case in which U and V are spin-$\frac{1}{2}$ particles, they imply that only unoriented directions in space are important. In the case in which photons are used, they entail that the tests are invariant with respect to rotations of $\pi/2$ of the polarizers.

12.6 DISCUSSION: THEORETICAL

Since the experiments carried out in this domain are not yet entirely conclusive, our discussion should in principle consider separately the different cases corresponding to all the experimental results that are a priori possible. On the other hand, (hypothetical) results that would bring to light a violation of quantum mechanics—in such simple phenomena as spin correlations—need hardly be discussed here, since their importance and many of their bearings are quite obvious. In this section let us, for this reason, assume as a working hypothesis that the present trend will be confirmed and that the experimental results will corroborate the quantum-mechanical predictions bearing on such correlations. In other words, let us assume that the set of concepts and assumptions of Section 12.2 will definitely be falsified, and let us call this, for short, the "extrapolated experimental result." We now investigate briefly some consequences of such an assumption.

1. Question of the significance of the non-Boolean logic. In the first chapters of this book, the basic principles of quantum mechanics are simply described as a set of rules that makes it possible to formulate observational predictions when the results of some previous observations are known. These rules are expressed by using ordinary language and, therefore, also ordinary, Boolean logic. On the other hand, in those chapters the question of whether these rules could be "translated" into a formulation that would incorporate propositions bearing on "the physical systems themselves" was left open. In the first stages of our inquiry we have merely discovered that this problem is far less trivial here than it is, for instance, in the classical case (in which case the answer is obviously "yes"). On the other hand, we know about the existence of axiomatics formulated directly in terms of propositions (see section 6.4 and 12.1). These axiomatics are somewhat more general than the one described in Chapter 3, which is based on Hilbert space. The propositions they deal with follow the rules of a non-Boolean calculus, but we notice that—at least as they are usually formulated—they *seem* to bear on properties of the physical microsystems themselves. This might look like an argument in favor of the conjecture already mentioned: that, perhaps, the non-Boolean character of the lattice of quantum propositions is just *a price we have to pay:* a price for being able to consider without contradiction that, when using this language, we speak of systems "as they really are," and not just of what *we* observe on them.

Of course, to say that a discourse speaks of systems "as they really are" goes anyway beyond what operationalism allows us to assert. But, at this stage, the question is not yet as follows: "Is it possible to show on a priori grounds that we are being preposterous when we assert such a knowledge to be possible"?* Rather, the question is this: "If we *assume* that this knowledge is possible, and if we take known facts into account, what is the maximal meaning that such an assumption can have?" To the question just formulated the foregoing analysis provides the answer that such a meaning is necessarily extremely limited—so much so that its very existence becomes doubtful. This is due to the fact that, if we want to keep Assumption 1 (the one that is at stake here), we must necessarily reject one at least of Assumptions 2, 3, and 4. Since, as shown below, it seems difficult to abandon either 3 or 4 we are led to give up Assumption 2. Hence, assuming that we can meaningfully describe a stable, isolated system S at some time t "as it really is," this does not allow us to believe that we could also describe it "as it really was" an instant before, even if S is known to remain isolated within a time interval incorporating the two times.

Let it be stressed once more that these conclusions are deduced from the "extrapolated experimental result" (see above) that inequalities of Bell's type are violated; in that sense they are in principle independent of the validity or nonvalidity of the postulates of quantum mechanics. It is worth no-

---

*Such questions are discussed in later chapters.

ticing that up to now they are not reversible: when S is isolated, it can be assumed without contradicting any previous result that to any proposition there corresponds a type of apparatus which will certainly give the answer "yes" to a *second* measurement made immediately after a first one, if this first one gave the answer "yes." In other words, it is quite possible that no manipulation made on any system other than S can "wipe off" the truth of a proposition a bearing on S, whenever S is isolated, whereas—if Assumption 2 is the (only) one that is violated—such a manipulation can, in some instances, *impart* to S the property that is described by proposition a. Even if we believe that the truth of a proposition can be inferred from the present to the (immediate) *future* whenever S is isolated, we remain with the impossibility of doing the same, under the same circumstances, from the present to the (immediate) *past*. Such an impossibility follows from giving up Assumption 2 and severely limits the meaning that can be given to the statement "such and such a set of propositions describes S as it really is."

2. *Question of the alternative assumptions.* If the usual notion of a system is retained, it seems rather artificial to give up Assumption 4 only. Indeed, it even seems doubtful that this could be done consistently.

In regard to Assumption 3, there exist some subtle ways of violating the principle to which it refers while keeping all the other ones and producing no *observable* effects of the future on the past: as mentioned in Section 11.1, deterministic theories can be found that violate none of the predictions of quantum mechanics; they are of the contextualistic [14], nonseparable variety. But their nonseparability (i) leads to no observable violation of the principle of finite velocity propagation of signals, and (ii) might be accounted for as a consequence of a retroactive effect: in the phenomenon described in Section 12.3, this effect would consist in a retroactive influence of the measurement made on U at time t_2 on the parameters describing the state of the system $U + V$ at time t_a [15.]. Up to now, however, the scientific community seems to be reluctant in regard to the idea that the future can act upon the past, even in a way that is not directly observable.

The idea (already discussed above) that Assumption 2 should be violated is somewhat less unattractive that those considered up to this point. At any rate, it does not sound completely unfamiliar to many of the theorists who have studied the foundations of quantum mechanics. But it partakes of similar difficulties. And the main question is, "Should it be abandoned *alone*?" If it is, a strange kind of irreversibility is thereby introduced into the fundamental facts of physics. Hence it may seem more natural to abandon also Assumption 1.

Finally, there are two main possible substitutions for the set of Assumptions 1 and 2. One of them introduces the idea of a nonseparability existing between the microsystem and the experimental arrangement (including the instruments with which the system will *later* interact). This seems to have been

Bohr's view and the essence of his answer to the E-P-R criticism; it is discussed here in Chapters 9 and 21. The other possible substitution for Assumptions 1 and 2 introduces a nonseparability between microsystems that have interacted in the past. This type of nonseparability is closely parallel to the nonseparability of the quantum-mechanical wave function. It is the (sometimes implicit) common feature of two or three otherwise different descriptions: the one that introduces hidden variables in a deterministic [16, 17] but contextualistic and nonseparable theory, the one that makes consciousness an active agent (see Chapter 22), and finally—if it can be proved that this does not reduce to one of the latter two—the description that makes objective the entire wave function of the universe (see Chapter 23).

Remark

The relationship between Assumption 2 and principle (*b*) of Section 8.2 is quite obvious. Note that in Section 8.2 the falsehood of a set of principles which included principle (*b*) as its "hard core" could be demonstrated only by postulating Assumption Q (absence of hidden variables) as well. Here, in what is substantially the same problem, we do not need Assumption Q. But this is so because the role of the latter is taken over by an experimental piece of information, namely, the violation (under suitable conditions) of Bell's inequalities.

12.7 CONCLUSIONS

In Chapter 11 we designated as the *principle of separability* the following principle, as formulated by Einstein. If S_1 and S_2 are two systems that have interacted in the past but are now arbitrarily distant, "the real, factual situation of system S_1 does not depend on what is done with system S_2, which is spatially separated from the former." The result of the foregoing analysis can then be summarized by saying that this principle cannot reasonably be considered any more as holding true.

Superficially, a conclusion such as the one above seems neither surprising nor new. After all, the nonlocality—in the general case—of the many-particle wave functions is quite obvious. It finds its best illustration in the Pauli principle, which also questions the possibility of individualizing systems in the way that separability would have it. Even classical physics admits of correlations between spatially separated events and hence of sudden changes in the probability distributions, induced by distant measurements. On the other hand, the very ease with which we find these apparent counterexamples to the separability principle should make us doubtful about their real validity as such. Obviously, none of the facts we have just listed was unknown to Einstein! That he could nevertheless give credence to the separability principle should induce us to try to be as critical in the use of our conceptual frameworks as we are accustomed to be in the use of our mathematical formalism. And, as soon as we make such an effort, we discover that of course Einstein was quite right. In his times, separability could be *questioned* but could not be *disproved*. For example, the fact alluded to above that distant correlations can take place even between spatially separated events (when influenced by some common anterior one) actually has nothing to do with the principle of separability, which refers, as just recalled, not to our knowledge but to "the real factual situation." More generally, the nonlocality of the many-particle wave function cannot be used straightaway as an argument against separability. For that purpose it must be associated with an *interpretation* of the wave function, and this leads to the long chain of somewhat subtle arguments that this book tries to disentangle.

Since in its principle the whole analysis of this chapter is completely independent of quantum mechanics, its conclusion against separability is much more direct. Its main defect is that the experimental results which should normally constitute its firm basis are recent, incomplete, and still somewhat controversial. We may be confident, however, that such experimental ambiguities will be resolved very soon.

REFERENCES

[1] G. Birkhoff and J. von Neumann, *Ann. Math.* **37**, 823 (1936).
[2] J. M. Jauch, *Foundations of Quantum Mechanics,* Addison-Wesley, Reading, Mass.
[3] C. F. von Weizsäcker, *Naturwissenschaften* **42**, 521 (1955); G. Mackey, *Mathematical Foundations of Quantum Mechanics,* New York, 1963.
[4] C. Piron, *Helv. Phys. Acta,* **37**, 439 (1964); J. M. Jauch and C. Piron, *Helv. Phys. Acta* **42**, 842 (1969); C. Piron, *Found. Phys.* **2**, 287 (1972).
[5] B. d'Espagnat, *Phys. Rev. D* **11**, 1424 (1975).
[6] A. Einstein, B. Podolsky, and N. Rosen, *Phys. Rev.* **47**, 777 (1935).
[7] J. S. Bell, *Physics* **1**, 195 (1964); J. F. Clauser, M. A. Horne, A. Shimony, and R. A. Holt, *Phys. Rev. Lett.* **23**, 880 (1969).
[8] *Foundations of Quantum Mechanics, Proceedings of the Enrico Fermi International Summer School,* Course IL, Academic Press, New York (esp. articles by J. M. Jauch, J. S. Bell, L. Kasday, A. Shimony, E. Wigner, B. De Witt).
[9] D. Gutkowski and G. Masotto, Catania Preprint.
[10] L. Kasday in Ref. [8].
[11] S. J. Freedman and J. F. Clauser, *Phys. Rev. Lett.* **28**, 938 (1972).
[12] R. A. Holt, Unpublished Harvard thesis (1973).
[13] G. Faraci, S. Notarrigo, A. Pennisi, and D. Gutkowski, *Boll. SIF* **93**, 39 (1972).
[14] A. Shimony in Ref. [8].
[15] O. Costa de Beauregard, Boston Preprint (1974).
[16] L. de Broglie, *J. Phys.* **5**, 225 (1927).
[17] D. Bohm, *Phys. Rev.* **85**, 166, 180 (1952); J. S. Bell, On the Hypothesis That the

Schrödinger Equation is Exact, Contribution to Pennsylvania State University Conference September 1971, TH 1424 CERN.

Further Reading

H. P. Stapp, *Phys. Rev.* **D3,** 1303 (1971).

CHAPTER 13

ASPECTS OF THE PROBLEM OF RETRODICTION

The problem of retrodiction is associated with basic questions, but for this very reason it is, unfortunately, quite difficult, even in classical physics. In quantum physics it is even more subtle. Hence it would not be appropriate to ascribe to it anything like a central role in our investigations in this book. Indeed, it seems advisable to avoid as much as possible using arguments based on retrodiction when discussing the problems of measurement theory or of general interpretations of the formalism. It may be pertinent to note that similar remarks also hold good in regard to information theory and its use. However, the difference is that we shall be able to discuss quite thoroughly the problems just mentioned above without applying anywhere the techniques of information theory. This is not exactly the case with respect to the concept of retrodiction. At a few stages in some of the discussions below we shall indeed find it necessary to use this concept, and this is the reason why it must be introduced and studied here [1]. On the other hand, we must keep in mind that its role in the investigations to be carried out is a somewhat marginal one that does not require a full examination of all the aspects of these questions. Nevertheless, the aspects that *are* necessary for our purposes happen to be somewhat subtle. Hence, although this chapter finds its proper place here on purely rational grounds, it is more difficult than the chapters that come immediately before and after it, and its general impact on the substance of our studies is smaller. For all these reasons it may be considered advisable to omit it on the first reading.

13.1 "INDEPENDENT TIME EVOLUTION" AND THE RETRODIC- TION PROBLEM

As a preliminary step for what follows, let us discuss the conditions under which it is possible to speak of an "independent time evolution" (our terminology) of a system S and to study what kinds of predictions can legitimately be made when these conditions are fulfilled.

149

In classical analytical mechanics these questions are easily settled. Between times t_1 and t_2, S has an independent time evolution if and only if a Hamiltonian function H can be found that correctly describes the said evolution and that is a function of the physical quantities (observables) of S alone (with the understanding, of course, that H can involve given time-independent potentials on which the reaction of S is zero). Then, if we know the state of S at any time t_a satisfying $t_1 < t_a < t_2$, we can calculate the corresponding state of S at any other time t_b also satisfying $t_1 < t_b < t_2$. The significance of such a calculation is that all the predictions of results of conceivable observations that we can deduce from the knowledge thus acquired, by using it in conjunction with the laws of (classical) mechanics, are fully correct. These considerations hold good irrespective of the sign of the difference $t_b - t_a$. Incidentally, when a system S interacts with another system T, it often happens that neither S nor T undergoes an independent evolution but that the composite system $S + T$ does.

In quantum mechanics, a consistent definition of what is called a "system" leads to problems that do not exist in classical physics, and correspondingly the answers to the questions raised are less straightforward. Indeed, too naive a transposition to quantum mechanics of what has just been recalled in regard to classical physics could easily lead to false experimental predictions, particularly in the case $t_b < t_a$: of this, the experimental situations that allow for the various formulations of the so-called Einstein-Podolsky-Rosen paradox offer clear-cut illustrations, which have been quite extensively discussed (see Chapters 8 and 12). Consequently, it is quite generally recognized that a radical difference exists in this respect between classical and quantum mechanics. On the other hand, in regard to the precise location of this difference some confusion apparently exists. Indeed, the idea seems to be present in the background of some discussion that, if the foregoing definition of an independent evolution is carried over to quantum mechanics, then the rules that allow us to calculate the state at time t_b if we know it somehow at time t_a (these rules are here embodied in the Schrödinger equation, of course) should be applicable only if $t_b > t_a$—in other words, they would provide either no information or false ones if applied when $t_b < t_a$.

Now, if such a drastic restriction had to be systematically associated with the Schrödinger equation, this would mean that the basic laws of quantum mechanics have a specific kind of basic irreversibility that has no counterpart in classical mechanics, and this quite independently of any supplementary considerations bearing on measurement, statistical theory, and so on. What—for later use—we want to stress in this section is merely that such a considerable change of outlook is not, offhand, necessary, and that it is even false, in the sense that, although when $t_b < t_a$ the rules alluded to are seldom applicable, they sometimes are, and then they lead to predictions that are correct. It is true, of course, that a radical difference exists between classical and quantum mechanics, which has a bearing on the problem of the independent time evolution of systems. But this difference is by no means constituted by an intrinsic irreversibility of the quantum laws of motion such as the one considered above. As already suggested, it bears exclusively on the greater relativity that the notion of a physical system presents in quantum mechanics compared to what is the case in classical mechanics.

The point is as follows. Let us consider again the well-known example of the spin-conserving decay, at a time t_0, of a spin-zero particle into two spin-$\frac{1}{2}$ particles, U and V. Let E be an ensemble of N such systems $U + V$, and let t_1 be a time at which an observation of the third component $S_3^{(U)}$ of the spin of U is made. We can legitimately assert that after t_1 each particle V has a definite spin component $S_3^{(V)}$ along $O\hat{z}$. On the other hand, between times $t = t_0 + 0$ and $t = \infty$ the particles V are free. A Hamiltonian describing the time evolution of free spin-$\frac{1}{2}$ particles is easily found and does not depend on observables not belonging to V (in other words, it is an operator in the Hilbert space of the V particles). Too naive a transposition of the definition used in classical mechanics might therefore induce us to assert without any supplementary qualification (a) that between times $t_0 + 0$ and ∞ the V particles "undergo an independent time evolution," and (b) that "therefore" we can deduce from our knowledge that each particle V has after t_1 a definite $S_3^{(V)}$ the "fact" that each of them "has" that same definite value also at any time t lying between t_0 and t_1. However, such a statement has consequences bearing on the spin correlation of U and V along directions other than $O\hat{z}$ that are obviously incorrect (see again Chapters 8 and 12).

These considerations about a possible measurement made on U at t_1 constitute an argument for "being careful" as soon as we attempt to retrodict. But let us introduce a set of K of "allowed" measurements; that is, let us adopt the convention that we shall never consider making any measurement other than those of that set. We can then, moreover, identify K with the set of all the experiments bearing on V alone. Then, for the measurements of this set, the retrodictive calculation described above is legitimate, at least in the sense that the description of the ensemble of system V at time t that it implies leads to no incorrect consequences in regard to the predictions bearing on the set K of measurements. Under these conditions we can try to develop a relative definition of the "independent time evolution" that would cover such a situation. For this purpose we first adopt the convention to use the word "system" in its usual sense, in spite of the fact that in order to take into account the phenomenon just described it is often convenient not to do so. For instance, we call particle V at any time $t > t_0$ a "system" in spite of its nonseparability from U as long as $t < t_1$. Such conventions can be adopted freely, provided, of course, that they are explicit and that the restrictions that quantum mechanics imposes on their use are kept carefully in mind. Here one such restriction is that we can use the definition given below only when we consider as "allowed" just the measurements that bear on the set (K) of the observable defined in V alone, plus, possibly, some carefully selected ones. Under these conditions we propose the following definition.

Definition

With the definition stated above of the word "system," let us consider all the systems S subject to known forces and to interactions of given (finite) range with other systems. We say that, in the interval of time (t_1, t_2), the S undergo an "independent time evolution in the restricted sense" or (briefly) an "independent time evolution," if it is possible to find a Hamiltonian H that operates on the Hilbert space of S alone and that—by means of the usual laws of quantum mechanics—makes it possible to establish for all the S's the same connections between "results of measurements" and "predictions of measurements that could be done" as are provided by the full Hamiltonian, whenever these results and predictions refer to measurements that (a) take place within the interval (t_1, t_2) and (b) bear on observables of S alone.

It should be noticed that the foregoing definition refers in fact to a whole *ensemble* of systems S, and that such an ensemble is determined by specifying the nature of these systems and the forces to which they are subjected, but *not* by specifying special "initial" (or "final" or "intermediate") conditions. This serves in discarding ensembles of systems that would be determined by specifying that they pass through the apertures of given screens; screens do not operate as "known forces," since their effects cannot be described by interaction Hamiltonians operating in the Hilbert space of the system. It *is* necessary to discard such cases, since otherwise, examples could easily be given of systems that would undergo an independent time evolution and for which it would nevertheless be faulty to "retrodict" in the way described above [i.e., by calculating $\psi(t_b)$ from $\psi(t_a)$ when $t_b < t_a$]. For instance, this would be the case if the system S were defined as being a polarized spin S particle that passes first a Stern-Gerlach magnet and then a hole in a screen, which has the effect of selecting only one of the possible emerging beams. According to the definition given above, such systems do not undergo an independent time evolution, for, even if a given system S passes through the hole (and hence suffers no interaction with the screen), this is

not, in general, the case for the whole ensemble of systems S to which the foregoing definition refers.

On the other hand, let S be a system, $H^{(S)}$ its Hilbert space, K the set of hermitean operators in $H^{(S)}$, let us assume that, within the time interval (t_1, t_2), the
time evolution of S is "independent" in the sense of the above stated definition
and let, again, the symbol H refer to the there mentioned hamitonian (which belongs to K of course). Let us consider an ensemble E of such systems and let us
assume that, in $H^{(S)}$, its density matrix $\rho(t_a)$ is somehow known at time t_a. We can
then compute $\rho(t_b)$ by means of the general formula $\rho(t_b) = U(t_b, t_a) \rho(t_a)$ where
$U(t_b, t_a)$ is the unitary evolution operator that corresponds to H. In the case $t_1 <
t_b < t_a < t_2$ this may be called a retrocalculation. As we already know, we must
then be careful when attempting to utilize $\rho(t_b)$ for uttering statements concerning observables not belonging to K, such as correlations between S and systems
with which S interacted before t_1. But, to repeat, as long as we are only interested
in statements concerning what can be observed on the elements S of E, retrocal-
culating this way–though hardly useful–is unobjectionable in principle as long as
the corresponding outcome is adequately used. Imagine, for instance, that a
physicist Paul who has, at time t_1, prepared E in a state described by $\rho(t_1)$ com-
putes $\rho(t_a) = U(t_a, t_1) \rho(t_1)$ and informs another physicist, Peter, that, in case no
measurement is performed on the systems S in between times t_1 and t_a, the state
of E at time t_a is to be this $\rho(t_a)$. With the help of the *sole* knowledge of $\rho(t_a)$,
Peter, who, actually, *has* decided to make some measurements immediately after
time t_b, can then retrocalculate, compute $\rho(t_b) = U(t_b, t_a) \rho(t_a)$, and use this $\rho(t_b)$
for calculating the probabilities that the measurements he is, in fact, going to
make yield such and such outcomes. These predictions will (quite obviously!) be
correct.

To sum up, we met here with conditions in which retrocalculation is entirely legitimate. Admittedly, in this case this procedure is also entirely re-
dundant since, in practice, there is no point in going over from $\rho(t_1)$ to $\rho(t_a)$
and then back to $\rho(t_b)$, instead of calculating $\rho(t_b)$ directly from $\rho(t_1)$. Nevertheless, since retrodiction is valid in this particular instance, its validity can
hardly be questioned in situations that, in regard to both the instrumental setup and the structure of the general argument, are identical to this one. And indeed, there are situations of such a type; to wit: situations in which the $\rho(t_a)$
that is used in the argument, far from being actually *measured* at time t_a, is the
$\rho(t_a)$ that would emerge in the (counterfactual) independent evolution case,
and whose only difference with the situation at hand is that, in them, instead
of being calculated from $\rho(t_1)$, this counterfactual $\rho(t_a)$ is assumed to be
(partly) known (so that, in such cases, retrodictive calculations are *not* redun-

dant). Such situations are met with as soon as it is assumed that S can obey the two conditions–admittedly hardly compatible in our Universe as it is!–of being a macrosystem and being isolated from its environment. The assumed, partial knowledge just alluded to consists in the fact that, on macrosystems, it is normally taken for granted that some physical quantities such as pointer coordinates have, in all conceivable–actual or virtual–circumstances, and in a kind of an "absolute" sense, definite values: for this assumption (essentially Assumption Z of Section 7.4) sets definite conditions on the structure of any conceivable ρ (t_a). In Section 17.2 we shall make use of the just described retroactive computation to investigate whether or not the assumed knowledge in question is really consistent with the standard quantum mechanical formalism.

13.2 CONNECTED PROBLEMS

Let it be stressed once more that the aspects of the retrodiction problems discussed above are rather special ones. The reason why they are investigated here in detail is just that they are made use of below. Other, quite important aspects are not discussed in this book because they have no direct connection with its main subject matter. These include, for instance, the role of the concepts of outgoing and incoming waves and the question of whether it is at all possible to consider a reduction of the wave packet that would work backwards in time [2].

It should also be noticed that, to some extent, the concept of retrodiction is related to the notions of time reversal and of irreversibility. As is well known, both these notions play a central role in physics. For this reason, extensive descriptions of their meanings and of their uses can be found in a great number of textbooks. Again, it seems unnecessary that new descriptions should be added here to the previously existing ones.

Under these conditions, it is appropriate to discuss here but one of these many questions. It is not, by far, the most important, but it should be mentioned at this stage, since otherwise misunderstandings might creep in as to

---

*Other cases exist. In particular the phenomena of correlation at distance studied in Chapter 8 can be used to develop other examples [1].

the legitimacy and meaning of some of the developments occurring in later chapters.

The point concerns the case in which two measurements are made in succession and in which, knowing the result of the second one, we want to infer from this knowledge the probabilities of the possible results of the first one. For simplicity, let us assume that these measurements are complete, ideal ones. Let

$$A\phi_i = a_i\phi_i \tag{13-1}$$

and

$$B\chi_\alpha = b_\alpha\chi_\alpha \tag{13-2}$$

be the eigenvalue equations of the observables A and B that are thus measured at times t_0 and t_1, respectively, on a system Σ or on an ensemble thereof. Let $U \equiv U(t_1, t_0)$ be the time evolution operator of Σ between these two times. Let $w_\alpha^{(i)}$ be the probability that the result $B = b_\alpha$ will be obtained at t_1 if the result $A = a_i$ has been obtained at t_0. Similarly, let $w_i'^{(\alpha)}$ be the probability that the result $A = a_i$ has been obtained at t_0 if the result $B = b_\alpha$ is obtained at t_1. Finally, let ψ be a state vector describing an ensemble E of N systems Σ before time t_0.

Let us ask first for the value of $w_\alpha^{(i)}$. The answer to this question is straightforward. According to the general rules of quantum mechanics $w_\alpha^{(i)}$ is independent of ψ and is given by

$$w_\alpha^{(i)} = |\langle \chi_\alpha | U | \phi_i \rangle|^2 \tag{13-3}$$

Now let us ask for the value of $w_i'^{(\alpha)}$. We might be tempted to argue that, since, at time t_1, B is known to have the value b_α on all the systems of the ensemble E_α that has to be considered in this problem, that ensemble is described at time t_1 by the state vector χ_α. Applying the argumentation developed in the foregoing section, we would then conclude that at time t_0 the state vector was $U^{-1}\chi_\alpha$ and that the value of $w_\alpha^{(i)}$ is correspondingly $|\langle \phi_i | U^{-1} | \chi_\alpha \rangle|^2 = w_\alpha^{(i)}$. However, that argumentation would be erroneous, since *in the present problem* each Σ interacts with an instrument at time t_1. Hence from the fact that E_α is described by χ_α at time $t_1 + 0$ we cannot deduce that it is also described by χ_α at time $t_1 - 0$. In general, therefore, there is no reason that it should be described by $U^{-1}\chi_\alpha$ at time t_0.

Hence some other method must be found. The correct method consists in remarking that the number $N_{i,\alpha}$ of elements of E on which both results are obtained is given either by

$$N_{ia} = N_i w_\alpha^{(i)} \tag{13-4}$$

or by

$$N_{i\alpha} = N_\alpha w_i'^{(\alpha)} \tag{13-5}$$

where N_i and N_α are the numbers of elements of E that lead to the results $A = a_i$ and $B = b_\alpha$, respectively.

Hence

$$w_i'^{(\alpha)} = \frac{w_\alpha^{(i)} N_i}{N_\alpha} \tag{13-6}$$

This formula constitutes the answer to our question, since N_i and N_α are easily obtained:

$$N_i = N |\langle \phi_i | \psi \rangle|^2 \tag{13-7}$$

$$N_\alpha = N \sum_j |\langle \chi_\alpha | U | \phi_j \rangle|^2 |\langle \phi_j | \psi \rangle|^2 \tag{13-8}$$

Incidentally, it may be noticed that, in general, N_α is not equal to $N |\langle \chi_\alpha | U | \psi \rangle|^2$. This is due to the fact that the presence of the instrument measuring A modifies the behavior of the systems.

It follows from (13–3), (13–6), (13–7), and (13–8) that $w_i'^{(\alpha)}$ is not equal to $|\langle \phi_i | U^{-1} | \chi_\alpha \rangle^2$. In other words, $w_i'^{(\alpha)}$ cannot be obtained by means of a retrodiction calculation applied to the state vector at $t_1 + 0$.

But this is obviously due to the fact that the system interacts at t_1 with another system. Hence the impossibility just mentioned cannot be used as an argument for rejecting such a calculation in cases in which no such interaction takes place and in which, nevertheless, the said state vector is known, by assumption or otherwise. These are precisely the cases that were studied in the foregoing section, where the conditions under which such calculations are valid were stated. The point that must be stressed here—in view of the content of later sections—is that, in fact, the formulas just written have no restrictive implications whatsoever in regard to the conditions of validity of the theory developed in Section 13.1 since they apply to another problem.

REFERENCES

[1] B. d'Espagnat, *Nuovo Cimento* **21B**, 233 (1974).
[2] Y. Aharonov, P. G. Bergmann, and J. L. Lebowitz, *Phys. Rev.* **134B**, 1410 (1974).

EXERCISES OF PART THREE

EXERCISE 1

Let two systems, I and II, interact, and let us consider them at a time t when they no longer interact. Let us, moreover, assume that at time t the state vector of the system can be written as

$$|\psi\rangle = \sum_j C_j |l_j\rangle |m_j\rangle$$

where $l_j \in H_I$ ($|m_j\rangle \in H_{II}$) is an eigenket of an operator $L(M)$ associated with an observable $L(M)$ of I (II). Let R be another observable belonging to I, and S another observable belonging to II. Show without using the density matrix formalism that the proper mixture containing a fraction $|C_j|^2$ of states defined by the values l_j of L and m_j of M gives (*a*) the same observable predictions as the pure state in regard to all measurements made on L or R alone, on M or S alone, and on correlations between L and S or between R and M; and (*b*) predictions that are in general different and incompatible in regard to the correlations between R and S (Furry). *Hint*; For question (*a*) use formula (3–3a).

EXERCISE 2

Use the results of Exercises 2 and 3 of Part Two to show that the pure case described by Eq. (8–8) is not equivalent to a proper mixture in which, for each element, the spin of U points in a definite direction n and the spin of V points in the opposite direction, where n is a unit vector isotropically distributed over the mixture. *Hint*: Evaluate the mean value of $[\mathbf{a}\cdot\sigma(U)][\mathbf{b}\cdot\sigma(V)]$ for the two cases, $\mathbf{a}$ and $\mathbf{b}$ being unit vectors.

EXERCISE 3

Discuss the validity of the "experimental proof" given at the very end of Section 8.3 that distant correlations are instantaneous.

EXERCISE 4

Use the remark at the end of Section 12.5 to develop tests of the assumptions considered in Section 12.2 that involve particles with nonzero rest masses and spins higher than 1.

EXERCISE 5

Imagine a deterministic, classical world in which any sufficiently precise measurement would modify the measured variable. Use the Bell-Clauser hidden variables model for spin-$\frac{1}{2}$ particles (Section 11.1) to develop an example showing that the set of the *operationally defined* propositions (corresponding to measurements that can actually be made in such a world) can constitute a non-Boolean lattice. Contrast this lattice with the Boolean lattice of the yes-no assertions bearing on the hidden variables themselves (assertions that only a Maxwell demon could consider as being operationally defined).

PART FOUR

MEASUREMENT THEORIES

We now enter the field of measurement theory, where a disturbing fact confronts us. There is not *one* measurement theory. Rather, several competing theories are in existence, and they are, as a rule, in conflict with one another. To do our best with this irksome situation we first review, in Chapter 14, the elements of measurement theory that meet with more or less universal approval. These include (i) the description of the measurement process as a simple correlation established between an observable of the system and an observable of the instrument, and (ii) a formalism based on the density matrix approach, which—although in any specific case its use could always be dispensed with—nevertheless gives an elegant synthesis of all the observable predictions. We then review some of the difficulties that are inherent in any attempt at proceeding further. By this we mean the difficulties that beset any tentative description of the instrument coordinate as a quantity that really *has* the value we observe. Two traditional but recently renewed approaches to this problem, whose purpose is to make such a description consistent with the superposition principle, are then critically examined. One of them (Chapter 15) is based on the concept of classical systems. The other (Chapter 16) is grounded on the macroscopic irreversibility of the measurement processes. The limitations of both views are shown.

When a quantum theory of the measurement process is developed, the important facts that the initial state of the instrument is not known exactly, and that most measurements are *nonideal* (or *of second type*), should be kept carefully in mind. The possibility that the difficulties mentioned above might have their origin in an insufficient consideration of these two points should therefore be investigated. This investigation is made in Chapters 17 and 18. In that respect our conclusion is negative. The overall content of this part does not, therefore, make it possible to deny the validity of Wigner's statement that "measurements which leave the system "object plus apparatus" in one of the states with a definite position of the pointer cannot be described by the linear laws of quantum mechanics."

Of course, here, as everywhere else in what follows (except when explicitly stated otherwise) the validity of Assumption Q (Section 4–2) is postulated. In particular, it is assumed that no hidden variable exists.

CHAPTER 14

ELEMENTS OF MEASUREMENT THEORY

In this Chapter, the reasons why measurement theory is a particularly controversial part of modern theoretical physics are first very briefly examined. This is followed by a review of the parts of the theory that are free from every such treacherous difficulty and therefore contribute a common basis to all the more elaborate attempts at understanding what takes place in a process of measurement.

14.1 SOME INTRODUCTORY REMARKS

The problem of measurement in quantum mechanics is considered as nonexistent or trivial by an impressive body of theoretical physicists and as presenting almost insurmountable difficulties by a somewhat lesser but steadily growing number of their colleagues. Before we embark on technicalities, it seems appropriate to survey rapidly the origins and causes of such an odd situation.

Let us, for this purpose, consider the most naive and elementary form of quantum measurement theory. The measurements with which we are concerned are not those of general quantities: the spin of a particle, the electric conductivity of a metal, and so on; they are exclusively those of transient particular properties of a system: position, momentum, spin component, and the like. Let $|\psi_1\rangle \ldots |\psi_n\rangle \ldots$ be the eigenvectors of such an observable L that pertains to a system S and whose value an experimentalist wants to measure by means of an apparatus A. For the sake of simplicity, let us assume that A can only be in one of a discrete set of eigenstates $\{|n, r\rangle\}$, where n labels the possible positions $G = g_n$ of a pointer, and r is a degeneracy index related to all the other variables in A. Presumably it is unnecessary to stress the fact that such a description of the apparatus is highly schematic and superficial; any conceptual difficulty to which it can lead will therefore have to be re-examined with care, within the framework of a more realistic formalism, before it can be said to be a true one. Nevertheless, this simple picture is useful in illustrating the problem.

Let us first consider the case in which S is initially in an eigenstate $|\psi_m\rangle$ of the quantity L to be measured:

$$L|\psi_m\rangle = l_m|\psi_m\rangle \tag{14-1}$$

and in which A simultaneously lies in one of the states $|0, r\rangle$, so that the co-ordinate G has the value zero. A is so constructed (this is a part of our assumptions) that, under such circumstances, the effect of the interaction is to bring it into one of the states $|m, s\rangle$ without changing the state of S. In other words, the state vector that describes the composite system $S + A$ undergoes the change described by

$$|\psi_m\rangle\,|0, r\rangle \rightarrow |\psi_m\rangle\,|m, s_{m,r}\rangle \tag{14-2}$$

If this is the case, after the interaction the pointer of A is clearly in the position $G = g_m$, which is in a one-to-one correspondence with the eigenvalue l_m of L.

Thus Eq. (14–2) can indeed be considered as a model describing a process of measurement of L by means of A.

Now, however, let us consider a different initial situation in which A is still described by $|0, r\rangle$ but in which S is in a superposition of states $|\psi_m\rangle$:

$$|\psi\rangle = \sum_m a_m|\psi_m\rangle \tag{14-3}$$

where the a_n are parameters; the superposition principle guarantees that Eq. (14–3) is indeed a possible initial state. Then, as a result of Eq. (14–2) and the linearity of the time-evolution law (Rule 3, Chapter 3), the final state of the compound system is *necessarily* described by

$$|\psi_f\rangle = \sum_m a_m|\psi_m\rangle\,|m, s_{m,r}\rangle \tag{14-4}$$

As shown below, a consequence of Eq. (14–4) is that, for an observer who measures G on A and finds g_m, system S subsequently behaves as if it were in state $|\psi_m\rangle$. This is just the wave packet reduction. However, at least if considered uncritically, Eq. (14–4) also seems to have the meaning that apparatus A is simultaneously in a superposition of states that correspond to different positions g_n of the pointer. Such a meaning, however, comes into direct conflict with what appears to us as a part of our most immediate experience, which is that macroscopic objects cannot occupy different positions at the same time. It seems, therefore, inherently absurd.

This, very briefly sketched, is the essence or, better, the origin and first aspect of the problem. Now, when confronted with it, the theorists respond in a great variety of ways. Some simply say—in a somewhat dogmatic fashion—that, since the pointer is a macroscopic object, it is *necessarily* in one definite place and that therefore the interpretation of Eq. (14–4) is straightforward: it simply "means" that the pointer *is* in one of the positions g_n with respective probabilities $|a_n|^2$. Others quite rightly stress the fact that the initial assumptions leading to Eq. (14–4) are much too crude and should be replaced by more realistic ones. A third group accepts the idea that, after all, even macroscopic objects can consistently be regarded as being in superpositions of macroscopically distinct states, but run across the difficulty that the observer himself, who is at the end of the chain, knows for sure that *he* is *not* in such a superposition, so that he cannot be simply identified with an ordinary macroscopic system.

All these points of view, theories, or statements will be discussed at length below. As will then appear, some of them go as far as to question the absolute validity of the general rules of quantum mechanics. Here, however, we must first give particular consideration to the standpoint of the physicists who simply assert that, for a priori reasons, there can be no real problem here, and thus, a fortiori no difficulty with quantum mechanics stemming from the point at issue.

The physicists who thus claim that the problem just alluded to is inherently a *false* one have indeed a position which appears very strong. They simply point out that the only legitimate purpose of science is to make predictions concerning the results of future observations from the knowledge of previous ones. Since, they say, quantum mechanics has hitherto not led to any prediction contradicted by experiment, there can, up to now, exist no valid reason for questioning any of its principles or any consequence of these principles.

In the particular case of quantum mechanics, this kind of argumentation can be developed as follows.

(*a*) Observe that quantum mechanics makes predictions only for ensembles of systems.

(*b*) Notice that, if an ensemble of identically prepared initial states is considered, the pure case of Eq. (14–4) which results from the interaction is in fact—because of our limited abilities in investigating complex correlations between atoms of a macroscopic system—practically undistinguishable from a mixture in which the pointer of the instrument occupies in each individual case a definite position g_n (with a relative frequency $|a_n|^2$).

(*c*) Observe that, if (*b*) were not true for a particular instrument, one should simply apply (*b*) to some instrument higher up in the "hierarchy" of successive instruments that culminates in the observer.

(*d*) Conclude that, since there is now no observable difference between the description in terms of (14–4) and the description by the corresponding mixture, we can use the one we like better (keeping in mind that this, of course, is of no importance, as is always the case when two alternative descriptions lead to the same observable results).

All contemporary scientists know the weight of such arguments. Indeed, everybody knows that science was encumbered in the past with many false problems; these originated mainly from the fact that the limitations which in recent times have been rather generally ascribed to the scientific purpose—that of correlating past and future observations and nothing more—had not yet been widely accepted or, indeed, even clearly stated. Nowadays, a salutary "reaction" operates. Hence any argument which points toward the nonobservability and therefore the absence of meaning of certain concepts or of particular descriptions is quite sure to receive, in the circles of contemporary physicists, all the attention it deserves. In spite of this, however, many theorists consider that, in the present instance, the argument described has to be examined particularly closely and critically.

The reason for this attitude can be sketched as follows (see Chapters 19 and 20 for a more thorough discussion). For the motivations alluded to above, it may be (and it even probably is) extremely wise to *define* science as being simply the body of rules that permits us to make predictions about future observations when we know the results of enough previous ones. We should then, however, argue consistently on this basis. In particular, we should then concede that any statement which does not bear, directly or indirectly, on observations, that is, which does not refer, in the last resort, to a human consciousness in its act of perception, is not truely a scientific one. If it were, science would be more than the simple body of rules mentioned above and the definition would not hold. Then, either we accept the extreme view that consists in not believing in any reality outside the human perceptions, or we somehow bow to the fact that truly scientific knowledge is extraordinarily limited, both in scope and in interest, since it gives no clue as to what the things *are*, independently of ourselves. Many of us who, for reasons of mental rigor, instinctively tend to agree with the definition of science given above are undoubtedly optimistic enough to believe that from the "proper" scientific knowledge we can nevertheless at least draw reasonable inferences as to some general traits of the reality around us. But then the purity of the doctrine is lost, and, as we shall see, difficulties may well creep in.

To state the same remark differently, it is of course quite true that both classical and quantum physics can be formulated simply as sets of rules which serve for predicting observations; we see, therefore, nothing particularly new or suprising in quantum mechanics when we choose that standpoint. However, it is also true that if we like we can reformulate classical physics in terms that refer to an independent reality (objects are then assumed to have such and such positions, etc., at least within some approximation). It is therefore a meaningful question to ask whether or not we can, if we like, consistently reformulate quantum mechanics along these same lines, if not in regard to microscopic (see Chapter 8), at least in regard to macroscopic objects.

Finally, let us observe that, even if it is argued that quantum physics needs the concepts derived from classical physics for its own formulation [1], the question posed above remains meaningful. It simply takes the following modified form: "In regard to macroscopic objects, do some observable predictions logically follow from the very introduction of these concepts as adequate tools for describing physical systems such as instruments? If so, are these predictions compatible with those given by an appropriate quantum-mechanical treatment of the overall systems, instruments being included?" Clearly, the foregoing questions bring down the problem from the heavens of the philosophical a priori to the more terrestrial and also more accessible regions of ordinary theory. Several of the later chapters deal with the problems thus raised and with the suggestions that have been made in order to solve them.

14.2   SURVEY OF IDEAL MEASUREMENT THEORY

The law of evolution in time described by expression (14–4) constitutes the essential part of the conventional measurement theory. In this theory, the existence of a Hamiltonian describing the instrument-system interaction and leading to Eq. (14–4) is postulated, not derived. In other words, this assumption is considered as a legitimate idealization of the experimental facts. A schematic but explicit example of such a Hamiltonian is given below.

As already mentioned, the reason why Eq. (14–4) can serve as a basis for a measurement theory is as follows. Let us assume that, at some time after the interaction between the system S and the instrument A, an observer measures the value of L on S *and* the value of G on A (for this purpose he uses other instruments whose mode of operation we refrain as yet from analyzing). If the observer carries on investigations of this kind on a whole ensemble of identically prepared $S + A$ systems, we know beforehand that he is bound to discover a 100 per cent correlation between the results of his L measurements and the results of his G measurements. From the rules of quantum mechanics, each time that he finds $G = g_n$ he must also find $L = l_n$, and conversely. In other words, as soon as our observer has measured G on A, he knows the value that he will find upon measurement of L on S. Under these conditions he can dispense with the latter operation. We say, therefore, that by measuring G on A *he has performed a measurement of* L.

Let us now imagine that the observer selects a subensemble of all the $S + A$ systems by keeping only those for which he finds the value $G = g_n$. Rule 10a of Chapter 3 then informs us that this subensemble is also described by a ket. In the ideal case, this ket is the projection (renormalized to unit norm) of the right-hand side of Eq. (14–4) on the subspace defined by $G = g_n$. The corresponding projection operator is

$$\sum_k |\psi_n\rangle \, |n, k\rangle \langle \psi_n| \, \langle n, k| \qquad (14\text{--}5)$$

so that the renormalized result is

$$|\psi_n\rangle \, |n, s_{n,r}\rangle \qquad (14\text{--}6)$$

Obviously, this product can be interpreted by saying that the subensemble of systems S is described by $|\psi_n\rangle$ and that of the systems A by $|n, s_{n,r}\rangle$. In other words, the wave packet reduction

$$|\psi\rangle \rightarrow |\psi_n\rangle$$

due to the interaction of S with A, instead of being postulated from the start, is here derived from a similar postulate, which bears, however, no longer on S but exclusively on the measurement made on A.

The foregoing description cannot be considered as a complete account of what happens during the measurement process, since it simply carries the problems over from the S–A interaction to the interaction of A with the apparatus that serves to measure G on A. This second process must now be analyzed as well. For this purpose the transition (14–2), which is at the basis of the ideal measurement theory, can be generalized as follows. Let us consider a process in two stages. In a first stage, the system S with which the observable L is associated interacts with an instrument A—whose pointer coordinate is here labeled A as well—according to scheme (14–2), which reads as

$$|\psi_m\rangle \, |a_0\rangle \rightarrow |\psi_m\rangle \, |a_m\rangle \qquad (14\text{--}7)$$

In a second stage, instrument A interacts with another instrument B which is so devised as to operate with respect to A in the same way as A operates with respect to S, that is,

$$|a_m\rangle \, |b_0\rangle \rightarrow |a_m\rangle \, |b_m\rangle \qquad (14\text{--}8)$$

The correct quantum-mechanical description for this succession of events is obtained by considering the complete state vector representing the whole system. As a consequence of the time-dependent Schrödinger equation, this state vector undergoes the changes symbolized by the arrows in the expression

$$|\psi_m\rangle|a_0\rangle|b_0\rangle \rightarrow |\psi_m\rangle|a_m\rangle|b_0\rangle \rightarrow |\psi_m\rangle|a_m\rangle|b_m\rangle \qquad (14\text{--}9)$$

(Notice that, for the sake of maximum simplicity, the degeneracy indices have been suppressed.) If, now, an initial state is considered in which S is not in an eigenstate of L:

$$|\psi\rangle = \sum_m c_m|\psi_m\rangle \qquad (14\text{–}10)$$

one sees by applying linearity in the usual way that the corresponding succession of events is described by the arrow in the expression

$$(\sum_m c_m|\psi_m\rangle)|a_0\rangle|b_0\rangle \rightarrow (\sum_m c_m|\psi_m\rangle|a_m\rangle)|b_0\rangle \rightarrow$$

$$\sum_m c_m|\psi_m\rangle|a_m\rangle|b_m\rangle \qquad (14\text{–}11)$$

Obviously, this generalization can be extended to an arbitrary number of instruments A, B, C, ... This is the principle of what is often called von Neumann's "hierarchy" or "chain." Formula (14–11) makes it quite clear that what propagates from S to A, then from A to B, etc., is a correlation: accroding to the rules of quantum mechanics, the information contained in (14–11), for instance, is that if, after the interaction between A and B has taken place, somebody happens to measure (by a procedure we do not analyze) L, A, and B on the corresponding ensemble, he will always obtain sets of three values labeled by the same index m, and he will find them with relative frequencies $|c_m|^2$.

Clearly, in regard to the observables L and A, this predictive content is identical to the one contained in the appropriate superposition of right-hand sides of evolution schemes such as (14–7). In other words, to calculate the probabilities that he will find the value l_m for L, the experimentalist who uses the "chain" of instruments A, B, . . . for his measurement (A may be his voltmeter, B his eye, etc.) can in principle use several equivalent methods.

First procedure: He considers S alone as being the "physical" system; he then writes a state vector that is in the Hilbert space of S alone and with it calculates the probabilities. In other words, he does not treat A, B, . . . as ordinary physical systems interacting with S; instead, he considers them as forming a part of the device by which the "further unanalyzed measurement procedure" is carried out.

Second procedure: He considers S plus A as constituting together the physical system; he then writes a state vector that is in the Hilbert space of S and A and makes his calculations with it. When he uses this method, he considers A just as he would consider any other physical system, but he still considers B, etc., as parts of the device by which the "further unanalyzed measurement procedure" is carried out.

Third procedure: He considers S plus A plus B as constituting together the physical system, and so forth. He knows that he will obtain the same result (e.g., $L = l_m$) if he applies his "unanalyzed measurement procedures" to a measurement of B and takes the aforementioned correlations into account as he would if he were to apply them "directly" to a measurement of L.

In practice L can, as in von Neumann's example [2], be the temperature of a liquid. Then A could be the thermometer; B, the light rays from the thermometer to the observer's eye; C, the observer's eye; D, his optical nerve; E, his cortical cells; and so forth. At one time, of course, this otherwise infinite enumeration must be terminated. It therefore must be said, "And finally the observer observes the quantity F on the corresponding instrument F (or G on G, etc.) and finds f_m (or g_m, etc.)." Since the theory does not analyze this last step, it must of course be considered as incomplete. It should nevertheless be recognized that, from a pragmatic standpoint at least, the theory adequately compensates for this shortcoming. This it does simply by pointing out the fact that, provided the chain *is* cut somewhere in the manner just described, the place where this is done—or, better, imagined to be done—is entirely irrelevant to the result.

14.3   AN EXPLICIT EXAMPLE

At this stage, a question of course comes to mind as to whether the notion of "ideal measurements" really represents adequately the physical process that occurs in a measurement. In this connection it is relevant to notice that a (highly schematized) system-apparatus interrelation which *does* lead to Eq. (14–2) or (14–4) can be constructed as follows [2, 3].

Let the observable L, to be measured on system S, be a function of the coordinate q and conjugate momentum p of that system exclusively. Let, moreover, the Hamiltonian describing the interaction between S and the instrument be a time-dependent one given by (with $\hbar = 1$)

$$H_{int} = \delta(t)\, L\left(q, \frac{1}{i}\frac{\partial}{\partial q}\right) \frac{1}{i}\frac{\partial}{\partial Q} \qquad (14\text{–}12)$$

where t is the time, and Q is the coordinate of the pointer (we work in the q, Q representation). Because of the impulsive character of H_{int}, the time-dependent Schrödinger equation is, between times 0_- and 0_+, the particular case of

$$i\frac{\partial \psi}{\partial t} = \frac{d\alpha}{dt}\, L\left(q, \frac{1}{i}\frac{\partial}{\partial q}\right)\frac{1}{i}\frac{\partial}{\partial Q}\, \psi \qquad (14\text{–}13)$$

that corresponds to $d\alpha/dt = \delta(t)$. Now (14–13) has as a general solution

$$\psi = \psi_m f[Q - \alpha(t)l_m] \qquad (14\text{–}14)$$

where f is an arbitrary function of its argument. In our problem $d\alpha/dt = \delta(t)$,

$\alpha(t)$ is the step-function. On the other hand, f is determined by the initial conditions. If, in particular, the instrument pointer initially is in a definite position, for instance $Q = 0$, the initial wave function of the composite system is the product

$$\psi_0 = \psi_m \, \delta(Q) \qquad (14\text{-}15)$$

Thus the function f is identical to the delta function, and the final state (at $t = 0_+$) is

$$\psi = \psi_m \, \delta(Q - l_m) \qquad (14\text{-}16)$$

because of (14–14). H_{int} therefore induces the transition

$$\psi_m \, \delta(Q) \rightarrow \psi_m \, \delta(Q - l_m) \qquad (14\text{-}17)$$

which is an example of process (14–2). Obviously, however, examples like this one stand rather far from the real complexity of the measurement processes actually occurring. Indeed, the "ideal measurement" is, as will turn out—see, in particular, Chapter 18—a process far too special to correspond to what really occurs in most cases. This is the reason why nonideal measurement processes are considered in several of the analyses that follow.

14.4 MEASUREMENTS AND THE DENSITY MATRIX FORMALISM

As the foregoing discussion shows, a wave packet reduction can be considered as a phenomenological procedure that, in many instances, can be applied in order to calculate the results of successive observations and the correlations between them without introducing into the formulas the complicated wave functions of the various instruments.

Let an ideal measurement be made on an ensemble of systems S, described by the density matrix ρ; that is, let an observable L be measured on each system S (this implies, of course, an interaction of each S with an instrument, but we refrain from analyzing this interaction). Then we can either:

(a) pick out the system S for which the value l_m has been found, or
(b) consider again the ensemble of *all* systems S, that is, mix up all the systems again.

In both cases, the question is, "What is the density matrix of the new ensemble?" Indeed, the desired operator can be expressed quite simply in terms of ρ. Let us derive this expression [4, 5] first in the case where all the eigenvalues of L are nondegenerate, and then in the case where some of them are degenerate.

1. Nondegenerate Eigenvalue

Case (*a*): The desired density matrix ρ_m is a projection operator on a one-dimensional vector space, namely, $|\psi_m\rangle$. It is therefore

$$\rho_m = P_m \equiv |\psi_m\rangle\langle\psi_m| \qquad (14\text{–}18)$$

Since

$$\text{Tr}(P_m\rho) = \langle\psi_m|\rho|\psi_m\rangle \qquad (14\text{–}19)$$

(this expression is the probability p_m that the value l_m is observed), equation (14–18) can be rewritten as

$$\rho_m = \frac{|\psi_m\rangle\langle|\psi_m|\rho|\psi_m\rangle\langle\psi_m|}{\text{Tr}(P_m\,\rho)} = \frac{P_m\rho P_m}{\text{Tr}(P_m\rho)} \qquad (14\text{–}20)$$

Case (*b*): The desired density matrix ρ' is the sum of the density matrices previously found, each multiplied by the relative number of systems in the subensemble it represents. This immediately gives

$$\rho' = \sum_m P_m\rho P_m \qquad (14\text{–}21)$$

2. Degenerate Eigenvalue

In this case simple expressions are obtained only if it is assumed, as was done in Chapter 3 (Rule 10a), that the measurement is not only "ideal" but also "moral," or, in other words, that it has the effect of simply projecting any state vector representative of a premeasurement state into the subspace of the Hilbert space that corresponds to the observed eigenvalue (14–2).

For the sake of the calculation, it is convenient here to express ρ as a linear combination:

$$\rho = \sum_k w_k\rho_k = \sum_k w_k|\phi_k\rangle\langle\phi_k| \qquad (14\text{–}22)$$

of projection operators ρ_k on mutually orthogonal states $|\phi_k\rangle$. This expansion is, as shown in Chapter 6, always possible, and the ρ_k are the density matrices of subensembles whose union constitutes a proper mixture having ρ as its density matrix. Of course, the w_k are the relative numbers of individual systems in all of these subensembles.

Case (a): Let P_m be the projection operator onto the subspace pertaining to eigenvalue l_m. Each subensemble described by a $|\phi_k\rangle$ is transformed by the measurement into its projection on that subspace renormalized to unity, that is, into

$$|\chi_{km}\rangle = \frac{P_m|\phi_k\rangle}{(\langle\phi_k|P_m|\phi_k\rangle)^{1/2}} \qquad (14\text{--}23)$$

The subensemble labeled k is thus transformed into a subensemble which has

$$|\chi_{km}\rangle\langle\chi_{km}|$$

as its density matrix. Recombining all the subensembles for a given m with weights n_{km}/n_m proportional to this relative population number, we obtain again after a straightforward calculation

$$\rho_m = \frac{\sum\limits_k w_k P_m|\phi_k\rangle\langle\phi_k|P_m}{\sum\limits_k w_k\langle\phi_k|P_m|\phi_k\rangle} = \frac{P_m\rho P_m}{\mathrm{Tr}(P_m\rho)} \qquad (14\text{--}24)$$

where we have used the relations

$$n_{km} = Nw_k\langle\phi_k|P_m|\phi_k\rangle; \qquad n_m = \sum_k n_{km}$$

in which N is the total population.

Case (b): The total probability that result l_m is obtained is obviously

$$p_m = \sum_k w_k\langle\phi_k|P_m|\phi_k\rangle \qquad (14\text{--}25)$$

so that the desired density matrix, obtained by combining the subensembles labeled m with weights proportional to their population number, is again

$$\rho' = \sum_m P_m\rho P_m \qquad (14\text{--}26)$$

14.5   AN ERRONEOUS ARGUMENT

In connection with expressions (14–21) and (14–26) derived above, a warning should be made [6] against a rather common but erroneous argument bearing on the problem of measurement. This reasoning (which has a mathematically neat appearance but is nonetheless fallacious) goes as follows. Since, as a rule, from Eq. (14–26) $\rho'^2 \neq \rho'$ even when $\rho^2 = \rho$, ρ' is, as a rule, the representative of a mixture even when the initial ensemble of systems is pure. Moreover, each separate term in the sum (14–26) corresponds to a given value of the measured quantity L. Thus, it is argued, the interaction of S with A changes the ensemble of systems S into a mixture of systems a given proportion of which *are* in states that have definite L values. The observer who looks at A then simply acquires the knowledge of this already established situation.

The reason why this argument is false is simply that the mixture of systems *S* that emerges from the interaction is an improper, not a proper, mixture, in the sense explained in Chapter 7. In other words, from the fact that the representative, ρ', of the ensemble of systems *S* has the form (14–21), it *cannot* be concluded that each system *S* has a well-defined *L* value. Indeed, in the case where the initial ensembles of systems *S* and of instruments *A* are both pure cases, we know already that such a conclusion would be false, since, as shown in Chapter 7, it would necessarily lead to some predictions that would conflict with those that are derivable from the state vector of *S* + *A*. These predictions bear on correlations between *S* and the instrument. A typical example of the errors that can be made by unduly attributing definite values to the observable *L* and/or *G* of the systems and of the instruments, respectively, is given in Exercise 2 on page 227 (see also Ref. [3]).

REFERENCES

[1] L. D. Landau and E. M. Lifshitz, *Quantum Mechanics,* Pergamon Press, London, 1958.
[2] J. von Neumann, *Mathematical Foundations of Quantum Mechanics,* Princeton University Press, Princeton, N.J., 1955.
[3] J. S. Bell and M. Nauenberg in *Preludes in Theoretical Physics: In Honor of V. F. Weisskopf,* A. De Shalit et al., Eds., North-Holland, Amsterdam, 1966.
[4] G. Lüders, *Ann. Phys.* **8**, 322 (1951).
[5] L. Furry, *Boulder Lectures in Theoretical Physics,* Vol. 8A (1965), University of Colorado Press, 1966.
[6] B. d'Espagnat in Ref. [3].

Further Reading

M. Born, *Natural Philosophy of Cause and Change,* Oxford University Press, London, 1949.

L. Tisza, *Rev. Mod. Phys.* **35**, 151 (1963).

L. de Broglie, *La Théorie de la Mesure en Mécanique Ondulatoire,* Gauthier-Villars, Paris, 1957.

CHAPTER 15

CLASSICAL PROPERTIES

If the measurement process is considered to be well represented by the "ideal measurement" scheme, and if the measuring instrument is described by a state vector, then, as shown in Chapter 14, the postulate of common sense that a macroscopic object such as a pointer necessarily is at an approximately definite place comes into contradiction with the linear laws of quantum mechanics.

In the present chapter, theories are reviewed which try to suppress this contradiction by taking into account the *role* of the instrument. These theories are centered on the notions of *classical* instruments and of classical measurements. Other theories, based on the concept of macroscopic properties, that is, on the essential complexity of the measuring apparatus, are reviewed in Chapter 16.

15.1 CLASSICAL MEASUREMENTS

One of the most distinctive features of quantum mechanics is the fact that it often assigns precise structures—or discrete "choices" among possible different structures—to the physical systems it describes. A hydrogen atom, for instance, has in its ground state a perfectly well-defined structure that makes it qualitatively different from a hydrogen atom lying in the first, second, . . . excited states. This is indeed the most startling difference—suitably emphasized by the very epithet "quantum"—between many systems described by quantum physics, or "quantum systems," and the hypothetical systems that would obey the laws of classical physics. "Hydrogen atoms" as described by classical physics would exhibit an infinite variety of orbits, radii, eccentricity, and so on. Presumably, if the parameters characterizing such an "atom" were just slightly changed, the properties of that system would be quantitatively but not qualitatively altered, so that the system would still be classified as an "atom" of the same type. Such features are, of course, not really specific to classical physics, since they are shared also by quantum systems when these are in states belonging to a continuum or at the limit of very large quantum numbers. However, in classical physics, they represent more or less the general case, so that one is naturally led to say (at least in a loose way) that a system has classical features if small modifications of its internal parameters are possible and do not qualitatively alter the "general appearance" of the system, or, in other words, its most noticeable properties.

Not all "large" systems exhibit predominantly classical features in this sense; some large living molecules can, for example, have such a degree of internal organization that "minor" changes in their internal parameters—a change in the position of a small group of atoms, for instance—can alter their properties in an extremely significant way. It turns out, however, that for several possible reasons (the detail of which is here irrelevant), most large systems—and instruments of measurement, in particular—do have the classical features sketchily described above. At least in regard to the use we make of them, minor modifications in the relative locations, velocities, and so forth of small groups of atoms inside these systems are obviously of negligible importance. The general guiding idea of the theories of measurement reviewed in this chapter (see, in particular, Refs. [1] and [2]) is to define more precisely and then to exploit general characteristics of this kind.

When we say that the precise location of such and such small groups of atoms inside the pointer of an instrument has no practical importance, what do we mean exactly? Obviously, what we have in mind is the fact that neither their precise location nor even the *very* precise location of the whole pointer has any importance, as a rule, in respect to the final result of the whole measurement. This, again, is a characteristic of quantum physics and, in particular, of the measurement of quantized observables; whereas in classical physics the measurement probe is usually small and delicate, so as to minimize the perturbation it might create on the object, the measurement systems in quantum physics are unavoidably immensely larger than the object under investigation and their precise coordinates are always, comparatively, poorly known. However, this fact has no undesirable consequences because, in particular, if an observable is known to be quantized, even a very imprecise knowledge of the value of the instrument coordinate is sufficient to eliminate all possibilities except one, and therefore to give a precise and exact value. This holds in regard to both general and particular properties.

As an instance of the former case, take the measurement of the spin of a type of particle by means of a Stern-Gerlach instrument. There, even a very poor statistical knowledge of mean positions is adequate to obtain an exact result, since it is sufficient to simply count the number of spots appearing on the screen. As an example of the particular property case, which incorporates, as already mentioned, the only kind of measurement that we really consider here, take the measurement, with the same instrument, of the z spin component of a silver atom, which is already known to be either in a state with pure $S_z = \frac{1}{2}$ or in a state with pure $S_z = -\frac{1}{2}$. Again, in order to get an exact result, it is sufficient to acquire but a rough knowledge of the z coordinate of the atom after it has traversed the instrument; as a matter of fact, a knowledge of only the sign of this quantity is sufficient.

The examples just described make it apparent that, in the type of theories considered in this section, the essential concept is *not* the complexity of the instrument. Rather, it is the irrelevance, in regard to the information to be obtained, of most of the finer details concerning the instrument variables, so that these variables can indeed be very few in number, and, correlatively, the instrument may be quite simple provided only that this property of irrelevance is satisfied. In other words, the fundamental notion in these theories, the one which we shall now have to define in a precise way, is not that of "a classical system" but that of a "classical measurement." Once we know what a classical measurement exactly consists of, it will of course be a trivial matter to choose an apparatus coordinate in such a way that:

(*a*) it is in correspondence with the physical quantity L to be measured, and

(*b*) it is such that a mere "classical measurement" of this coordinate is sufficient to provide an exact value for the quantity L of interest.

In order to arrive at a useful definition of classical measurement, it is advisable to introduce new quantities A', B', C', ..., which:

(*a*) are approximations to the observables A, B, C, . . . under consideration,

(*b*) have a discrete spectrum and can therefore be exactly measured, and

(*c*) commute with each other. How this can be done is shown explicitly by von Neumann for the case of two observables, P and Q, such as momentum and position, whose commutator, $-i\hbar$, is a number. The procedure consists in first looking for a complete system of functions [defined, e.g., in configuration space: $\psi = \psi(q)$] whose amplitudes are appreciably different from zero only in some small segment of the q axis and whose Fourier transforms $\Psi(p)$ are appreciably different from zero only in some small segment of the p axis. A system of functions that possesses these properties is the following (labeled by two independent indices $\mu, \nu = 0, \pm1, \pm2, . . .$):

$$\psi_{\mu,\nu}(q) = \left(\frac{\gamma}{\hbar\pi}\right)^{1/4} \exp\left\{-\frac{\gamma}{2\hbar}\left[q - \mu(2\pi\hbar\gamma)^{1/2}\right]^2 + i\left(\frac{2\pi}{\gamma\hbar}\right)^{1/2}\nu q\right\} \quad (15\text{–}1)$$

This system of functions can be orthogonalized by applying Schmidt's method. Let $\phi_{\mu,\nu}(q)$ be the resulting system of functions and $\Phi_{\mu,\nu}(p)$ be their Fourier transforms. The corresponding kets are, of course,

$$|\mu, \nu\rangle = \int dq \, \phi_{\mu,\nu}(q)|q\rangle \quad (15\text{–}2)$$

The operators

$$Q' = \sum_{\mu,\nu} (2\pi \, \hbar\gamma)^{1/2} \, \mu \, |\mu, \nu \rangle \langle \mu, \nu| \qquad (15\text{--}3)$$

and

$$P' = \sum_{\mu,\nu} \left(\frac{2\pi\hbar}{\gamma}\right)^{1/2} \nu \, |\mu, \nu \rangle \langle \mu, \nu| \qquad (15\text{--}4)$$

evidently commute with each other. They have as eigenvalues $q_\mu' = (2\pi\hbar\gamma)^{1/2} \, \mu$ and $p_\nu' = (2\pi\hbar/\gamma)^{1/2} \, \nu$, respectively. On the other hand, the mean square deviation of Q with respect to Q' is, in state $|\mu, \nu \rangle$,

$$\sigma_Q^2 = \langle \mu, \nu | (Q - Q')^2 | \mu, \nu \rangle$$
$$= \|(Q - Q')|\mu, \nu > \|^2 \qquad (15\text{--}5)$$

Now

$$(Q - Q')|\mu, \nu \rangle$$
$$= \int dq \, \phi_{\mu,\nu}(q)[q - (2\pi\hbar\gamma)^{1/2}\mu]|q \rangle \qquad (15\text{--}6)$$

and therefore, with $\langle q|q' \rangle = \delta(q - q')$,

$$\|(Q - Q')|\mu, \nu \rangle \|^2$$
$$= \int dq \, |\phi_{\mu,\nu}(q)|^2 [q - (2\pi\hbar\gamma)^{1/2}\mu]^2 \qquad (15\text{--}7)$$

Similarly, remembering that $\langle p|q \rangle = (2\pi\hbar)^{-1/2} \, e^{-ipq/\hbar}$ [$|p \rangle$ is the ket corresponding to a definite value of P; $\langle q|p \rangle$ is therefore the plane wave $(2\pi\hbar)^{-1/2} \, e^{ipq/\hbar}$],

$$P|\mu, \nu \rangle = \int \int dq \, dp \, \phi_{\mu,\nu}(q) \, P|p \rangle \langle p|q \rangle$$
$$= \int dp \, \Phi_{\mu,\nu}(p) \, p|p \rangle \qquad (15\text{--}8)$$

$$P'|\mu, \nu \rangle = \left(\frac{2\pi\hbar}{\gamma}\right)^{1/2} \nu |\mu, \nu \rangle$$
$$= \left(\frac{2\pi\hbar}{\gamma}\right)^{1/2} \nu \int dq \, dp \, \phi_{\mu,\nu}(q)|p \rangle \langle p|q \rangle \qquad (15\text{--}9)$$
$$= \left(\frac{2\pi\hbar}{\gamma}\right)^{1/2} \nu \int dp \, \Phi_{\mu,\nu}(p)|p \rangle \qquad (15\text{--}10)$$

and therefore the mean square deviation of P with respect to P' in state $|\mu, \nu \rangle$ is

$$\sigma_p{}^2 = \langle\, \mu, \nu\,|(P - P')^2|\,\mu, \nu\,\rangle = \|(P - P')|\,\mu, \nu\,\rangle\|^2$$

$$= \int dp\, |\,\Phi_{\mu,\nu}(p)|^2\left[p - \nu\left(\frac{2\pi\hbar}{\gamma}\right)^{1/2}\right]^2 \tag{15–11}$$

Under these conditions von Neumann has verified the following inequalities:

$$\sigma_P < 60\left(\frac{\hbar\gamma}{2}\right)^{1/2}; \qquad \sigma_Q < 60\left(\frac{\hbar}{2\gamma}\right)^{1/2} \tag{15–12}$$

where, he says, the factor 60 could probably be greatly reduced. At any rate, its presence does not matter, since the product of σ_Q and σ_P is an extremely small number.

Inequalities (15–12) show that, with any reasonable choice for γ (which is still an arbitrary parameter), P' and Q' are indeed very good approximations for P and Q. Moreover, P' and Q' are, as already pointed out, compatible observables.

According to the result just obtained, it is legitimate to assume that, if two quantities Q and R with a continuous spectrum do not commute, two other quantities Q' and R' that are approximations to Q and to R, respectively and that commute with each other can be constructed. As a consequence, Q' and R' can be measured independently in any order, and in such a way that a measurement of one of these quantities does not affect the other one. These properties, which can, of course, be generalized to more than two variables, are precisely those that characterize the observables of classical physics. They imply, in particular, the obvious and well-known consequence that definite values for these observables can, without inconsistency, be attributed to the system itself, independently of the observer, or, in other words, of any further measurement which is planned on that system. Thus we can now define precisely what we mean by a set of "classical measurements": it is one that can formally be reduced to a set of measurements of quantities of the type Q', R', and so forth, which are themselves called "classical observables."

Clearly Q', R', can be chosen in such a way that it is possible to construct a great many—in fact, an infinity of—states that are quantum mechanically different from each other, and that nevertheless correspond to the same values for the quantities Q', R', and so on. If only quantities of the latter type can be measured, then all these states, called "microstates" in what follows, cannot be distinguished from one another. Following Jauch [1], it is convenient to use the word "macrostate" to designate the whole set of microstates on which all the classical observables forming a definite complete set have definite values (in the language in which the word "state" is associated with ensembles, these values cannot differ from one element of the ensemble to the next). If a system exists on which only classical measurements are possible, then obviously *only macrostates have a physical significance for that system: microstates have none.*

15.2   A PROPOSITION BEARING ON CLASSICAL OBSERVABLES

Now that classical measurements have been precisely defined, the next step is to apply this definition to the theory of measurement. For that purpose let us again consider the quantum system S on which a quantity L is to be measured. Following Jauch [1], what we want to show is that, if only classical measurements are possible *on the instrument*, there is no inconsistency [even in the general case considered in Chapter 14, where S is described by (14–3)] in attributing, after the interaction has taken place, definite values both to the instrument variable and to the observable L of S that the experimental setup is fitted for measuring. What is to be shown is, in other words, that under these conditions there is (contrary to the case in Chapter 14) no difference between the observable predictions of the complete state vector describing instrument *plus* quantum system and those of an appropriate mixture of states characterized by definite values of the above-mentioned quantities.

To prove this, it is not necessary to describe the initial state of the instrument by means of a mixture or density matrix (as we shall do in Chapter 17); a simple-minded description using a state vector is indeed sufficient (if valid for that case, the result is of course also valid for a mixture). Let us thus again consider (14–4), which is the state vector that describes the complex system after the interaction has taken place, and let us first rewrite it as

$$\sum_{n,r} b_{n,r} | \psi_n \rangle \, | n, r \rangle \tag{15–13}$$

where

$$| n,s \rangle = \sum_r | n, r \rangle \, \alpha_{n,s,r} \tag{15–14}$$

and

$$b_{n,r} = a_n \alpha_{n,s,r} \tag{15–15}$$

In going from (14–4) to (15–13) we have exploited the possibility of changing the basis in the subspace with fixed n of the Hilbert space of the instrument. Consequently, the $\alpha_{n,s,r}$ are arbitrary parameters that determine the choice of the $\{| n, r \rangle\}$. Now let us compare the predictions of the complete state vector (15–13) with those of *any* of the proper mixtures defined as follows:

"a proportion $| b_{n,r} |^2$ of composite

systems in each $| \psi_n \rangle \, | n, r \rangle$" $\tag{15–16}$

If, considering the sequence of events that led to (15–13) or (15–16) as a mere interaction between systems, we imagine for a moment that an observer will now make measurements of his choice on the composite systems described by (15–13) or (15–16), we immediately see (e.g., by considering the mean values) that (15–13) and (15–16) give the same statistical predictions as regards (a) the quantities belonging to system S alone and (b) the correlation between L and the "instrument coordinate," the latter being, by definition the observable, hereafter called G, that obeys

$$G|n, r\rangle = n|n, r\rangle \qquad (15\text{--}17)$$

In general, however, if the whole set of the observables pertaining to the composite system is considered, there exists no mixture of type (15–16) that gives the same statistical predictions as the pure case (15–13); this well-known result, which is at the origin of our troubles, is just a special case of the general result, shown in Chapter 6, that there exists no mixture that is equivalent to a pure case, except that pure case itself. If, for instance, K and T, with eigenvectors $|k, z\rangle$ and $|t, u\rangle$, denote, respectively, an observable belonging to S and an observable belonging to the instrument, then:

(a) the mean value of T is

$$\sum_{n,r,r'} b_{n,r}^* \, b_{n,r'} \, \langle n, r|T|n, r'\rangle \qquad (15\text{--}18)$$

for the pure case (15–13), whereas it is

$$\sum_{n,r} |b_{n,r}|^2 \, \langle n, r|T|n, r\rangle \qquad (15\text{--}19)$$

for the mixture (15–16), and

(b) the probability of finding simultaneously a value k for K and a value t for T is

$$\sum_{z,u} |\sum_{n,r} b_{n,r} \langle k, z|\psi_n\rangle \langle t, u|n, r\rangle|^2 \qquad (15\text{--}20)$$

if (15–13) is used, whereas it is

$$\sum_{n,r} |b_{nr}|^2 \sum_{z,u} |\langle k, z|\psi_n\rangle \langle t, u|n, r\rangle|^2 \qquad (15\text{--}21)$$

if (15–16) is used.

Now it is at this stage that the assumption made in this section, namely, that the measurements that can be made on the instrument are all of the "classical" type defined above, comes into play. What we have to show is that, if this restriction is made, there indeed exists at least one mixture of type (15–16) that gives the same statistical predictions as the pure case (15–13).

The restriction means that all the observables pertaining to the instrument are in fact quantities of the primed (i.e., Q') type and that therefore G and all the possible T commute with each other. If this is the case, there exists a system $\{n, t, v\}$ of orthogonal vectors, in the Hilbert space of the instrument, that are simultaneous eigenvectors of all these observables (the index t stands here for a set of indices that label the eigenvalues of the set of observables T). Let us choose the $|n, r\rangle$ that appear in (15–13) from among this set of vectors. It should perhaps be stressed again that this is always possible, since a transformation of type (15–14) can always be performed on the state vectors of A that correspond to a definite value of n, so as to bring them into the desired form. With this choice (15–13) reads as

$$\sum_{n,t',v} b_{n,t',v} |\psi_n\rangle |n, t', v\rangle \tag{15–22}$$

so that expression (15–18) of the mean value of T takes the form

$$\sum_{n,t',v,t'',v'} b^*_{n,t',v} b_{n,t'',v'} \langle n,t', v | T | n, t'', v'\rangle$$

$$= \sum_{n,t',v,t'',v'} b^*_{n,t',v} b_{n,t'',v'} t' \delta_{t't''} \delta_{vv'}$$

$$= \sum_{n,t',v} |b_{n,t',v}|^2 t'$$

which is identical to the form taken by expression (15–19) when the basis $\{|n, t', v\rangle\}$ is chosen. Similarly, expression (15–20) for the probability of finding a value k for K and a value t for T now reads as

$$\sum_{z,n'',v''} \sum_{n,t',v,n',t'',v'} b^*_{n,t',v} b_{n',t'',v'} \langle \psi_n | k, z\rangle$$

$$\langle k, z | \psi_{n'}\rangle \langle n, t', v | n'', t, v''\rangle \langle n'', t, v'' | n', t'', v'\rangle$$

$$= \sum_{z,n'',v''} |b_{n'', t, v''}|^2 |\langle k, z | \psi_{n''}\rangle|^2$$

Again, this is precisely the form taken by expression (15–21) when the basis $\{|n, t', v\rangle\}$ is chosen.

These are precisely the results that we wanted to prove; they show that, if only measurements of the classical type can be made on the instrument, then, indeed, a description of the final configurations, in which both the instrument coordinate and the quantity L are assumed to have definite (though unknown) values, cannot be contradicted by any measurement.

15.3 SCOPE AND CRITICISM OF THE THEORY

The arguments developed in Section 15.1 and 15.2 show that, when the instrument is of a certain complexity, it is extremely difficult to devise measurements that will contradict the description in terms of a mixture. We should therefore not be surprised—this is the main lesson of the theory described here—that our usual description of common physical systems as "entities" having in all circumstances well-defined positions and shapes works extremely well in practice.

It should, however, be carefully noted that the "impossibility" of nonclassical measurements alluded to above is of only a practical—and thus of an approximate—nature. It is not in any way a consequence rigorously derived from first principles. Indeed, we made it plausible by referring to the complexity, to the "macroscopic" nature of the instrument, an argument which, in the present context, lacks rigor since the classical measurements were defined in Section 15.1 without introducing such notions, and since these notions were, as a matter of fact, not defined in a precise way.*

To show more clearly that the foregoing remarks do indeed restrict rather severely the general scope (or "philosophical importance," if one prefers) of the theory under discussion, let us introduce a simple example. Following Bohm [3] and Wigner [4], we choose the Stern-Gerlach experiment.

Let a beam of spin-$\frac{1}{2}$ particles that propagate along the $O\hat{y}$ axis traverse a magnetic field localized in a small region around O and presenting in that region a gradient directed in the $O\hat{z}$ direction. Let $|u_\pm\rangle$ be the eigenkets corresponding to the eigenvalues $\pm\frac{1}{2}$ of the z component of the particle spin, that is, let

$$\sigma_z|u_\pm\rangle = \pm|u_\pm\rangle \tag{15-23}$$

and let f_0 be the initial wave packet describing the particles before they enter the magnet. It can be shown [3] that to a sufficient approximation the propagation of the wave packet along the y direction can be treated classically, so that it is sufficient to consider the wave packet at any time, and f_0 in particular, as a function of z alone.

Similarly, a sufficient approximation to the real behavior of the state vector is obtained by assuming that, during the finite time ΔT that the particle spends inside the field, the latter is so strong that the free Hamiltonian can be neglected as compared to the interaction Hamiltonian. The behavior of the wave

---

*It is true that the limitations on the precision of the measurements of two noncommuting observables arising from the definition of "classical measurements" given in Section 15.1 greatly resemble the uncertainty relations, which of course *are* derived from first principles. This resemblance is purely formal, however. Heisenberg's uncertainty relations bear on the dispersions of ensembles as a whole, whereas the limitations that serve to define "classical measurements" essentially concern our possibilities of actually performing measurements in different elements of our ensemble. Thus from the uncertainty relations alone one can certainly not derive the conclusion that nonclassical measurements are impossible; indeed, nothing in the general principles of quantum theory forbids one from making an arbitrarily precise measurement of a position coordinate on one system and, simultaneously, an arbitrarily precise measurement of the corresponding momentum on some *other* system belonging to the same ensemble.

packet $f(z, t)$ at all times was explicitly calculated by Bohm [3], using these approximations. The result is that initial particles described (at $t < 0$) by

$$f_0(z)|u_\pm \rangle \tag{15-24}$$

are, for sufficiently large t, described by

$$f_\pm(z)|u_\pm \rangle \tag{15-25}$$

where $f_\pm(z, t)$ are wave packets that are centered at the points

$$z_\pm(t) = \mp \ \mu m^{-1} H' \varDelta T \cdot t \tag{15-26}$$

and whose width is much smaller than $z_\pm$. In (15–26), m is the mass of the particle, μ its magnetic moment, and H' the gradient of the magnetic field, so that (15–26) is, in fact, just the classical expression for the quantity under investigation. The corresponding wave packet in momentum space also turns out to be well separated.

For simplicity, let us first consider a case in which we know that the incoming particles have either $\sigma_z = +1$ or $\sigma_z = -1$ before they enter the magnet. A particle that has $\sigma_z = +1$ goes in the "upper" (i.e., $+$) beam; one that has $\sigma_z = -1$, in the "lower" (i.e., $-$) beam. Thus we can consider that the quantum system on which the measurement is made is *the spin* of the particle (rather than the particle itself), and that correspondingly the instrument coordinate (labeled G above) is the *ordinate z of the particle*. More generally, we can consider that the instrument coordinate is any quantity that is markedly different, when the particle is in the upper beam, from what it is when the particle is in the lower beam. As already stressed, a very poor knowledge of either the position or the momentum of the particle is sufficient to give exactly and with certainty the value of the quantity σ_z. In the language used in this Chapter, we can say, therefore, that classical measurements made on "the instrument" (i.e., on the position of the particle in this case) are sufficient to provide the desired result. Indeed, all the measurements that are made on particle positions in the usual setups associated with Stern-Gerlach magnets are of this type.

Thus the $S + A$ system, in which S is the spin of the particle and A its position, is a clear illustration of the fact that the concept of a classical measurement is quite independent of the complexity of the instrument used. Here the instrument is simple (it is just one particle), and yet all the measurements that must be (and generally are) done on it in order to obtain the desired result are of the classical type. Of course, that first instrument is itself observed by means of another instrument—a counter, for instance—which is in turn observed by means of a third instrument (a tape recorder, or the eye of a physicist), and so on. Here, however, these are not the relevant facts. The relevant fact is that the types of theories we are now discussing make it legitimate to consider the z coordinate of the particle as being itself an instrument coordinate, instead of associating it with the system S on which the measurement is made.

Under these circumstances, the general proposition proved above applies of course to this particular example. If, instead of (15–24), the initial state is a superposition such as

$$2^{-1/2} f_0(z) (|u_+\rangle + |u_-\rangle) \qquad (15\text{–}27)$$

that is, if the incident beam is polarized in the $O\hat{x}$ direction, the final state is, because of the linearity principle, a superposition of (15–25) with the same coefficients as in (15–27), that is,

$$2^{-1/2}[f_+(z)|u_+\rangle + f_-(z)|u_-\rangle] \qquad (15\text{–}28)$$

However, because of the proposition just referred to, if only classical measurements (here, rough measurements of z or p) are possible, (15–28) has the same predictive content as that of a mixture of states

$$f_+(z)|u_+\rangle \qquad (15\text{–}29a)$$

and

$$f_-(z)|u_-\rangle \qquad (15\text{–}29b)$$

in the proportions $\frac{1}{2}$ and $\frac{1}{2}$. For this reason, the theories under discussion can consistently claim that there is no physical difference between an ensemble of N systems described by (15–28) and a mixture consisting of $N/2$ systems described by (15–29a) plus $N/2$ systems described by (15–29b). They conclude that, after the traversal of the magnet, half the particles *are* in the upper (lower) beam and *have* $\sigma_z = +1 (-1)$. In the language in which the word "state" is associated with ensembles, this can also be expressed by stating that (15–29ab) and (15–28) describe one and the same "macrostate," which is then considered as nonhomogeneous.

Now the main interest of the special example under discussion here is to show in a quite elementary way that the conclusion reached above can definitely *not* be true. To this end it is sufficient to imagine that the two beams, after being suitably bent by appropriate external fields, are merged again by means of a second Stern-Gerlach magnetic field that is a replica of the first one (see Figure 15–1).

A detailed description of this *gedanken experiment* and of the experimental setup it requires can be found in Groenevold [5], for instance. It is shown there that, in particular, a considerable screening of the emerging beam is necessary if the experiment is to succeed. However, in principle this is not a

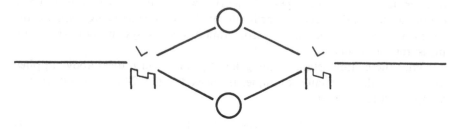

Figure 15–1.

difficulty. The complete calculation of the evolution in time of the state vector initially described by expression (15–27) shows that after the traversal of the second Stern-Gerlach magnet this state vector becomes

$$2^{-1/2}f(z) \, (|u_+ \rangle + |u_- \rangle)$$

This shows that, if the component S_x along $O\hat{x}$ of the spin of the particle is then measured on an ensemble, the result $+\frac{1}{2}$ will be obtained in all the cases, and the result $-\frac{1}{2}$ will never occur. This, however, is not at all the result that is predicted by the mixture (15–29a) and (15–29b). Indeed, if the ensemble were described by such a mixture at some place between the magnets, it is easy to show that a measurement of S_x carried out after the traversal of the second magnet by the particle would necessarily yield $+\frac{1}{2}$ in half the cases and $-\frac{1}{2}$ in the other half. Hence we must now conclude that the statement which asserts that in between the magnets the particle *is* in the upper beam with $S_z = +1/2$ in half the cases and in the lower beam with $S_z = -1/2$ in the other half is simply not true, since it leads to verifiable predictions that conflict with those of quantum mechanics.

   Of course, such a result is not a paradox. It simply recalls to us that, along with the classical measurements that are customary (and are easily done), nonclassical measurements can also in principle be performed.*

   This simple fact is sufficient, however, to show that the assertion that macrostates only have physical significance is not generally valid. If the quantum rules were correct, and if the instrument variables nevertheless really *had* definite values for every system, all the observable consequences of these facts would agree in every circumstance with the correct theoretical predictions.

---

*In particular, measurements that are not "classical" according to our terminology can turn out to be practically feasible in one important case. This is the one in which the time evolution of a pure case such as (15–28) and that of the "equivalent" mixture (15–29) are such that after some time these ensembles are no longer equivalent (in the sense that they do not lead to the same predictions even in regard to the restricted set of observables considered in this chapter).

The assertion is often made that some systems *A*, called "classical systems," are such that after they have interacted with a quantum system *S* some observables of *A*, called "classical observables," always have definite values. At least to some good approximation, measuring instruments and instrument coordinates are then said to be "classical systems" and "classical observables," respectively. However, the definition that is then given of a classical observable, or, more precisely, of a set of such observables, is usually just that it corresponds to a set of classical measurements (as defined in Section 15.1). Moreover in the theories presently analyzed, a classical system is defined as one on which *we can* measure classical observables only. One weakness of such definitions is, of course, that they make use in a decisive way of the limitations of our abilities as human beings in order to define a few essential concepts: a few concepts which (at least in the conventional description of microphysics) play a decisive role in the laws that ultimately the whole universe, including man himself, is supposed to obey. Even if we reconcile ourselves to this idea, however, there is another, serious weakness of these definitions: by extrapolating the argument given above, we may legitimately fear (at least until otherwise shown) that they define an empty class. Both defects would be cured at once if we could define a class of systems by their composition or structure and *show* that these systems are classical in the sense that (*a*) some observables pertaining to them can consistently be said in all cases to possess definite values (and obey classical evolution laws), and (*b*) no other observable leads to difficulties. This program constitutes the motivation that underlies some of the attempts described in the next chapter. In the absence of such a definition as the one just specified and of the corresponding theorem, we must concede that the theories described here are interesting but essentially incomplete.

REFERENCES

[1] J. M. Jauch, *Helv. Phys. Acta* **37**, 293 (1964).
[2] P. K. Feyerabend in *Observation and Interpretation, Proceedings of the 9th Symposium of the Colston Research Society,* S. Korner, Ed., Butterworth, London.
[3] D. Bohm, *Quantum Mechanics,* Prentice-Hall, Englewood Cliffs, N. J., 1951.
[4] E. P. Wigner, *Am. J. Phys.* **31**, 6 (1963).
[5] H. J. Groenevold, *Foundations of Quantum Theory, Proceedings of the Saltzburg Colloquium on Philosophy of Science, 1968,* International Union of History and Philosophy of Science.

Further Reading

V. A. Fok, *Czech. J. Phys.* **7**, 643 (1957).

A. Lande, *Foundations of Quantum Theory,* Yale University Press, New Haven, Conn., 1955.

M. Renninger, *Z. Phys.* **158**, 417 (1960).

CHAPTER 16

MACROSCOPIC INSTRUMENTS

Instruments are necessarily of macroscopic size because our sense organs —which can be considered as instruments occupying a certain rank in the von Neumann hierarchy—are of macroscopic size. It is only natural, therefore, that physicists should have attempted to make use of this property for constructing quantum measurement theories. Moreover, one of the remarkable properties possessed by most macroscopic systems, including all instruments, but lacking in microscopic systems is irreversibility. This simple observation suggests very strongly that the phenomenon of irreversibility and the phenomenon of "wave packet reduction," which is associated in some way or other with the measurement process, should somehow be correlated.

16.1 IRREVERSIBILITY: DUALISTIC APPROACH

Before engaging in a perusal—even a succinct one—of these ideas, it is necessary to point out that there are two broad, general directions in which we can a priori attempt to direct them.

1. We can consider *both* principles of time evolution (Schrödinger-like evolution and reduction of wave packet) as distinct, inequivalent, but still equally valid principles: it is then necessary to formulate a rule, which, without falling into contradiction with any data, exactly specifies when each of these principles should be applied. In the present context the general idea is then to say that the Schrödinger time-dependent equation, or its generalizations, applies for *reversible* phenomena and the wave packet reduction for irreversible ones. The difficulty with this kind of approach is, of course, that it is not easy to separate the phenomena into two classes, the reversible and the irreversible processes, in a perfectly objective fashion. This is even more true in the present case, where the class of irreversible processes must of necessity not contain any of the simple or moderately simple phenomena in which experiment shows that phase correlations and so forth persist.

2. We can consider that only *one* principle of time-evolution is fundamental, namely, the one connected with the Schrödinger time-dependent equation. The wave packet reduction—or, to be more precise, the phenomena for which the wave packet reduction offers a simple "phenomenological" description—should then appear as a consequence of the interaction of the investigated system S with other systems, and with instruments in particular. The difficulty with such an approach is, as shown in Chapter 14, that in the most general case that the superposition principle makes physically possible it becomes difficult to make these views consistent with the idea that the pointer of the instrument necessarily is in some macroscopically well-defined position. The hope is that consideration of the irreversible evolution processes undergone by the instrument might suppress this difficulty. This standpoint is discussed in Sections 16.2 and 16.3.

Let us first look at approach 1. Since we do not, for the moment, want to ground our theory on the influences of the "remote stars," we can consider the composite system (measured system plus instrument) as practically isolated. It suffers irreversible evolution, so that what is actually implied in this approach is that *some* isolated systems should evolve in time differently from what is predicted by the time-dependent Schrödinger equation or by its generalizations. Such a theory, supposing that it could be given a coherent form not in conflict with any data, would at any rate be a modification of quantum mechanics. As such, it falls outside our general program. The interested reader can find a description of theoretical attempts running (in a broad sense) in this direction in Ludwig's works (see especially Refs. [1] and [2]).

In spite of the above remark, it is interesting to consider the question whether quantum physics does or does not provide an "objective" distinction between reversible and irreversible processes. This of course is a vast subject, which has distinctly philosophical aspects, since the very definition of the word "objective" is not the same for a "realist" and for a "positivist" (see Chapter 20). Here we make the question more definite by specifying that we consider the word only in the "realistic" sense ("objective" then means "independent of the existence of any observer"). Moreover, we are satisfied with simply pointing out one or two elementary facts in a rather cursory way. The first of these, which was stressed by von Neumann [3], is that a quantum-mechanical analogue of the classical entropy can be defined, and that this quantity has at least the appearance of an objectively defined quantity, in the sense specified above. This can be seen from its explicit expression: for a mixture of systems of which Np_n "are" in state n, von Neumann's entropy is, by definition,

$$Y = -N\mathcal{N} \sum_n p_n \log p_n = -N\mathcal{N} \operatorname{Tr}(\rho \log \rho) \qquad (16\text{–}1)$$

where $\mathcal{N}$ is Boltzmann's constant, and ρ is the density matrix that describes the ensemble. If this ensemble is an improper mixture, expression (16–1), contrary to the corresponding classical expression for entropy in terms of probabilities, is a definition which, at first sight at least, does not refer in any way to the fact that the knowledge of the observers is not as great as it could theoretically be. There is some motivation, therefore, for stating that it is "objective," much in the same sense as the quantum-mechanical probabilities are said to be "objective." On the other hand, such a statement is made rather unconvincing by the observation that forming an improper mixture by disregarding part of a composite system is in a way a subjective act.

The second fact to which attention should be drawn is that, although Y serves as a useful definition of entropy in many problems (e.g., it is maximal at thermal equilibrium), it lacks the property of the (coarse-grained) entropy that is of main interest to us here. As is immediately seen by applying in the usual way the time evolution operator to its constituent operators ρ and log ρ, Y does not change as long as the systems are isolated, even in cases where the classical analogues of these systems would undergo irreversible evolutions.

This precludes the hope of constructing a "purely objective" distinction between reversible and irreversible processes simply by using definition (16–1) of Y. It is true that von Neumann, for instance, has constructed another acceptable form for quantum entropy, which does not present the above defect. This however, he could do only by introducing the concept that in Chapter 15 was given the name of "classical measurement." Since the definition of classical measurements explicitly refers to some inabilities of human observers, an approach to our problem based on this new definition of entropy* would be no more "purely objective" (in the sense in which the latter word is defined above) than is, for instance, the coarse-grained classical one, or than the theory of measurement described in Chapter 14 actually is.

Recently, important progress has been made toward more "objective"—in our sense—descriptions of the irreversible approach to equilibrium, particularly by van Hove [4] and by Prigogine [5] and his school: in these theories coarse graining is eliminated, in a way; but this is accomplished at the expense of considering only infinite systems and of postulating a rule for the order in which the limits $t \rightarrow \infty$ and $V \rightarrow \infty$ should be taken (t is the time and V is the volume of the system). It is not inconceivable, of course, that these or similar ideas may one day be helpful in constructing a measurement theory based on irreversibility. However, in the general approach investigated in this Section, they are not directly helpful, because of the fact that they establish irreversibility for (infinite) systems obeying the Schrödinger time-

---

*Which can be called "coarse-grained quantum entropy" in analogy with "coarse-grained classical entropy."

dependent equation, whereas what we are aiming at is (let it be recalled) a theory in which irreversible systems would *not* obey this equation.

Finally, let it be noted for future reference that the expression for Y gives zero for a pure case and a positive value for any mixture different from a pure state. It thus increases when a "measurement without selection" in the sense of Chapter 14 is made on the pure case. It can be shown [3] that the same is true also if such a measurement is made on a mixture.

16.2 THE ONE-PRINCIPLE APPROACH: STATISTICAL PHASE CANCELLATION

The theories still to be investigated are those utilizing approach 2 of Section 16.1, in which only one time evolution principle is assumed, namely, that associated with the Schrödinger equation. In the case that several systems interact, this equation is of course to be applied to the composite system. In particular, a measurement made on a microscopic system S with an apparatus A is simply an interaction of these two systems, and therefore is also to be described by the Schrödinger equation [it has, moreover, the special characteristics of inducing correlations, as depicted, e.g., by (14–2)]. The problem is to show that it can *also* (even in the more general case where S is not in an eigenstate of the measured quantity L) be described, at least approximately, in terms of a wave packet reduction. If, for the sake of clarity, we prefer to short-circuit the latter expression, we must show at least that, in all circumstances, the observable predictions which result from that "Schrödinger" description of evolution are not at variance with those implied by the assumption that the macroscopic parts of A are always in macroscopically definite places.

In this section, a simple attempt at such a theory is discussed. It is based on the idea that the interaction between S and A induces phases which, because of the complexity of the apparatus, should be distributed at random.

Let us, for simplicity, assume that L has only two eigenvalues, l_1 and l_2, with eigenvectors ϕ_1 and ϕ_2, respectively, and that the system S initially is in state

$$\psi = C_1\phi_1 + C_2\phi_2 \tag{16-2}$$

or, to be more precise, that the ensemble of systems S is initially the pure case described by ψ. The theory asserts that, after the interaction with A, the ensemble of systems S can, *for every purpose*, be considered as an ensemble of systems distributed over the states with state vectors

$$C_1\phi_1 e^{i\alpha_1} + C_2\phi_2\, e^{i\alpha_2} \tag{16-3}$$

where α_1, α_2 are numbers that depend in a crucial way on the initial internal coordinates of the instrument A. These numbers must therefore be considered as distributed at random, and they are correspondingly to be averaged over when mean values over the ensemble are calculated.

It is easy to see that this description of the final ensemble leads to the correct predictions in regard to the results of future measurements that would be carried out on systems S alone. The argument goes as follows. Let B be any observable on S: here, since l_1 and l_2 are nondegenerate, S can be considered as being the spin of a spin-$\frac{1}{2}$ particle and B as the spin component along an arbitrary direction. The mean value $\bar{B}$ of B on the ensemble of systems S is, according to (16–3), given by the expression

$$
\begin{aligned}
\bar{B} = &\ |C_1|^2 \langle \phi_1 | B | \phi_1 \rangle + |C_2|^2 \langle \phi_2 | B | \phi_2 \rangle \\
&+ 2 \operatorname{Re} [C_1{}^* C_2 \langle \phi_1 | B | \phi_2 \rangle] \cos (\alpha_1 - \alpha_2) \\
&+ 2 \operatorname{Im} [C_1{}^* C_2 \langle \phi_1 | B | \phi_2 \rangle] \sin (\alpha_1 - \alpha_2)
\end{aligned}
\tag{16–4}
$$

averaged over the phases α_1, α_2. However, the mean values of the cosine and sine functions being zero, $\bar{B}$ reduces to the two first terms. On the other hand, the sum of these first two terms is precisely the result that would be obtained if, instead of the ensemble considered here, a proper mixture consisting of proportions $|C_i|^2$ of systems in states ϕ_i ($i = 1, 2$) were considered. In other words, as long as the only observables that are taken into consideration are those belonging to S, there is no observable difference between an ensemble of N systems, distributed at random over the states with state vectors (16–3), and a collection of $N|C_1|^2$ systems having the value $L = l_1$ for L and of $N|C_2|^2$ systems having the value $L = l_2$ for this same observable quantity. Under these conditions it seems legitimate to assert that (16–3) really corresponds to such a collection. Hence the "paradox" of measurement theory appears as solved.

This, again, is the answer provided by the elementary theory of phase cancellation considered in the present section. Let us now investigate what is the true significance of such an answer. The theory under discussion claims that the ensemble E of the systems S can for *every purpose* be considered as the union of several ensembles $E(\alpha_1, \alpha_2)$, each of which is described by one of the state vectors (16–3). However, for such a statement to be true a necessary condition is of course that none of its observable consequences be in contradiction with known facts. In particular, none of them should contradict the assumption that a correlation exists between L and the instrument variable G that serves to measure L, for in the absence of such a correlation the instrument would simply *not* serve the purpose for which it has been introduced in the first place. Now, let g_1 and g_2 be the values that G takes when L takes the values l_1 and l_2, respectively; let P_1 (P_2) be the projection operator corresponding to g_1 (g_2) in the Hilbert space of the instrument; and let $N(\alpha_1, \alpha_2)$ be the number of elements in each $E(\alpha_1, \alpha_2)$ [$\sum_{\alpha_1, \alpha_2} N(\alpha_1, \alpha_2) = N$]. To evaluate this correlation, we must introduce an ensemble of N instruments each of which is associated with one of the N systems S. Let $\rho(\alpha_1, \alpha_2)$ be the density matrix that describes the ensemble of the $N(\alpha_1, \alpha_2)$ instruments corresponding to the elements of $E(\alpha_1, \alpha_2)$. From Assumption Q (Section 4.2) it is known that $\rho(\alpha_1, \alpha_2)$ exists. Then the probability that upon simultaneous measurement of L and G (made immediately after the interaction by means of some other, unanalyzed instruments) the results $L = l_i$ and $G = g_j$ $(i, j = 1, 2)$ should be found is

$$P_r(L = l_i \text{ and } G = g_j) = \text{Tr}\left[(|\phi_i\rangle \langle \phi_i| \otimes P_j) \cdot M\right] \qquad (16\text{–}5)$$

where

$$M = \sum_{\alpha_1, \alpha_2} \frac{N(\alpha_1, \alpha_2)}{N} M(\alpha_1, \alpha_2) \qquad (16\text{–}6)$$

and $M(\alpha_1, \alpha_2)$ is the density matrix that describes the ensemble of the composite systems S + instruments, the S part of which belongs to $E(\alpha_1, \alpha_2)$. Since $E(\alpha_1, \alpha_2)$ is a pure case, it is known from a result of Section 6.6 (Proposition 1) that the most general form of $M(\alpha_1, \alpha_2)$ is

$$M(\alpha_1, \alpha_2) = |C_1\phi_1 e^{i\alpha_1} + C_2\phi_2 e^{i\alpha_2}\rangle \langle C_1\phi_1 e^{i\alpha_1} + C_2\phi_2 e^{i\alpha_2}| \otimes \rho(\alpha_1, \alpha_2) \qquad (16\text{–}7)$$

with the help of Eqs. (16–6) and (16–7), (16–5) gives

$$P_r(L = l_i \text{ and } G = g_j) = \sum_{\alpha_1, \alpha_2} \frac{N(\alpha_1, \alpha_2)}{N} |C_i|^2 \, \text{Tr}\left[P_j \, \rho(\alpha_1, \alpha_2)\right] \qquad (16\text{–}8)$$

But eq. (16–8) implies, for example, that

$$\frac{P_r(L = l_2 \text{ and } G = g_1)}{P_r(L = l_1 \text{ and } G = g_1)} = \frac{|C_2|^2}{|C_1|^2} \qquad (16\text{–}9)$$

and this relation is not compatible with the strict correlation mentioned above, since the latter means that the left-hand side of eq. (16–9) should be zero.

Since the result obtained is negative, we must conclude that the elementary theory of statistical phase cancellation does not solve the difficulties of the quantum theory of measurement.

16.3   MORE REFINED CONSIDERATIONS

Interesting detailed analyses of the S, A interaction have been proposed

[6, 7] that describe in a thorough way the irreversible processes occurring in the instrument after the interaction with system S has taken place. For the obvious reason that only "macroscopic observables" can in practice be observed on the instrument, these investigations focus on such quantities (or, by extension, on the quantities that commute with the macroscopic observables).

In these theories macroscopic observables are defined by considering several mutually orthogonal subspaces V_1, V_2, ..., V_r, ... of the Hilbert space of the instrument. Here the index r really represents a set of indices a, k, v which specify the values that all the macroscopic observables of the instrument have when the instrument is in any of the states belonging to the subspace V_r.

This way of introducing the macroscopic observables obviously implies, as it should, that these observables have at least one common system of eigenvectors (in fact, they have several) and are therefore mutually commuting. The set of observables considered here thus essentially coincides with particular choices of classical observables of the type introduced in Chapter 15. The general proof given in Section 15.2 therefore applies to the present case: if a system S interacts with an instrument, there always exists a proper mixture that provides the correct predictions in regard to all the quantities which it is possible to measure and that is composed of states for which each macroscopic observable can be said to have some definite value. As one sees (we diverge here from some statements appearing in Refs. [6] and [7]; see discussion below), this result is really completely independent of the time evolution of the instrument after its interaction with system S, and, in particular, of its irreversible character. In other words, the disappearance of all cross terms in the expressions for probabilities and mean values is just as effective at time $0 +$ as it is at time t, if $t = 0$ is the time when the interaction takes place.

If this is the case, then what, if any, is the role of irreversibility in the measurement process? The answer to this question may be given as follows. In the general case, the fact that at any given time t a mixture M exists which gives the same predictions as a pure case P for a selected set $\{B\}$ of observables does not by itself imply that at a time $t' > t$ the predictions derived from the transformed expressions of M and P through the Schrödinger time evolution operator still coincide, even in regard to only the observables of set $\{B\}$. On the contrary, in the case in which the instruments are of such a complexity as to obey the ergodic conditions usually associated with irreversibility, the investigations described in Ref. [6] and particularly the subsequent developments due to Prosperi [8] tend to show that there exists a mixture (endowed with all the desired properties) which is practically equivalent to the pure state *for all times* greater than some lower bound. Here "equivalent" is of course to be understood in a restricted sense, since the two ensembles give the same predictions only when the instrument observables are restricted to those which commute with the macroscopic observables (as was already the case in regard to the equivalence at one time only). But the persistence of the equivalence is a new and important fact.*

To understand better the scope and limitations of these analyses, we may consider an example. Let instrument A be a Stern-Gerlach apparatus equipped with Geiger counters on each of the outgoing beams, let S be the spin of an atom polarized in an arbitrary direction, and let us consider an ensemble of identical composite systems $S + A$. The elementary considerations of Section 15.2 then suffice to show that, at any time t subsequent to the interaction time, a mixture can always be found that has two properties: (a) it is constituted of systems $S + A$, each of which is such that one and one only Geiger counter can be said to have been triggered, and (b) it gives the same predictions as the correct overall state vector in regard to any measurement made at time t on any observable that commutes with the set of the macroscopic observables. To this result (which, again, is quite independent of the detailed structure of the instrument) the investigations reviewed here tend to add a further and most welcome piece of information, since their purpose is to show *in a very general* way that if the instrument is ergodic a mixture of the above-described type exists which is suitable at a time t and *remains suitable* at all times subsequent to t.

In spite of their unquestionable interest, the technical aspects of the works whose results we are now reviewing need not be described here because of two limitations in the scope of these theories. The first one is that, although these developments describe adequately most instruments, it seems doubtful that they can be applied to instruments of *any* type. The second one is that, at the present stage at least, they cannot be used to define classical systems in a purely "objective" way, if the word "objective" is meant to imply in particular "that makes no reference to the inabilities of the observers." The reason for this is that the mixtures these theories propose give the same predictions as the pure case *only for a limited set of observables* (those which commute with the macroscopic ones). It is true, of course, that our instruments are macroscopic. Nevertheless, given any macroscopic instrument, we can always, in principle at least, manufacture other macroscopic instruments that will enable us to measure microscopic quantities in the first one (maybe that first instrument will be destroyed by the operation, but such an outcome, which often occurs in physics, is irrelevant to the argument). Up to now, at least, there is in the theories presently reviewed neither a principle nor a theorem showing that such experiments are impossible. Under these circumstances, if we tried to define classical systems by asserting that on such systems no quantity is measurable that does not commute with the macroscopic observables, it is clear that our assertion would indeed define but an empty class.

To conclude, the theories reviewed in this Chapter, and principally the observations of the present section, show that it is *in practice* always possible

*More recent studies by other groups [9, 10] lead to conclusions pointing in the same direction.

to attribute definite macroscopic properties to the measuring instruments. The danger that the error we thus make will ever be detected has been proved to be vanishingly small. These theories do *not* show, however, that the instruments can, without contradiction, be said really to have these properties, since they do not show that all the consequences of such an assumption which are in principle observable (and which therefore have physical significance) actually come out right. In other words, only by explicitly referring to human inabilities (to measure quantities other than macroscopic on macroscopic instruments) could it be claimed that these theories make their point.

Rather recently, Hepp [11] advanced a theory of the quantum measurement process that shows some similarities to the ones reviewed above, as well as some differences. Since this theory is formulated in the language of the algebra of observables, and since the corresponding formalism is not described in this book, it is not possible to describe here the technical aspects of Hepp's theory. On the other hand, readers who already know that theory may be interested in a comparison of its conceptual basis with those of the theories reviewed so far.

As just mentioned, there are both similarities and differences. One similarity is to be found in the assumption, common to all these theories, that when a system is "large" in some sense not all the Hermitean operators in the Hilbert space of the system are observable, and that the nonobservable ones have no physical significance whatsoever. The main difference is that in the theory presently analyzed the criterion for specifying the set Σ of the operators that *do* correspond to observables is not based on the human inability to take direct cognizance of quantities other than macroscopic. Let us not describe in detail the criterion for the specification of the set Σ that *is* used in this theory. To begin with, we merely note that it is introduced by considering infinite systems (systems having an infinite number of degrees of freedom). Coherent states of a quantum system S coupled to an infinite apparatus A are then shown, under special conditions, to suffer changes that destroy the phase relations for the operators of the set Σ (belonging to the composite system $S + A$).* It is concluded that the ensemble of the $S + A$ systems is then

---

*Hence a pure case evolves into an ensemble that, from then on, cannot be distinguished from a mixture. For short, we can say that "a pure case evolves into a mixture." In a way, this does not come here as a surprise. Let us stress again that, even in the case of (suitably chosen) systems having a *finite* number of degrees of freedom, such a phenomenon can also occur (as is shown above) when it is assumed that a sufficiently large set of Hermitean operators does not correspond to observables. On the other hand, it must be emphasized that, in the case of systems having an *infinite* number of degrees of freedom, the choice of the Hermitean operators that do correspond to observables can be made in a way that is distinctly neater, both from the mathematical and from the operational points of view. This is Hepp's mathematically important result. The question of whether or not this particular type of "transition from a pure case to a mixture" (in our simplified terminology) is actually significant in regard to the process of measurement is discussed below.

constituted of elements that *have* some desired properties (e.g., locations of pointers) which could not otherwise be attributed consistently to them.

This is the result sought. On the other hand, it turns out, upon inspection, that the specification of the set Σ used in this theory implies the exclusion from the set of the observables of some operators, the nonlocality of which exceeds the dimensions of Λ, where Λ is a sufficiently large but finite region of space. It can be argued that in the context of the theory this requirement of finiteness is the hidden way in which a reference to the limited abilities of human observers does in fact reappear. Moreover (and even more significantly, although the two facts are related), the special conditions alluded to above include the requisite that *either* the evolution is "catastrophic" (no Hamiltonian) *or* the time needed for the phases to vanish is infinite. Hence, in general, "the composite system $S + A$ can really be said to *be* in a state corresponding to some position of the pointer" only after an *infinite* time has elapsed. As Bell showed, for any finite time t, elaborate measurements could be invented (in accordance with the formalism) the results of which would contradict the assertion* inside the quotation marks.

If what is looked for is but a pragmatical description of the measurement process, the preceding feature creates no difficulty. In particular, for t finite but large (compared with the appropriate relaxation times), it can be shown that such measurements are not feasible in practice. Hence we can assert that the statement according to which the pointer *is* in some definite graduation interval is *approximately true*. In so doing, we adopt the convention that a statement can be "approximately true" even in cases in which some particular measurement would show it to be false, provided only that this latter measurement is not feasible in *practice*, so that the error will never come to light. The use of statements that are approximately true in the above sense is a common practice in physics; and they are often employed, as here, in connection with the idealization that "t large" and "t infinite" are "the same thing."

This practice is quite obviously unobjectionable in most cases. For instance, when (in some idealized classical theory) two Gaussian wave packets that propagate in opposite directions without spreading are considered, it is proper to say that for t finite but sufficiently large these packets "do not interfere." However, it would be even more proper to assert that their interference at time t is smaller than some number $M(t)$. The latter statement has all the correct information content of the first one plus the advantage that no measurement can falsify it! More generally, we usually consider (except, perhaps, if we are die-hard positivists or pragmatists) that an approximate statement S about a system T should be replaceable by a true statement S' having two properties: (*a*) it should have all the correct information content of S (this implies, in particular, that it should bear specifically on T), and (*b*) it should be

---

*For a proof of this see Ref. [12] or Exercise 6, p. 229.

such that no conceivable measurement made on that particular T could falsify it. Unfortunately, Hepp's theory is no better in that respect than any other theory of measurement presented so far. If we consider an ensemble of N measurements, and if we call $T_1, \ldots, T_i, \ldots, T_N$ the individual pointers, there is *no* true statement S'_i endowed with properties (*a*) and (*b*) with respect to the approximate statement S_i that at some large but finite t the pointer T_i is in some definite graduation interval (for any value of i). In fact, all the true statements that might at first sight appear as acceptable candidates apply to the ensemble of the T's, viewed as one great "system," instead of applying to individual pointers, as they should according to (*a*). It is hardly necessary to point out that the same kind of objection can be made also in regard to the theories of the measurement process reviewed earlier in this chapter.

REFERENCES

[1] G. Ludwig, *Z. Phys.* **152**, 98 (1958).
[2] G. Ludwig in *Werner Heisenberg und die Physik unserer Zeit*, Braunschweig, Vieweg, 1961.
[3] J. von Neumann, *Mathematical Foundations of Quantum Mechanics*, Princeton University Press, Princeton, N. J., 1955.
[4] L. van Hove, *Physica* **21**, 517 (1955); **25**, 268 (1959).
[5] I. Prigogine, *Non-Equilibrium Statistical Mechanics*, Interscience Publishers, New York, 1962.
[6] A. Daneri, A. Loinger, and G. M. Prosperi, *Nucl. Phys.* **33**, 297 (1962).
[7] L. Rosenfeld, *Progr. Theor. Phys. (Kyoto)*, Suppl. 222 (1965).
[8] G. M. Prosperi in *Foundations of Quantum Mechanics, Proceedings of the Enrico Fermi International Summer School*, Course IL, Academic Press, New York, 1971.
[9] L. Lanz, G. M. Prosperi, and A. Sabbadini, *Nuovo Cimento* **2B**, 184 (1971).
[10] C. George, I. Prigogine, and L. Rosenfeld, *Nature* **240**, 25 (1972).
[11] K. Hepp, *Helv. Phys. Acta* **45**, 237 (1972).
[12] J. S. Bell, *Helv. Phys. Acta* **48**, 93 (1975).

Further Reading

A. Daneri, A. Loinger, and G. M. Prosperi, *Nuovo Cimento* **44B**, 199 (1966).

J. Jauch, E. Wigner, and M. Yanase, *Nuovo Cimento* **48**, (1966).

L. Rosenfeld, *Nucl. Phys.* **A108**, 241 (1968).

A. Loinger, *Nucl. Phys.* **A108**, 245 (1968).

A. Peres and N. Rosen, *Phys. Rev.* **135B** 1486 (1964).

E. Wigner and M. Yanase, *Proc. Natl. Acad. Sci.* **49**, 910 (1963).

G. F. Chew, *Phys. Rev.* **4D**, 2330 (1971).

CHAPTER 17

OF CATS AND POINTERS

Up to now, the initial state of the instrument has been ascribed a state vector. This convention can be criticized, however, either by observing that the instrument is a complicated system which is therefore imperfectly known, or, more adequately still (this is again the difference between proper and improper mixtures), by noting that the instrument has interacted with many other systems before the time when it is used for measuring L, and that consequently it can be associated, not with any definite state vector, but only with a density matrix (in other words, an ensemble of such instruments identically prepared would be an improper mixture). We may wonder whether this remark offers a clue for solving the difficulty encountered at the end of Chapter 14. There it was shown that, if the ensemble E of instruments is initially a pure case, it is in general impossible to consider that every instrument coordinate has some definite eigenvalue after the interaction with the system has taken place. Let us investigate whether or not this is still the case when ensemble E is a mixture.

17.1 AN APPLICATION OF THE CONVENTIONAL FORMALISM

Since we do not propose to study any correlation involving the systems with which the instrument interacted in its past history (i.e., before it is used for measuring L), it is legitimate for us to describe the ensemble E by means of a density matrix $\rho_0{}^A$. Furthermore, it is legitimate, for the purpose thus limited, to replace an initial improper mixture of instruments by any proper mixture with the same density matrix. Following Wigner [1], let us therefore diagonalize $\rho_0{}^A$. Let p_q and $A_0{}^{(q)}$ be the resulting eigenvalues and a set of corresponding eigenfunctions, respectively (some of the p may be equal, in which case the choice of the corresponding $A_0{}^{(q)}$ is not unique; this, however, is unimportant for our purpose). We shall work with the proper mixture consisting of composite systems (S plus instrument A) in which a fraction p_q of the instruments A are initially in state $A_0{}^{(q)}$.

Let us first consider the simple case in which the system S on which the measurement is to be carried out is initially in one of the eigenstates of the quantity L to be measured, and let ψ_m be that eigenstate. Then, since we still assume that the measurement is an ideal one in the sense previously defined (see Chapter 3), the effect of the unitary operator $U(t, t_0)$ that describes the evolution of the compound system during all the time the interaction takes place is represented by the arrow in the expression

$$\psi_m A_0^{(q)} \rightarrow \psi_m F_m^{(q)} \tag{17-1}$$

In Eq. (17–1) the $F_m^{(q)}$ are state vectors in the Hilbert space attached to A and describe a situation in which the pointer of A is in graduation interval m. (Initially we may think, e.g., that the pointer was in graduation interval 0, in which case the $F_0^{(q)}$ are linear combinations of the $A_0^{(q)}$.) Since the $F_n^{(q)}$ with different values of the index n correspond to macroscopically distinct positions of the pointer, they are mutually orthogonal. Let us also notice at this stage that two $F_n^{(q)}$ with the same n but with different q indices are also orthogonal to each other (and therefore linearly independent). This is due to the fact that two $A_0^{(q)}$ with different values of q are mutually orthogonal by construction and to the property of unitary transformations such as (17–1) of preserving orthogonality. The system of the $F_n^{(q)}$ is, in general, not complete, but we can always complete it by introducing vectors $F_n^{(q)}$ with other q values and by setting the corresponding p_q equal to zero. After the interaction has taken place, a relative proportion p_q of the composite systems (S plus A) is, according to the above analysis, in the state described by the right-hand side of Eq. (17–1).

After these preliminaries let us now consider, just as in Chapter 14, the more general case in which system S is initially in a quantum superposition of states ψ_n given by

$$\psi = \sum_n a_n \psi_n \tag{17-2}$$

Again, it follows from the linearity of the time-dependent Schrödinger equation (or of its possible generalizations) that, for all the $N p_q$ cases in which the instrument may be considered as being initially in state $A_0^{(q)}$ (N is the total number of composite systems in the ensemble we consider), the state of the composite system after the interaction has taken place is necessarily described by

$$Y^{(q)} = \sum_n a_n \psi_n F_n^{(q)} \tag{17-3}$$

On the other hand, let us assume that no superselection rule is at work in the description of our instrument. Then no principle is at present known that would set limits on the observability of the mean values which serve to characterize the density matrix. A mixture M' of states of the composite systems can therefore be a correct description of the final state of affairs—that is, lead to no false observable predictions—only if its density matrix is identical with the density matrix corresponding to an ensemble M of composite systems, a proportion p_q of which is in each state (17–3). Of course this does not imply that the proper mixture M' should, in regard to its composition in terms of definite state vectors, be identical with M, since to a given density matrix there correspond several mixtures. Indeed, M' need not even be made of mutually orthogonal states. A condition that its component states must necessarily satisfy, however, is (as shown in Section 6.5) that they be linear combinations of the states (17–3) that compose M.

As a result of this analysis the only states that we are allowed to use for constructing the ensemble M' of final states of the composite systems are linear combinations of the $Y^{(q)}$. On the other hand, if we accept the postulate of "common sense"—hereafter called "Postulate P"—that in any individual composite system the pointer of the instrument is in some definite graduation interval, we must require that M' be a union of ensembles M'_m, each corresponding to one such interval. We must therefore require in turn that M' be describable by means of a density matrix which should be a linear combination of density matrices describing the M'_m. But, according to our assumption above, a given ensemble is describable by means of one density matrix only. Hence finally we must require that the component states of M' have the general form

$$Z_m{}^{(q')} = \sum_q x_m{}^{(q',q)}\, \psi_m\, F_m{}^{(q)} \qquad (17\text{–}4)$$

This follows from the general content of Section 7.4 (see, in particular, Proposition S and the corollary to Condition R), since states (17–4) are the most general ones in which (a) the measured quantity L has a definite value l_m and (b) the pointer is in the definite graduation interval that, according to Eq. (17–1), corresponds to l_m (the coefficients $x_m{}^{(q',q)}$ are arbitrary parameters, to be chosen at convenience). From all the discussion above, it thus follows that Postulate P is tenable only if the $Z_m{}^{(q')}$ are expressible as linear combinations of the $Y^{(q)}$.

From now on, let the collection of all the $Z_m{}^{(q)}$ that have the same value of m be referred to as the macrostate m. We say that macrostate m is empty if no instrument of the ensemble is in any $Z_m{}^{(q)}$ with that particular value of m (or in any superposition of these). Then the above condition implies that sets of numbers $b_m{}^{(q',q)}$, $x_m{}^{(q',q)}$ should exist such that

$$Z_m{}^{(q')} = \sum_q b_m{}^{(q',q)}\, Y^{(q)} \qquad (17\text{–}5)$$

for all macrostates m that are not empty.

Since the $F_n{}^{(q)}$ are linearly independent, (17–5) can in turn be written as

$$a_n b_m{}^{(q',q)} = \delta_{nm} x_m{}^{(q',q)} \qquad (17\text{–}6)$$

Now, if only one of the parameters a—for example, a_1—is different from zero, such a set obviously exists, namely,

$$x_1{}^{(q',q)} = a_1; \qquad b_1{}^{(q',q)} = 1;$$

$$x_n{}^{(q',q)} = b_n{}^{(q',q)} = 0 \quad \text{for } n \neq 1 \qquad (17\text{–}7)$$

but, if two or more a_n are different from zero, for example, a_1 and a_2, such sets no longer exist, as the following applications of Eq. (17–6) to particular cases clearly show:

$$n = 1; \qquad b_m{}^{(q',q)} = 0 \quad \text{for } m \neq 1$$

$$n = 2; \qquad b_m{}^{(q',q)} = 0 \quad \text{for } m \neq 2$$

that is, in particular, for $m = 1$, so that

$$b_m{}^{(q',q)} = 0 \quad \text{for all } m$$

Since, in the general case (no superselection rule for systems S) (17–2) is, with any choice of the parameters a_n, a physically possible initial state, the conclusion is that the simultaneous validity of description (14–1) of ideal measurements *and* of Postulate P is not compatible with the linearity of the law of evolution with time of quantum mechanics.

17.2   TWO GENERALIZATIONS

The argument that has just been described is based on some assumptions, several of which are admittedly idealizations. It is convenient to defer to chapter 18 the examination of one of the most important of these, namely, the hypothesis that the measurement process is an ideal one in the sense previously defined. Here, however, we can discuss two objections that are frequently made to two other simplifying assumptions used above.

The first of these criticisms is based on the fully correct observation that systems never have completely well-defined properties since, strictly speaking, completely isolated systems do not exist. This remark is true also in classical physics, where a system can never be said to have, for instance, a perfectly well-defined energy, since it always shares some amount of energy with the systems with which it, ever so weakly, interacts. In quantum theory the same is true, of course, and this is certainly a valid reason (one could easily invent a few more) for rejecting the idea that a macroscopic system (such as an instrument pointer) should necessarily be in some macroscopically strictly well-defined state. Without making any revolutionary change in our concepts —indeed, by just extending in a natural way to quantum physics what has always been known in classical physics—we can thus very well accept the following idea. At any one moment in its history a pointer, for instance, is not to be described by one (or a mixture of) wave function(s) all of which are strictly zero except in *one* graduation interval. Instead, we make the more realistic assumption that the pointer can be described by one (or a mixture of) wave function(s) all of which are large in that interval and very small in all the other intervals.

To take this assumption into account [2], let us replace expression (17–4) for $Z_m{}^{(q)}$ by the following one:

$$Z_m{}^{(q)} = \sum_{q'} x_m{}^{(q,q')}\, \psi_m F_m{}^{(q')} + \sum_{q',l,n \neq m} y_{n,m,l}^{(q,q')}\, \psi_l F_n{}^{(q')} \qquad (17\text{–}8)$$

with

$$|y_{n,m,l}^{(q,q')}| \ll |x_m{}^{(q,q')}| \qquad (17\text{–}9)$$

Again, the $Z_m{}^{(q)}$ must be linear combinations of the $Y^{(q)}$:

$$Z_m{}^{(q)} = \sum_{q'} b_m{}^{(q,q')} Y^{(q')} \qquad (17\text{–}10)$$

Since the $F_n{}^{(q)}$ are linearly independent, this condition gives

$$\left. \begin{array}{l} b_m{}^{(q,q')}\, a_m = x_m{}^{(q,q')} \\[6pt] b_m{}^{(q,q')}\, a_n = y_{n,m,n}^{(q,q')} \end{array} \right\} \quad (n \neq m) \qquad \begin{array}{r} (17\text{–}11) \\[6pt] (17\text{–}12) \end{array}$$

However, these relations imply that, if $a_m \neq 0$,

$$y_{n,m,n}^{(q,q')} = \frac{a_n}{a_m}\, x_m{}^{(q,q')} \qquad (17\text{–}13)$$

Now, we can always consider a case in which the quantum system S that is brought into interaction with the instrument is in a state so chosen that $a_1 \approx a_2$, for instance. In this case, Eq. (17–13) contradicts inequality (17–9). We must conclude, therefore, that the enlargement of the class of acceptable Z states considered here is not a sufficient improvement to remove the difficulty of the measurement theory.

However, a second objection should also be considered. To justify our claim that the density matrices that describe the mixtures M and M' must coincide, we had to postulate that no superselection rule is present. This assumption can be criticized as being unrealistic. Let us therefore show that even without this assumption the difficulty remains.

To this end, it suffices to write the density matrices of the instrument A and of the composite system $S + A$ in a basis whose elements are simultaneous eigenvectors of the "superselection charges" Q that generate the superselection rules in question (see Section 7.1). Then, since no system can be found in a state of different, superposed Q values, the nonzero part of the density matrix for the $S + A$ system is made of a set of smaller square matrices that lie along its first diagonal. This structure is preserved during the interaction time, for the superselection charges are constants in time. Since all these superselection charges have been taken into account, the submatrices must now be considered (by assumption) as measurable. Hence the arguments developed above can be applied separately to each of these submatrices. Therefore the conclusions reached in the foregoing sections hold also in this case.

Although the very simple proof just presented suppresses the difficulties specifically connected with the possible existence of some superselection rules, it remains insufficient if we want the arguments previously developed to have a degree of generality that can be considered as fully satisfactory in every respect. The reason for this is that some elements of the density matrix could conceivably be nonmeasurable quite independently of the existence of the superselection rules.

To study this case, let us consider the specific example in which S is the spin of a spin-$\frac{1}{2}$ particle. Let the ψ_n $(n = 1, 2)$ be the eigenvectors of the z component S_z of S, and let us choose the ensemble of initial systems so that it constitutes a completely polarized beam along the direction $O\hat{x}$. This means that

$$\psi = 2^{-1/2}(\psi_1 + \psi_2) \qquad (17\text{–}14)$$

Moreover, let us assume that the instrument never alters the spin state of a particle whose spin has a definite value along $O\hat{z}$. To be quite specific, let us consider the case in which the instrument pointer can lie in one of three intervals, $m = 0, 1, 2$. Initially it lies in interval 0. It goes over into interval n $(n = 1, 2)$ as a result of its interaction with a particle whose spin is described by ψ_n.

Let $H^{(A)}$ be the Hilbert space of the instrument A, and let $A_0^{(r)}$ be an orthonormal basis of the subspace of $H^{(A)}$ that corresponds to a situation in which the pointer is in the scale interval 0. If $\mathscr{U}$ is the operator that describes the evolution between the initial and the final time (before and after the interaction), then by assumption we have for any $A_0^{(r)}$ the relation

$$\mathscr{U}\psi_n A_0^{(r)} = \psi_n F_n^{(r)} \qquad (17\text{–}15)$$

where $F_n^{(r)}$ describes a state in which the pointer lies in graduation interval n $(n = 1, 2)$. The $F_n^{(r)}$ are mutually orthogonal since the $A_0^{(r)}$ have this property (see above). In general, the $F_n^{(r)}$ do not constitute a basis for the subspace $H_n^{(A)}$ of $H^{(A)}$ that corresponds to a situation in which the pointer lies in graduation interval n. Let us therefore introduce other orthogonal vectors, $F_n^{(s)}$, in $H_n^{(A)}$, that are orthogonal to the $F_n^{(r)}$ and that are such that the set $\{F_n^{(t)}\} = \{F_n^{(r)}\} \cup \{F_n^{(s)}\}$ is a basis of $H_n^{(A)}$. Finally, we can, of course, supplement if necessary the set $\{A_0^{(r)}\} \cup \{F_1^{(t)}\} \cup \{F_2^{(t)}\}$ with a set of vectors of $H^{(A)}$ that, together with the foregoing ones, constitute an orthogonal basis of $H^{(A)}$.

Let us define

$$\Phi_n^{(s)} = \mathcal{U}^{-1} \psi_n F_n^{(s)} \tag{17-16}$$

Denoting by $\langle \, | \, \rangle$ the scalar product, we have then

$$\langle \psi_m A_0^{(r)} | \Phi_n^{(s)} \rangle = \psi_m A_0^{(r)} | \mathcal{U}^\dagger \mathcal{U} | \Phi_n^{(s)} \rangle$$
$$= \langle \psi_m F_m^{(r)} | \psi_n F_n^{(s)} \rangle = 0 \tag{17-17}$$

The most general state of $S + A$ that corresponds to a pointer lying in graduation interval n ($n = 1, 2$) and to a spin state of S being correspondingly ψ_n is

$$Z_n^{(t)} = \psi_n \sum_{t'} x_n^{(t,t')} F_n^{(t')} \tag{17-18}$$

as is easily shown, using again Eq. (7-17) and Proposition S.

Therefore, if, after the interaction has taken place, the pointer of each individual instrument is to lie in one definite graduation interval, as we assume to be the case, then the ensemble of the $S + A$ systems must, after the interaction, be describable as a mixture of the $Z_n^{(t)}$ (let the corresponding weights, or probabilities, be $p_n^{(t)}$). Can we deduce from this fact some general conditions that this same ensemble of $S + A$ systems should satisfy at time $t = 0$, that is, before S and A interacted, if the fact in question is true? To show that the answer to this question is positive, let us observe that in the idealized model which constitutes our main field of study in this chapter all the composite systems $S + A$ undergo an "independent time evolution" in the sense defined in Section 13.1. Let us note also that the only observables of interest are those which belong to the systems $S + A$ themselves, since obviously such observables as the correlations between the systems $S + A$ and the other systems with which they have interacted in the past will never be measured (in other words, we shall never measure any observable associated with any element of the set K' of the Hermitean operators that do not belong to the Hilbert space of $S + A$). Hence the arguments presented at the end of Section 13.1* can be ap-

---

*Some thinkers (see Ref. [3]) question the validity of a retrodictive calculation in the cases considered here. For this purpose they make use of one or several of three arguments, (a), (b), and (c). Argument (a) is simply the observation that, in practice, the composite system $S + A$ can never be completely isolated. This argument is correct, but it merely pushes the

plied to the composite systems $S + A$. It follows that the ensemble considered above must be describable at time $t = 0$ as a mixture M, with weights
$p_n{}^{(t)}$, of the "initial" states $Z_{n,\text{in}}^{(t)}$, which are deduced from the $Z_n{}^{(t)}$ by applying the inverse time evolution operator $\mathcal{U}^{-1}$:

$$Z_{n,\text{in}}^{(t)} \equiv \mathcal{U}^{-1} Z_n{}^{(t)} = \sum_{r'} x_n{}^{(t,r')} \psi_n A_0{}^{(r')} + \sum_{s'} x_n{}^{(t,s')} \Phi_n{}^{(s')} \quad (17\text{--}19)$$

However, let us consider the system $S + A$ as an ordinary physical system, on
which we can measure at time t_0 an observable of our choice, and let us compute with the help of M the probabilities $P_\pm$ that at time t_0 the spin component
S_x should be found with a value equal to $\pm\frac{1}{2}$, the pointer of the instrument
being simultaneously found lying in graduation interval 0, as it should. According to the general rules of quantum mechanics, these probabilities are
necessarily equal to

$$P_\pm = \sum_{n,t} p_n{}^{(t)} \sum_r |M_{n,\pm}^{(t,r)}|^2 \quad (17\text{--}20)$$

with

$$M_{n,\pm}^{(t,r)} = \langle A_0{}^{(r)} \, 2^{-1/2} \, (\psi_1 \pm \psi_2) | Z_{n,\text{in}}^{(t)} \rangle \quad (17\text{--}21)$$

Now, from Eq. (17–19),

$$M_{n+}^{(t,r)} = 2^{-1/2} \, x_n{}^{(t,r)}$$
$$M_{n,-}^{(t,r)} = 2^{-1/2} \, (-1)^{n+1} \, x_n{}^{(t,r)} \quad (17\text{--}22)$$

since all the terms involving the $\Phi_n{}^{(s')}$ vanish because of Eq. (17–17); therefore

$$P_\pm = \frac{\sum_{n,t,r} p_n{}^{(t)} |x_n{}^{(t,r)}|^2}{2} \quad (17\text{--}23)$$

---

cont'd from p. 203
problem one degree of complexity forward (see Chapter 24), so that, taken all by itself, it
does not constitute a satisfactory answer to the questions considered here. Argument (*b*)
is the observation that a measurement process should not really be called a measurement
until indelible records have been created. But what are indelible records? Would they be
conceivable if the Schrödinger equation were exact? The material reported in Chapter 16
shows that such questions admit of no easy answers. Hence, again, argument (*b*) simply
shifts the problem at hand, without really contributing an answer. Argument (*c*) consists in
applying to the present problem the argumentation reported in Section 13.2. However, the
latter argumentation applies only to a different case, namely, the one in which the system
(here $S + A$) is actually subjected to an interaction with some instrument A' which provides
some information on the state of $S + A$ at time t_2. This is not the case here, since we did not
introduce any A'. Hence argument (*c*) is trivially erroneous.

In other words, if the assumption considered in this section were correct, the probabilities P_+ and P_- would be equal. However, since the S are completely polarized along $+O\hat{x}$, and since the probabilities $P_\pm$ are computed at a time before the interaction with A, these probabilities are obviously 0 and 1. This shows that there exists no choice of the $x_n^{(t,r')}$ and of the $p_n^{(t)}$ that can fulfill the necessary requirements.

Unless the nonideality of the measurement should bring some radically new element into the picture, we are thus led to understand quite literally Bohr's contention that instruments should not be considered simply as physical systems and that their *role as instruments* is essential in the theory: for, if the $S + A$ systems considered before the interaction were just physical systems obeying the rules of quantum mechanics, any ensemble of them should be describable as a mixture, and no rule would then forbid us to argue as above and to reach the foregoing paradoxical conclusions. In other words, if we follow Bohr in asserting that the pointers are always in definite graduation intervals, then (apparently) we must also follow him when he insists that, because of its finality, an instrument is something more than just a (macroscopic) physical system.

In the early days of quantum mechanics, Schrödinger [4] imagined the following experiment. A cat and a flask of poison are enclosed together in a container well isolated from the outside. When the flask is broken, the poison kills the cat. The breaking is triggered by the discharge of a Geiger counter placed in one of the beams of a two-hole experimental setup such as the one described in Chapter 2. The beam intensity is very low, so that only one electron is emitted during the whole experiment. Schrödinger then pointed out that under these conditions, if the cat is considered as a quantum system and described by a wave function, its final state is necessarily a quantum superposition of the two states "cat alive" and "cat dead," so that it should in no case be considered *either* as being alive *or* as being dead. This state of affairs continues until an "outside observer" opens the container and looks inside, for it is only through the latter process that the wave packet is reduced and that, in one case out of two, the cat is thereby truly killed. Owners of cats will, however, generally agree with Schrödinger that this description is paradoxical, since even before the outside observer comes into action there is at any rate one individual for whom the observable "cat alive" or "cat dead" has a quite definite value, in one half of the cases at least—this individual being, of course, the cat.

When this paradox is not just smiled away, as is usually the case, it is sometimes pointed out that the argument does not really hold because the cat (or, in more proper language, an ensemble of similar cats) should be described by a mixture. The content of the present chapter shows that this remark does not yet provide a satisfactory solution to Schrödinger's difficulty.

REFERENCES

[1] E. P. Wigner, *Am. J. Phys.* **31**, 6 (1963).
[2] B. d'Espagnat, *Nuovo Cimento*, Supplement 4, 828 (1966).
[3] F. Belinfante *Measurement and Time Reversal in Objective Quantum Theory*, Pergamon, Oxford, 1975.
[4] E. Schrödinger, *Naturwissenschaften* **23**, 807 (1935).

CHAPTER 18

NONIDEAL MEASUREMENTS:
SMALL INFLUENCE OF OUTSIDE WORLD

Interesting as they are in many respects, the theories reviewed in Chapters 16 and 17 fail, as we have seen, to provide us with a satisfactory description of the general physical processes that are usually considered to take place when a quantum system interacts with an instrument. At any rate this is true if by "satisfactory description" is meant a description that makes use only of physical concepts, and if, in turn, by "physical concepts" are meant concepts whose definition nowhere necessitates the consideration of human faculties and of their practical limitations.

Should, then, this requirement be considered as too stringent? Or, in other words, is it really incompatible with the linear laws of quantum mechanics to assert that in all circumstances a macroscopic object such as the pointer of an instrument *does* lie in a definite graduation interval, quite independently of us, and of our knowledge or abilities? In Chapter 17 we gave one answer to this question. That answer cannot yet be taken as final, however, since we investigated only the case in which the measurement is of the ideal type. As a matter of fact, it turns out that most realistic measurements cannot be ideal ones, so that it is necessary to extend the considerations of Chapter 17 also to the non-ideal case.

In this chapter, several reasons why most realistic measurements are non-ideal are investigated and the considerations of Chapter 17 are extended to nonideal measurements.

18.1 PRELIMINARIES

Examples of nonideal measurements are easily found. Such is, as a rule, the (often discussed) measurement of the momentum p_a of a particle along some axis a and with an uncertainty $\Delta p'_a$. The instrument often modifies the value of the measured quantity, irrespective of what the state of the particle initially is. Correlatively, after a result lying between p'_a and $p'_a + \Delta p'_a$ has been obtained upon measurement of p_a, the particle or the ensemble of particles cannot be described by means of eigenvectors of p_a, all corresponding to values of p_a lying in, the range p'_a, $p'_a + \Delta p'_a$.

An obvious reformulation of this statement is simply that the result of the measurement refers to the state of the system *before*, not *after*, the interaction. If this remark is associated with our previous considerations bearing on ideal measurements, it is easily seen that four categories of measurements must be distinguished from one another. These are as follows.

1. The ideal measurements in which the system S on which the measurement is made at $t = 0$ is initially in one of the eigenstates of the quantity L to be measured. In this case, the result of the measurement can be considered as being the value that the quantity L *has* both before and after $t = 0$.

2. The ideal measurements in which initially S is not in one of the eigenstates of L. In this case, the result of the measurement can be considered as being the value the quantity L has *after* $t = 0$.

3. The nonideal measurements in which initially S is in an eigenstate of L. The result of the measurement can then be considered as being the value L had *before* $t = 0$.

4. The nonideal measurements in which initially S is not in an eigenstate of L. The result of an individual measurement cannot then be given any interpretation in terms of an attribute possessed by S (although it often enables us to calculate the value L takes for $t > 0$). But, of course, if the measurement is repeated on an ensemble E of systems, the collection of the results gives significant information regarding E.

Statements 1 to 4 must be corrected to some extent if Bohr's philosophy is adhered to. Since it suggests that the instrument bestows, in a sense, physical properties on the system with which it interacts, we can say that, quite generally, the result of the measurement is the value L had before time $t = 0$. But we must then, of course, always remember that we are using the verb "to have" in a quite esoteric sense, since, as we know, L cannot be considered as having the value in question independently of the instrument with which it will later interact. Very formally we could even say that the values of several noncommuting observables can be known in principle with infinite precision. For instance, if we measure at time $t = 0$ the position x of a particle of mass m whose momentum has previously been exactly determined, we can assert that we know the exact values that both p and x had at any time $t = 0$: we know p by assumption, and we also know x since we interpret here the x measurement as giving the x value immediately *before* $t = 0$ (the value at any $t < 0$ can then be calculated since $v = p/m$ is known for $t < 0$). Obviously, however, those two items of knowledge are of different natures. The uncertainty relations remain unaffected.

18.2 GENERALITY OF NONIDEAL MEASUREMENTS

The need for a measurement theory that takes into account the possible nonideality of the actual measurement processes acquires its full importance in the light of a remark first made by Wigner [1] and then generalized by Araki and Yanase [2, 3], which is the subject of this paragraph.

Examples of nonideal measurements, that is, of measurement processes that do not conform to the pattern of (14–2), are of course quite easily found. One might a priori believe that instances of ideal ones conforming to (14–2) are developed just as easily. However, this is, strictly speaking, not really true, and a large class of phenomena commonly believed to be in complete agreement with (14–2) turns out on closer inspection not to follow exactly this pattern of evolution in time, except in the limit where the instrument has infinite "complexity."

To show this, let us consider again the effect on particles of spin $\frac{1}{2}$ of a Stern-Gerlach instrument designed for measuring the z component S_z of that spin. If the phenomenon followed exactly the pattern of (14–2), as has been assumed until now, it could be rigorously described by the two expressions

$$Iu_+ \rightarrow A_+u_+ \qquad (18\text{–}1)$$

$$Iu_- \rightarrow A_-u_- \qquad (18\text{–}2)$$

where the arrows symbolize the evolution in time from the state *before* to the state *after* magnet traversal, where $u_\pm$ are the two eigenfunctions of S_z, and where I and A are now normalized eigenvectors that describe everything but the spin of the considered particle S (I and A refer to the initial and to the final states, respectively).

Let the particles S initially propagate along $O\hat{y}$, and let us now imagine two experiments. In one of them, the incoming particles are completely polarized along $O\hat{x}$ in the positive direction. In the other one, the particles are polarized along this same axis in the opposite direction. Because of the linearity of the quantum laws of time evolution applied to expressions (18–1) and (18–2), these two experiments are necessarily described by the two expressions

$$Iv_+ = I(u_+ + u_-)2^{-1/2} \rightarrow 2^{-1/2}(A_+u_+ + A_-u_-)$$
$$= \frac{(A_+ + A_-)v_+ + (A_+ - A_-)v_-}{2} \qquad (18\text{–}3)$$

$$Iv_- = I(u_+ - u_-)2^{-1/2} \rightarrow 2^{-1/2}(A_+u_+ - A_-u_-)$$
$$= \frac{(A_+ - A_-)v_+ + (A_+ + A_-)v_-}{2} \qquad (18\text{–}4)$$

where the $v_\pm$ are the eigenfunctions of S_x.

Let L_x be the x component of the total angular momentum of everything, excepting only the particle spin. The x component M_x of the total angular momentum of the composite system is

$$M_x = L_x + S_x$$

and the same relation holds for mean values

$$\bar{M}_x = \bar{L}_x + \bar{S}_x \tag{18-5}$$

Used in conjunction with (18-3) and (18-4), Eq. (18-5) gives, because of the orthonormality of the $v_\pm$ and of the $A_\pm$,

$$\bar{M}_x = 2^{-2}[\langle A_+ + A_- | L_x | A_+ + A_- \rangle$$
$$+ \langle A_+ - A_- | L_x | A_+ - A_- \rangle] + 0 \tag{18-6}$$

in regard to the right-hand side of (18-3), and

$$\bar{M}_x = 2^{-2}[\langle A_- - A_+ | L_x | A_- - A_+ \rangle$$
$$+ \langle A_+ + A_- | L_x | A_+ + A_- \rangle] + 0 \tag{18-6a}$$

in regard to the right-hand side of (18-4). These two values are seen to be equal, so that the mean values of M_x are predicted to be the same for both final states. They are obviously different, however, for the two initial states, since the two L_x are then the same whereas the S_x are different. Thus these results are in contradiction to the principle of conservation of the total angular momentum, applied to the totally isolated system constituted by the particle plus the instrument. Consequently, the basic assumptions (18-3) and (18-4) cannot be considered as strictly correct.

The phenomenon described here is a fairly general one. It can indeed be shown [2] that no observable that does not commute with the additive conserved quantities described by bounded operators (such as total angular momentum along an axis and so forth) can be subjected to an ideal measurement.*

The solution to this difficulty was also pointed out by Wigner [1]. Its principle is to consider that the measurement is not an ideal one, and that instead of being described by (18-1) and (18-2) it is described by

$$Iu_+ \to A_+u_+ + \varepsilon_+u_-; \quad Iu_- \to A_-u_- + \varepsilon_-u_+ \tag{18-7}$$

*For an extension to quantities described by unbounded operators, see Stein and Shimony [4]. For an extension to measurements of the type $I \cdot u_m \to A_m \cdot v_m$, where the v_m are eigenstates of some observable pertaining to the object, see Wigner and Yanase [5].

where ε_+, ε_- also refer to the instrument but have small norms. Expressions (18–6) and (18–6a) are then replaced by more complicated expressions that are no longer equal to one another. As a matter of fact, their difference is of first order in ε_+. Since the difference between the left-hand sides of (18–6) and (18–6a) is unity, this is possible only if the matrix elements of L_x between various states of the instruments are normally very large compared with $\hbar$, so as to compensate for the smallness of ε_+. With instruments of macroscopic size and complexity this is, of course, the case (for "optimal size" of instruments, see Ref. [3]).

This analysis shows that for most observables the notion of ideal measurements is an idealization indeed, in the sense that it is a limiting case, which real measurements can never exactly reproduce. However, it shows at the same time that this limiting case can be approached in principle with an arbitrarily small error, provided that the instruments used are large enough. It shows, therefore, that the requirement that the instrument be a macroscopic system is an important feature of the theory of measurement.

For the reason just given, the discussion presented in Chapter 15, which was based on a Stern-Gerlach measurement of a spin component, remains valid, in spite of the fact that the measurement in question was there treated as an ideal one, strictly following the (14–2) pattern. Indeed, due to the enormous comparative size and complexity of the apparatus (here necessarily including the magnet, coils, and so forth), it can be considered that the ideal measurement pattern of (14–2) is followed during these processes to a very good approximation. This is all that is needed for the discussions in question to be correct. These discussions might, however, be inconclusive, if the ideal measurements turned out to be no measurement at all; that is, if all the "burden" of the reduction of the wave packet could possibly be transferred to the "nonideal" measurements, of whose considerable importance we are now better aware. This is the question that should now be investigated.

18.3 NONIDEAL MEASUREMENTS WITH INITIAL STATE AS MIXTURE

The main features that a theory of nonideal measurement should possess have been outlined by Landau and Lifshitz [6]. In order to describe them in the simplest possible way, these authors introduce at the start the simplifying assumption (which of course is not essential) that the spectrum of the possible value g_n of the instrument coordinate g (e.g., the position of the end of a pointer on a graduated scale) is both discrete and nondegenerate. The eigenvector corresponding to g_n is denoted by F_n. The set of all the orthonormal vectors F_n spans the Hilbert space $H^{(A)}$ corresponding to the instrument. Similarly, a set of orthonormal vectors ψ_n exist, which are in a one-to-one correspondence with the set F_n and span the Hilbert space $H^{(S)}$ associated with the system S, which is subject to a measurement. Finally, the instrument, A, is so constructed that, if initially A is in state F_0 and S in state ψ_n, then, after the interaction has taken place, A is in state F_n, so that it can be said that a value g_n is registered by the instrument. The "nonideality" of the measurement is then reflected in the fact that the interaction with the instrument does, here, perturb the system. Consequently, in the final state, system S is no longer in the state ψ_n. If ϕ_n denotes the final state of S, the whole evolution just described can be summarized by the arrow in the expression

$$\psi_n F_0 \rightarrow \phi_n F_n \qquad (18\text{--}8)$$

The difference between (18–8) and (14–2) is that the ϕ_n are not identical to the ψ_n and that they are not even necessarily orthogonal to each other. Hence, the ϕ_n cannot, in general, be considered as being the eigenvectors of any observable.* The ψ_n can, however, and we thus can say that, by definition, the instrument measures the value that a certain observable L, with eigenvectors ψ_n, had *before* the interaction.

The rest of the argumentation of Section 14.1 can immediately be transposed here: if initially the system S is in a superposition such as

$$\sum_n a_n \psi_n$$

the linear laws of quantum mechanics have the consequence that, after the interaction has taken place, the composite system is necessarily describable by

$$\sum_n a_n \phi_n F_n \qquad (18\text{--}9)$$

It is here, however, that Landau and Lifshitz take a very important step. Since, they say, system A is macroscopic, it is necessarily, at any time, in a state corresponding to a definite position, g_n, of the pointer, that is, in a state described by a definite F_n. In order that the correlation introduced by expression (18–9) should be respected, it is then necessary to say that the final composite system is in one of the states described by

$$\phi_n F_n \qquad (18\text{--}10)$$

Finally, let us consider an ensemble of identically prepared initial states: in order that the statistical prediction of Eq. (18–9) concerning g_n should also be respected, it is necessary to say that, as a result of the interaction, the ensemble initially described by

---

*In special cases it may happen that $\varphi_n = U\psi_n$, with $U \cdot U^+ = U^+ \cdot U = 1$. However, these cases should be discarded here since the difficulty connected with ideal measurements and discussed in the preceding section is present in them also; see Ref. [5].

$$F_0 \sum_n a_n \psi_n \qquad (18\text{--}11)$$

is transformed into a (proper) mixture constituted by

$$\text{``a proportion } |a_n|^2 \text{ of composite}$$
$$\text{systems in each state } \phi_n F_n\text{''} \qquad (18\text{--}11a)$$

However, as we already know, this position is not tenable as it stands; an ensemble cannot at the same time be represented adequately by a pure state such as (18–9) and by a proper mixture such as (18–11a) *unless parts of the density matrix are not observable* [in the language of the density matrix, one can similarly observe that a density matrix which is a projection operator, such as the one corresponding to (18–11), cannot be transformed by its mere evolution in time into a density matrix that does not have this property]. Here, however, as already explained, we do not want to resort to explanatory attempts based on the alleged "inobservability" of parts of the density matrix (see Section 18–6).

There remains, therefore, only one obvious possibility to be explored. This is to try to generalize to the present case of nonideal measurements the considerations discussed in Chapter 17. There the description of the initial state as a mixture did not resolve the difficulty of the measurement theory, but it may be hoped that with the introduction of nonideal measurement things will turn out to be different.

The generalization that we are looking for can easily be obtained as follows [7, 8].* First, the simplifying assumption mentioned above that the g_n are nondegenerate should, of course, be dropped. By convention g has, from now on, the value g_n if the end of the pointer lies in the nth interval of the graduated scale. The spectrum of the g_n is thus, indeed, discrete, but each g_n now has a considerable degeneracy. As in Chapter 17, let the symbol q stand for the whole set of the corresponding degeneracy indices. In complete rigor the spectrum of the q's should be taken as partly continuous; in order to alleviate the notations we can, however, without real loss in generality treat it as a very densely populated discrete one. Also, for simplicity, we consider in what follows the measurement of an observable L that has a nondegenerate spectrum.

Under these conditions, let us replace (17–1) by the more general form

$$\psi_n A_0^{(q)} \rightarrow \sum_{\hat{q}} \phi_{n,q,\hat{q}} \, F_n^{(\hat{q})} \qquad (18\text{--}12)$$

In (18–12) the meaning of the symbols $A_0^{(q)}$ is the same as in (17–1), and all the comments developed after Eq. (17–1) are therefore also valid here (notice that, if the $F_n^{(\hat{q})}$ with same n and different $\hat{q}$ were not orthogonal to start

---

*For a more compact proof of the result of this section, see Shimony and Stein [4].

with, we could always make them orthogonal by redefining the $\phi_{n,q,\hat{q}}$). Just as the $A_n{}^{(q)}$, the $F_n{}^{(q)}$ belong, of course, to the Hilbert space $H^{(A)}$. As for the $\phi_{n,q,\hat{q}}$, which are vectors in $H^{(S)}$, according to the considerations above, they need not be orthogonal to each other and they are not chosen as normalized. However, it should be noted that, as a consequence of the linear independence of the $A_0{}^{(q)}$ and of the linearity of the time evolution operator that induces (18–12), any linear relation such as

$$\sum_q C_{n,q}\, \phi_{n,q,\hat{q}} = 0 \qquad\qquad (18\text{–}13)$$

in which the coefficients $C_{n,q}$ are independent of $\hat{q}$, implies that

$$C_{n,q} = 0 \qquad\qquad (18\text{–}13a)$$

for all values of q [to prove this, multiply (18–13) by $F_n{}^{(\hat{q})}$, sum over $\hat{q}$, apply the inverse time evolution operator, and use the linear independence of the $A_0{}^{(q)}$].

These conventions being made, the whole argumentation can be transposed, right from the beginning and almost literally, from Section 17.1. The only noticeable changes are that expression (17–3) for $Y^{(q)}$ should be replaced by

$$Y^{(q)} = \sum_{n,\hat{q}} a_n\phi_{n,q,\hat{q}}\, F_n{}^{(\hat{q})} \qquad\qquad (18\text{–}14)$$

and that expression (17–4) for $Z_m{}^{(q)}$ should be replaced by

$$Z_m{}^{(q)} = \sum_{\hat{q}} \lambda_m{}^{(q,\ \hat{q})}\, F_m{}^{(\hat{q})} \qquad\qquad (18\text{–}15)$$

where the $\lambda_m{}^{(q,\hat{q})}$ are now just vectors in $H^{(S)}$. Equation (18–15) then obviously gives the most general vector in $H^{(S)} \otimes H^{(A)}$ that corresponds to a definite value of g (namely, g_m). Identification (17–5), which is necessitated here by exactly the same considerations as in Chapter 17, then gives, for any nonempty macrostate m,

$$\sum_{\hat{q},n} \Big(\sum_{q'} b_m^{(q,q')}\, a_n\phi_{n,q',\hat{q}} - \delta_{nm}\lambda_n^{(q,\hat{q})}\Big) F_n^{(\hat{q})} = 0 \qquad\qquad (18\text{–}16)$$

that is,

$$\sum_{q'} b_m^{(q,q')} a_n\phi_{n,q',\hat{q}} = \delta_{nm}\lambda_n^{(q,\hat{q})} \qquad\qquad (18\text{–}17)$$

because of the linear independence of the vectors $F_n^{(\hat{q})}$ in $H^{(A)}$.

Here, again, sets of numbers $b_m^{(q,q')}$ and of vectors $\lambda_m^{(q,\hat{q})}$ must be found such that (18–17) holds. If, however, two or more of the a's, for example, a_1 and a_2, are nonzero, then, again, such sets do not exist since:

(*a*) for $m \neq 1$, (18–17) gives

$$\sum_{q'} b_m^{(q \cdot q')} a_1 \phi_{1,q',\hat{q}} = 0$$

and therefore [Since we have shown that any relation such as (18–13) implies the corresponding relation (18–13a)]

$$b_m^{(q \cdot q')} = 0 \qquad \text{for } m \neq 1 \text{ and nonempty}$$

(*b*) for $m \neq 2$,

$$\sum_{q'} b_m^{(q \cdot q')} a_2 \phi_{2,q',\hat{q}} = 0$$

and thus, for the same reason,

$$b_m^{(q \cdot q')} = 0 \qquad \text{for } m \neq 2 \text{ and nonempty}$$

so that finally   (18–18)

$$b_m^{(q \cdot q')} = 0 \tag{18-19}$$

for every m labeling a nonempty macrostate.

The result of this section is thus negative: the consideration of system-instrument interaction processes of type (18–12), that is, of "nonideal" measurements, does not by itself solve the difficulties of measurement theory.

The generalizations carried out by Fine [9] and by Fehrs and Shimony [10] provide further support for this conclusion.

18.4   A GENERALIZATION

The nonideal measurements considered in the foregoing section can still be generalized to some extent [11]. This is done simply by *not* assuming the law of time evolution (18–12). In other words, even in the simple case in which the observable L to be measured is initially equal to one of the eigenvalues l_n of the corresponding operator, we do *not* demand any more that, after the measurement, the value of the instrument variable should be g_n: it is still sharply defined—that much we assume—that is, it is equal to one of the eigenvalues of the operator G. But nonvanishing probabilities $W_m^{(n)}$ exist that this eigenvalue should be some g_m differing from g_n. We merely assume (i) that

$$\sum_{m \neq n} W_m^{(n)} = \varepsilon_n \ll 1 \tag{18-20}$$

and (ii) that after the measurement the instrument variable should be sharply defined, even in the general case in which the object is initially described by

$$\psi = a_1|\psi_1\rangle + a_2|\psi_2\rangle \qquad (18\text{--}21)$$

with

$$L|\psi_i\rangle = l_i|\psi_i\rangle \qquad (18\text{--}22)$$

and

$$a_1, a_2 \neq 0$$

Even with these very mild requirements, it can be shown that the problem of measurement as formulated above admits of no solution as long as the density matrix is considered to be observable.

As in Chapter 17, let us diagonalize the density matrix $\rho_0{}^A$ that describes the instrument before the interaction. Let p_q and $A_0^{(q)}$ be its eigenvalues and its eigenvectors, respectively. The proof is particularly simple in the case in which $\rho_0{}^A$ has no degenerate eigenvalues, that is, in the case in which all the p_q are different. In this case the density matrix of the system $S + A$ can be written initially, with obvious notations, as

$$W_{\text{in}} \equiv |\psi\rangle\langle\psi|\otimes\rho_0{}^A = \sum_q p_q|\psi A_0^{(q)}\rangle\langle\psi A_0^{(q)}| \qquad (18\text{--}23)$$

Hence the corresponding density matrix W *after* the interaction is

$$W = UW_{\text{in}}U^{-1} = \sum_q p_q|\Omega^{(q)}\rangle\langle\Omega^{(q)}| \qquad (18\text{--}24)$$

where

$$\Omega^{(q)}\rangle = U|\psi A_0^{(q)}\rangle = a_1|\Omega_1{}^{(q)}\rangle + a_2|\Omega_1{}^{(q)}\rangle \qquad (18\text{--}25)$$

with

$$|\Omega_i^{(q)}\rangle = U|\psi_i A_0^{(q)}\rangle, \qquad i = 1, 2 \qquad (18\text{--}26)$$

Since the eigenvalues of a tensor product of operators are the products of the eigenvalues of the two factors, the nonvanishing eigenvalues of W_{in} are the p_q. The (nonvanishing) eigenvalues of W are therefore also the p_q. They are nondegenerate. Therefore (18–24) is the only possible expansion of W as a linear combination of all projection operators. According to our assumption (ii), every ket $|\Omega^{(q)}\rangle$ must therefore be an eigenvector of $I \otimes G$. However, according to our other assumptions, this is also the case in regard to $|\Omega_1^{(q)}\rangle$ and $|\Omega_2^{(q)}\rangle$. Since a_1 and a_2 are both different from zero, this is possible only if $|\Omega_1^{(q)}\rangle$ and $|\Omega_2^{(q)}\rangle$ correspond to the *same* eigenvalue of $I \otimes G$. Such an argument holds for every value of the index q. It follows that the value of the instrument variable *after* the measurement has taken place is independent of whether the initial state of the object is $|\psi_1\rangle$ or $|\psi_2\rangle$. Under these conditions the instrument does not discriminate between the values l_1 and l_2 of L: hence it does not play its role and must be rejected. Quantitatively, the results we have derived imply that $W_m^{(1)} = W_m^{(2)}$. They show therefore that inequalities (18-20) are not compatible with the other assumptions made.

In the more realistic case in which some of the p_q are degenerate, the expression analogous to (18-24) for W is not unique. It can be shown that even in this case the assumptions stated above are not all compatible with one another. It follows that these assumptions do not constitute the basis of an acceptable theory of measurement. Here the proof [11] is more subtle than in the nondegenerate case. See Exercise 3 on page 228.

18.5 DISCUSSION OF THE GREEN MODEL FOR MEASUREMENT*

A model for measurement with an apparatus has been put forward by Green [12] and discussed by Furry [13]. It describes a type of measurement process which, as is, for instance, the case in a Stern-Gerlach experiment, can quite naturally be described as taking place in two steps. The first of these is a deflection—for example, by a magnet—which, in the general case where the initial quantum system S (particle) to which the measured observable belongs is not in an eigenstate of that observable, splits the beam into several distinct parts. We can consider the case (e.g., spin $\frac{1}{2}$) where this quantity has only two eigenvalues; here the two distinct parts of the beam are

$$|\psi_\pm\rangle = |u_\pm\rangle f_\pm(z) \qquad (18\text{-}27)$$

Then if, for instance, the initial state is completely polarized along $O\hat{x}$, the complete state vector of the particle after the magnet traversal is

$$|\psi\rangle = 2^{-1/2}[|u_+\rangle f_+(z) + |u_-\rangle f_-(z)] \qquad (18\text{-}28)$$

as already described in Chapter 15. The second step in the process under investigation consists in the interaction of system S with a detector, which is a macroscopic object, with many degrees of freedom. In the experiment investigated there are in fact two distinct detectors, D_+ (D_-), located in the region where f_+ (f_-) is different from zero. It is assumed that these detectors leave the spin of S unaltered. For the reasons discussed earlier in this section, it is appropriate not to describe these detectors by means of definite state vectors. Rather, an ensemble of elementary processes, in each of which one individual particle traverses the magnet and interacts with the two-detector system, should be introduced, and the ensembles of detectors should accordingly be described by density matrices ρ_+ and ρ_-, respectively. Since the density matrix (or "statistical operator") of the particle is $|\psi\rangle\langle\psi|$, the complete density matrix before the particle-detectors interaction is the tensor product

---

*The content of this section is not essential to comprehension of the general argument.

$$M_0 = |\psi\rangle\langle\psi|\rho_+\rho_- \tag{18-29}$$

Now let V be the interaction Hamiltonian between the particle and the systems of detectors. Since each detector interacts with one "beam" only,

$$V|\psi_\pm\rangle = V_\pm|\psi_\pm\rangle \tag{18-30}$$

so that the effect of the particle-detectors interaction is to convert M_0 into*

$$M = e^{-iVt}|\psi\rangle\rho_+\rho_-\langle\psi|e^{iVt} \tag{18-31}$$

$$= P_+ + P_- + P_{+-} + P_{-+} \tag{18-32}$$

with, in Furry's notation,

$$2P_+ = |\psi_+\rangle\langle\psi_+|e^{-iV+t}\rho_+\,e^{iV+t}\,\rho_- \tag{18-33a}$$

$$2P_- = |\psi_-\rangle\langle\psi_-|\rho_+\,e^{-iV-t}\,\rho_-\,e^{iV-t} \tag{18-33b}$$

$$2P_{+-} = |\psi_+\rangle\langle\psi_-|e^{-iV+t}\rho_+\rho_-\,e^{iV-t} \tag{18-33c}$$

$$2P_{-+} = |\psi_-\rangle\langle\psi_+|\rho_+\,e^{iV+t}\,e^{-iV-t}\,\rho_- \tag{18-33d}$$

The important feature of Eq. (18-32) is that the sum $P_+ + P_-$ is just the density matrix describing an ensemble in which half the particles would *be* in the upper $(= +)$ beam (the detector D_+ *being* correspondingly in its "detection: yes" final state, and the detector D_- in its "detection: no" final state), and the other half of the particles would similarly *be* in the lower $(= -)$ beam (the detectors D_+, D_- simultaneously *being* in the reverse-to-the-above situations). If, therefore, we could show that the two other terms P_{+-}, P_{-+} are small, we could formulate the conclusion that to a very good approximation the detectors D_+ and D_- can in all respects be said to *be*, after the whole process has taken place, in definite macroscopic situations (or "macrostates"). This conclusion —it is hardly necessary to stress this again—would be generally welcome in that it would bring the general body of quantum physics into accordance with the general postulates of a naive realism, applied to macroscopic systems.

As a matter of fact, it is sometimes claimed (see, in particular, Ref. [12]) that, at least for some special forms of $V_\pm$ and for particular structures of the detectors, a proof of the smallness of P_{+-} and P_{-+} can indeed be given. This claim, however, is erroneous, as has been shown convincingly by Furry [13]. This author has pointed out some weak points in the alleged derivation of Ref. [12]; what is more important, however, he shows from quite general arguments that P_{+-} and P_{-+} cannot be small. His proof is simply to calculate the

---

*For simplicity we assume here that the contribution of the free Hamiltonian to the evolution operator during the interaction time can be neglected.

matrix products such as P_{+-} P_{-+} which intervene in P^2. This gives, for instance,

$$4P_+^2 = (\langle\psi_+|\psi_+\rangle)|\psi_+\rangle\langle\psi_+|e^{-iV+t}\,\rho_+^2\,e^{iV+t}\,\rho_-^2$$

$$4P_{+-}P_{-+} = (\langle\psi_-|\psi_-\rangle)|\psi_+\rangle\langle\psi_+|e^{-iV+t}\,\rho_+^2\,e^{iV+t}\,\rho_-^2$$

wherefrom it is apparent that the terms P_{+-}, P_{-+} cannot be regarded as smaller than the term P_+.

Thus the conclusions to be drawn from a study of the Green model coincide, as was to be expected, with those of the general theory of Section 18.3. They even make these conclusions more nearly complete on several points. One of these is that, in the considerations of the present section, contrary to what was the case in those of Section 18.3, a formal replacement of the continuous spectra of eigenvalues by discrete though densely populated ones does not even appear necessary. Another one is that we now know the existence of at least some *important* differences between the true and the alleged density matrices. Thus, if density matrices are measurable quantities, we know for sure that we could in principle develop experiments that would sharply contradict the too naive classical description of the instruments. What prevents us from doing so in practice is only the existence of acute technological limitations (for the size and complexity of the necessary instruments are, in fact, incredibly large).

18.6   GENERALIZATIONS

On several points the arguments developed in the preceding sections are not yet completely general.

First of all, all these arguments are based on the postulate that the density matrix is in principle measurable. This is a highly questionable assumption, especially in view of the fact that some superselection rules can hold with respect to the instrument. Nevertheless, the specific problem induced by the possible existence of superselection rules is not, in fact, a source of serious difficulties. Indeed, the case in which the density matrix is partly nonmeasurable because of the presence of some superselection rules can easily be reduced to the one in which the density matrix is measurable; in this respect the argument developed in Section 17.2 is so general that it obviously applies also to nonideal measurements.

On the other hand, some of the elements of the density matrix could conceivably be nonmeasurable independently of the existence of the superselection rules, as already pointed out. Several methods can be thought of for overcoming this objection. Here let us describe one that is fairly general and at the same time entirely straightforward, in spite of some lengthy calculations. It is a generalization to "quasi-ideal" measurements (see definition below) of an argument already developed in Section 17.2. For the sake of definiteness, we describe it, as in Section 17.2, on a specific example, namely, the interaction of an instrument A with the spin S of a spin-$\frac{1}{2}$ particle. The instrument is designed to measure the z component of the spin, and the particle beam is polarized along $O\hat{x}$. We denote by ψ_n $(n = 1, 2)$ the eigenfunctions of S_z, by $H^{(A)}$ the Hilbert space of A (which includes also the spatial coordinates of the particles), by $H^{(S+A)}$ the Hilbert space of the composite system $S + A$, and by $A_0^{(r)}$ an orthonormal basis of the subspace of the space $H^{(A)}$ that describes a pointer localized in graduation interval 0. Let $\mathscr{U}$ be the evolution operator between the two times t_0 and t_1 that are prior and subsequent to the interaction, respectively. Let

$$\mathscr{U}\psi_n A_0^{(r)} = \Omega_n^{(r)} \tag{18-34}$$

$(n = 1, 2)$. Let H_n be the subspace of $H^{(S+A)}$ that corresponds to a localization of the pointer inside graduation interval n $(n = 1, 2)$ *and* to a value of S_z equal to $\pm\frac{1}{2}$ according to whether $n = 1$ or 2. Contrary to the assumption made in Section 17.2, we no longer assume that the $\Omega_n^{(r)}$ are *strictly* in $H^{(n)}$. Instead, we use the following procedure. Let $\tilde{H}_n$ be the subspace of $H^{(S+A)}$ that is spanned by the $\Omega_n^{(r)}$, let Π_n $(\tilde{\Pi}_n)$ be the projection operator onto $H_n(\tilde{H}_n)$, and let $|a_n\rangle$ $(|\tilde{a}_n\rangle)$ be any normalized vector of $H_n(\tilde{H}_n)$. Then we can formulate in a precise way the assumption that the measurement is a quasi-ideal one by requiring

$$\|\Pi_n |\tilde{a}_n\rangle\| \approx 1$$

and

$$\|(1 - \Pi_n)|\tilde{a}_n\rangle\| \ll 1 \tag{18-35}$$

These relations express the fact that $\tilde{H}_n$ is "nearly" embedded in H_n and have as a consequence the fact that, if an $S + A$ system is in a state described by a linear combination of the Ω_n, the probability of observing on this system a value of S_z, or a pointer position, corresponding to the other value of n is very small.

From (18-35) and from the orthogonality of the H_n, it follows that for $m \neq n$

$$\|\Pi_m |\tilde{a}_n\rangle\| \ll 1$$

Now let us consider

$$\Pi_1 \tilde{\Pi}_2 |a_1\rangle$$

Since $\tilde{\Pi}_2|a_1\rangle$ is in $\tilde{H}_2$ and has a magnitude no larger than unity, $\|\Pi_1\tilde{\Pi}_2|a_1\rangle\| \ll 1$ so that from the Schwarz inequality

$$|\langle a_1|\Pi_1\tilde{\Pi}_2|a_1\rangle| \leq \|a_1\| \, \|\Pi_1\tilde{\Pi}_2|a_1\rangle\| \ll 1$$

But the left hand side is just

$$\|\tilde{\Pi}_2|a_1\rangle\|^2$$

Hence

$$\|\tilde{\Pi}_2|a_1\rangle\| = \varepsilon' \ll 1 \tag{18–36a}$$

and, by exchanging the role of the indices,

$$\|\tilde{\Pi}_1|a_2\rangle\| \ll 1 \tag{18–36b}$$

These relations will be useful in what follows.

Let now H'_n be the subspace of H_n that is spanned by the vectors $\Pi_n\Omega_n^{(r)}$. Let H''_n be the subspace complementary to H'_n with respect to H_n, that is, the set of all vectors of H_n that are orthogonal to H'_n. Let $\psi_n F_n^{(s)}$ be an orthonormal basis in H''_n ($s \neq r$). Finally, let us define

$$\Phi_n^{(s)} = \mathscr{U}^{-1}\,\psi_n F_n^{(s)}$$

and

$$\varepsilon_n^{(r)} = (1 - \Pi_n)\Omega_n^{(r)}$$

Denoting by $\langle\,|\,\rangle$ the scalar product, we have

$$\langle \Phi_n^{(s)}|\psi_m A_0^{(r)}\rangle = \langle \psi_n F_n^{(s)}|\Omega_m^{(r)}\rangle$$
$$= \langle \psi_n F_n^{(s)}|\Pi_m\Omega_m^{(r)}\rangle + \langle \psi_n F_n^{(s)}|(1 - \Pi_m)\Omega_m^{(r)}\rangle$$

the first term in the last member is zero because it is a scalar product of two vectors that lie in mutually orthogonal spaces (H''_n and H'_n if $m = n$, H_n and H_m if $m \neq n$). Thus

$$\langle \psi_m A_0^{(r)}|\Phi_n^{(s)}\rangle = \langle \varepsilon_m^{(r)}|\psi_n F_n^{(s)}\rangle \tag{18–37}$$

(In fact, it can easily be shown that this expression is zero for $m = n$, but this is irrelevant to what follows.)

Our purpose is to investigate whether or not it is possible to describe the final state of $S + A$ as a mixture, with proportions $p_n^{(t)}$ to be determined, of systems each of which would have at least approximately well defined values, both for S_z and for the number that labels the graduation interval on which the pointer lies. If $Z_n^{(t)}$ is a (normalized) state vector of an $S + A$ system that possesses these properties, it should (in particular*) be possible to express $Z_n^{(t)}$ as

---

*Any $Z_n^{(t)}$ that satisfies the prescribed conditions can indeed be written in that form (see Exercise 4 on page 229).

$$Z_n^{(t)} = \sum_r x_n^{(t,r)} \, \Omega_n^{(r)} + \sum_s x_n^{(t,s)} \, \psi_n F_n^{(s)} + \sum_{i t'} y_{n,i}^{(t,t')} \, G_{n,i}^{(t')}$$

where the $G_{n,i}^{(t')}$ are orthonormal state vectors of $H^{(S+A)}$, and where

$$\sum_r |x_n^{(t,r)}|^2 + \sum_s |x_n^{(t,s)}|^2 \approx 1$$

$$\sum_{i,t'} |y_{n,i}^{(t,t')}|^2 \ll 1 \qquad\qquad (18\text{--}38)$$

Making use of the arguments already described in Section 17.2, we can then assert that, at a time t_0 prior to the interaction, the ensemble of $S + A$ systems should be describable as a mixture, with probabilities $p_n^{(t)}$, of the states

$$Z_{n,\mathrm{in}}^{(t)} = \mathcal{U}^{-1} Z_n^{(t)} = \sum_r x_n^{(t,r)} \, \psi_n A_0^{(r)} + \sum_s x_n^{(t,s)} \, \Phi_n^{(s)} + \sum_{i,t'} y_{n,i}^{(t,t')} \, K_{n,i}^{(t')}$$

where

$$K_{n,i}^{(t')} = \mathcal{U}^{-1} \, G_{n,i}^{(t')}$$

With the help of $Z_{n,\mathrm{in}}^{(t)}$ let us now evaluate the probability $P_+ \, (P_-)$ that if, at time t_0, a joint measurement of S_z and of the position of the pointer is made, a value $S_x = \frac{1}{2} \, (-\frac{1}{2})$ is found, together with a pointer lying in graduation interval 0. These probabilities are given by

$$P_\pm = \sum_{n,t} P_n^{(t)} \sum_r |M_{n,\pm}^{(t,r)}|^2$$

where

$$M_{n,\pm}^{(t,r)} = \langle \, 2^{-1/2} \, (\psi_1 \pm \psi_2) \, A_0^{(r)} \, | \, Z_{n,\mathrm{in}}^{(t)} \, \rangle$$

that is,

$$M_{1,\pm}^{(t,r)} = 2^{-1/2} \, (x_1^{(t,r)} + \zeta_{1,\pm}^{(t,r)})$$

$$M_{2,\pm}^{(t,r)} = 2^{-1/2} \, (\pm x_2^{(t,r)} + \zeta_{2,\pm}^{(t,r)})$$

with

$$\zeta_{n,\pm}^{(t,r)} = \langle \, (\varepsilon_1^{(r)} \pm \varepsilon_2^{(r)}) \, | \sum_s x_n^{(t,s)} \, \psi_n F_n^{(s)} \, \rangle$$

$$\qquad\qquad + \sum_{i,t'} y_{n,i}^{(t,t')} \, \langle \, (\psi_1 \pm \psi_2) \, A_0^{(r)} \, | \, K_{n,i}^{(t')} \, \rangle$$

[because of Eq. (18–30)].

Our choice of the initial conditions entails $P_- = 0$ and $P_+ = 1$. The former condition implies that $M_{n,-}^{(t,r)} = 0$, that is,

$$x_1^{(t,r)} = -\zeta_{1,-}^{(t,r)}; \qquad x_2^{(t,r)} = \zeta_{2,-}^{(t,r)}$$

whenever $p_n^{(t)}$ is $\neq 0$. Carrying these relations into P_+, we find that

$$2P_+ = \sum_t \{p_1^{(t)} \sum_r |\zeta_{1,+}^{(t,r)} - \zeta_{1,-}^{(t,r)}|^2 + p_2^{(t)} \sum_r |\zeta_{2,+}^{(t,r)} + \zeta_{2,-}^{(t,r)}|^2\}$$

Together with the inequality (valid for any pair of complex numbers a and b)

$$|a + b|^2 \leqslant 2(|a|^2 + |b|^2)$$

this gives

$$2P_+ \leqslant 8 \sum_t p_1^{(t)} \{\sum_r |\langle \varepsilon_2^{(r)} | \sum_s x_1^{(t,s)} \psi_1 F_1^{(s)} \rangle|^2$$

$$+ \sum_r |\langle \psi_2 A_0^{(r)} | \sum_{i,t'} y_{1,i}^{(t,t')} K_{1,i}^{(t')} \rangle|^2 \}$$

$$+ 8 \sum_t p_2^{(t)} \{\sum_r |\langle \varepsilon_1^{(r)} | \sum_s x_2^{(t,s)} \psi_2 F_2^{(s)} \rangle|^2$$

$$+ \sum_r |\langle \psi_1 A_0^{(r)} | \sum_{i,t'} y_{2,i}^{(t,t')} K_{2,i}^{(t')} \rangle|^2 \}$$

Let $\Psi_n = \sum_s x_n^{(t,s)} \psi_n F^{(s)}$. The first term in each curly bracket in the expression above then has the form

$$\sum_r \langle \psi_n | (1 - \Pi_m) | \Omega_m^{(r)} \rangle \langle \Omega_m^{(r)} | (1 - \Pi_m) | \psi_n \rangle$$

$$= \langle \psi_n | (1 - \Pi_m) \tilde{\Pi}_m (1 - \Pi_m) | \psi_n \rangle$$

with, moreover, $m \neq n$. Since $\psi_n \in H_n$, $\Pi_m \psi_n = 0$ and the expression above reduces to

$$\|\tilde{\Pi}_m \| \psi_n \rangle \|^2$$

which is much smaller than one because of Eqs. (18–36a) and (18–36b).

As to the second term in each curly bracket, it can be written as

$$\langle \sum_{i,t'} y_{n,i}^{(t,t')} K_{n,i}^{(t')} | \Pi_0 | \sum_{i,t'} y_{n,i}^{(t,t')} K_{n,i}^{(t')} \rangle$$

where Π_0 is the projection operator into the space spanned by the $\psi_n A_0^{(r)}$. This quantity is therefore smaller than the length of the vector $\sum_{i,t'} y_{n,i}^{(t,t')} K_{n,i}^{(t')}$, which is just $\sum_{i,t'} |y_{n,i}^{(t,t')}|^2$ because of the orthonormality of the $K_{n,i}^{(t')}$ (which follows from that of the $G_{n,i}^{(t')}$). However, this sum is also much smaller than one because of assumption (18–38).

Thus we find that P_+ has the general form

$$P_+ = \sum_t (p_1^{(t)} \alpha_1^{(t)} + p_2^{(t)} \alpha_2^{(t)})$$

where $\alpha_1^{(t)}$ and $\alpha_2^{(t)}$ are both much smaller than unity. Since $\sum_t (p_1^{(t)} + p_2^{(t)})$ must be equal to unity, this implies that P_+ cannot be equal to unity. We therefore arrive at a contradiction. In other words, the introduction of the concept of quasi-ideal measurements, as defined by conditions (18–35), (18–36), and (18–38), is not sufficient to resolve the difficulty pointed out above within the framework of a theory of purely ideal measurements.

At first sight, it may seem that the argument used in this section is connected with the well-known paradox of Loschmidt, which bears on statistical irreversibility. On closer inspection, however, it turns out that there is only a rather superficial similarity between the two ideas. In Loschmidt's case it is necessary to reverse precisely all the velocities in order to obtain a significant result; in other words, it is necessary to select one very particular state among a very large number of possible states. Here, on the contrary, we considered all the possible states of the instrument and system that could be accepted according to very general "commonsense" requirements. Moreover, in Loschmidt's case the very particular final state that corresponds reversibly to the very particular initial state really exists. Here *none* of the final states that can be accepted according to the commonsense requirements survives our consistency test. See Ref. [14] for a more elaborate discussion.

18.7 CONCLUSION

In a way, the results obtained could serve as an illustration of the proper manner in which the Copenhagen interpretation of quantum mechanics should be understood, because they show on an example that we can hardly hope to solve the conceptual problems raised by quantum measurement theory just with the help of the usual quantum-mechanical *techniques*, even if we supplement these by some assumptions bearing on the inobservability of a class of Hermitean operators. On the other hand, if we try to understand the results described above with the help of the Copenhagen interpretation, it seems that we can do so only if we are willing to make full use of one of its central principles, which we encountered in Chapter 9 and which is epistemological rather than physical. This principle, further analyzed in Chapter 21, states that no physical property should ever be attributed to a quantum system (or to an ensemble of quantum systems) with the sole possible exception of very particular ones: if A is the instrument that the system *will* interact with, these physical properties are exclusively those that A has been designed to measure. If we apply this principle to our example, we are led to assert that, under the given experimental conditions, and *if A is considered as a measuring instrument*, the incoming beam should indeed be considered as a mixture of $S_z = \pm\frac{1}{2}$ states (giving $P_+ = P_- = 1/2$), in spite of the fact that this same beam is considered by the experimenter who *produces* it to be fully polarized along $O\hat{x}$. As a matter of fact, under the strict Copenhagen orthodoxy, it would seem that we could assert that our beam *is* polarized along $O\hat{x}$ (and that we could give a meaning to $P_{\pm}$) only if we brought a second instrument into the picture, which would be oriented so as to measure S_x. But then the whole setup would be changed, and the calculations made in the text would not apply.

It is certainly very strange that the privilege of being able to possess (and impart) intrinsic physical properties should be attributed to measuring *instruments* and not, for instance, to beams or other physical systems. It is also quite surprising that this power of the instruments should "operate," so to speak, backward in the past. Such views, if seriously held, would necessitate very careful, restrictive definitions of what is meant by "instrument," of what is meant by "causality," and ultimately of what is meant by "reality." Since none of these steps is easy, it is advisable to defer their discussion to the next two chapters.

The possibility remains, of course, to give up the assumption we started with: that the pointer of the physical system that we called "instrument" always lies, at least approximately, on some definite graduation interval. In the spirit of the interpretation of the Copenhagen viewpoint discussed above, this is an entirely admissible step, since, as soon as we consider the probabilities P_+ and P_- for the observation of $S_x = \pm\frac{1}{2}$, this implies that we deprive A of its dignity as the *instrument* of observation. Then A is only an ordinary physical system; and, just like several of the diaphragms that Bohr introduced in his *gedanken* experiments, it must be treated quantum mechanically although it is macroscopic: nothing in the Copenhagen orthodoxy then forbids us to think of it as being in a superposition of macroscopically distinct states. Alternatively, we can also renounce the views of Copenhagen altogether and simply assert that A is in such a superposition independently of whether or not we choose to call it an instrument. Indeed there exists a theory, which Everett [15] called the "relativity of state theory," that assumes just this. It is reviewed in Chapter 23.

A final comment is perhaps appropriate here. It is stated above that we can *hardly* hope to solve the problems referred to by means of the usual quantum-mechanical techniques. Can this judgment be made sharper? In particular, did we examine all the possible generalizations of the concept of measurement, so as to prove that none of them is consistent? Admittedly, this is not really the case. To take only a specific example, we did not investigate here the possibility of associating the generalization studied in Section 18.4 and the one studied in Section 18.6. It turns out, however (see Ref. [14]), that this idea also leads to some inconsistencies.

More generally, the study of the field in question is far from being complete. A considerable amount of elaborate investigation remains to be done and should indeed be done before we can pronounce final judgments. On the other hand, the work already carried out is certainly sufficient to give a very strong indication. And this indication, as we have seen, is negative, unless we keep to an explicitly pragmatical conception of physics.

REFERENCES

[1] E. P. Wigner, *Z. Phys.* **133**, 101 (1952).
[2] H. Araki and M. Yanase, *Phys. Rev.* **120**, 622 (1961).
[3] M. Yanase, *Phys. Rev.* **123**, 666 (1961).
[4] H. Stein and A. Shimony, *in Foundations of Quantum Mechanics, Proceedings of the Enrico Fermi International Summer School*, Course IL, Academic Press, New York, 1971.
[5] E. P. Wigner and M. M. Yanase, *Ann. Jap. Assoc. Philos. Sci.* **4**, 171 (1973).
[6] L. D. Landau and E. M. Lifshitz, *Quantum Mechanics*, Pergamon, London, 1958.
[7] B. d'Espagnat, *Nuovo Cimento*, Supplement **4**, 828 (1966).
[8] J. Earman and A. Shimony, *Nuovo Cimento* **54B**, 322 (1968).
[9] A. Fine, *Phys. Rev.* **D2**, 2783 (1970).
[10] M. H. Fehrs and A. Shimony, *Phys. Rev.* **D9**, 2317 (1974).
[11] A. Shimony, *Phys. Rev.* **D9**, 2421 (1974).
[12] H. S. Green, *Nuovo Cimento* **9**, 880 (1958).
[13] W. H. Furry, *Boulder Lectures in Theoretical Physics*, Vol. 8A (1965), University of Colorado Press, 1966.
[14] B. d'Espagnat, *Nuovo Cimento* **21B**, 233 (1974).
[15] H. Everett III, *Rev. Mod. Phys.* **29**, 454 (1957).

EXERCISES OF PART FOUR

EXERCISE 1

Consider a measuring instrument that provides only "yes" or "no" information and that responds "yes" with a probability c_i when the system is in state $|\phi_i\rangle$. Write the probability that the result of the measurement is "yes" when the system is described by a density matrix ρ. Express this probability as the trace of the product of ρ and of another matrix called the "efficiency matrix."

(Wigner)

EXERCISE 2

Consider again the example of an interaction between a quantum system S and an instrument A that is described in Section 14.3. Now, however, assume that the initial state of the quantum system is

$$\psi = \sum_m c_m \psi_m$$

where ψ_m is an eigenvector of the measured observable L, which, for simplicity, is assumed to correspond to the eigenvalue $l_m = m$

$$L\psi_m = m\psi_m$$

(a) Use Eq. (14–17) and the linearity of the laws of evolution to write the final state of the $S + A$ system.

(b) Assume that the measuring interaction is switched on again at times τ and 2τ (this implies adding to (14–12) the extra term

$$[\delta(t - \tau) + \delta(t - 2\tau)] L \cdot \frac{1}{i}\frac{\partial}{\partial Q})$$

and during the period τ suppose that each eigenstate ψ_m evolves into a combination

$$\psi_m \rightarrow \sum_n b_{nm}\psi_n$$

while Q is a constant of motion during these time intervals. Then write the total state vector of the $S + A$ system just after time 2τ.

(c) Use the result obtained in (b) to write the probability of finding a value Q_0 for the pointer position and simultaneously a value l_0 for L. Eliminate n from the expression.

(d) Using the result obtained in (c), write the value of the probability for the pointer to be found in position Q_0 just after time 2τ irrespective of the value of L.

(*e*) Calculate now the probability considered in (*d*) by assuming that the pointer takes a definite (though unknown) position just after times $t = 0$ and $t = \tau$, that is, calculate the probabilities of these various possible intermediate steps and use them for evaluating the final result. Verify that the result thus obtained does not agree with the result obtained in (*d*) and should therefore be considered as incorrect.

(Bell and Nauenberg)

EXERCISE 3

Some of the eigenvalues p_q of the density matrix $\rho_0{}^A$ considered in Section 18.4 are now degenerate of order N_q.

Show that, according to the assumption made in this section, W should be describable as a linear combination $W = \Sigma_q p_q W_q$ of projection operators W_q, none of which can transform a ket belonging to a given eigen subspace of $1 \otimes G$ into a ket belonging to another of these subspaces.

Let $|\Omega_i{}^{(q,r)}\rangle$ $(i = 1, 2)$ and $|\Omega^{(q,r)}\rangle$ be the kets corresponding to the $|\Omega_i{}^{(q)}\rangle$ and $|\Omega^{(q)}\rangle$ defined in Section 18.4 (r is a degeneracy index); let $\{|\xi_{qs}^{(i)}\rangle\}$ be bases of the density matrices $W_i{}^{(q)} = \Sigma_q p_q |\Omega_i{}^{(q)}\rangle \langle \Omega_i{}^{(q)}| (i = 1, 2)$; and let

$$|\Omega_i{}^{(q,r)}\rangle = \sum_s C_{q,r,s}^{(i)} |\xi_{q,s}^{(i)}\rangle$$

Show that for any value of the index q

$$W_q = \sum_{i,j} a_i a_j^* \sum_{ss'} \sum_r C_{q,r,s}^{(i)} C_{q,r,s'}^{(j)*} |\xi_{q,s}^{(i)}\rangle \langle \xi_{q,s'}^{(j)}|$$

Show that, according to the assumptions of Section 18.4, there must exist a choice of the $\{|\xi_{q,s}^{(i)}\rangle\}$ such that each element of these sets belongs to some definite eigenspace H_n of $1 \otimes G$. Show also that there must be at least one value k of q such that the number n_1 of the kets $|\xi_{k,s}^{(1)}\rangle$ that belong to a given H_n is different from the number n_2 of kets $|\xi_{k,s}^{(2)}\rangle$ that belong to this same H_n. Assume that $n_1 > n_2$. Relabel the $|\xi_{k,s}^{(i)}\rangle (i = 1, 2)|$ so that those belonging to H_n should be the first n_i ones. Show that

$$\sum_{r=1}^{N_k} C_{r,s}^{(1)} C_{r,s'}^{(2)*} = 0 \quad \text{for } s \leq n_1, \; s' > n_2$$

Show that the N_k-tuples $C_{r,s}^{(1)}$ (fixed s) and $C_{r,s'}^{(2)}$ (fixed s'), $r = 1, \ldots, N_k$, can be considered as vectors in an N_k-dimensional complex vector space with an appropriate scalar product and that all the N_k-tuples such that $s \leq n_1$ and $s' > n_2$ are orthogonal to one another in this space. Show that their number is larger than N_k, and then explain why this is impossible. Conclude that the assumptions made in Section 18.4 are incompatible also when the p_q are degenerate.

(Shimony)

EXERCISE 4

The notation is that of Section 18.6. Let $|\alpha\rangle \in \tilde{H}_n$ and $|\beta\rangle \in H_n''$, and let $\||\alpha\rangle\|^2 + \||\beta\rangle\|^2 \approx 1$. If $|Z\rangle$ is any given vector, it is always possible to define $|\gamma\rangle$ as

$$|\gamma\rangle = |Z\rangle - |\alpha\rangle - |\beta\rangle$$

Show that the conditions $\||Z\rangle\| = 1$ and $\|\Pi_n|Z\rangle\| \approx 1$ then entail $\||\gamma\rangle\| \ll 1$.

EXERCISE 5

Compare Appendix C of the text quoted in the footnote appearing p. 203 with the content of Ref. [14] (Chapter 18). Try to clarify the problem.

EXERCISE 6

The z component $\sigma_z^{(U)}$ of the spin of a spin $\frac{1}{2}$ particle U propagating along Ox is measured by means of an instrument made of a semi-infinite linear array of spins $\sigma^{(n)}$ fixed at positions $x = n = 1, 2, \ldots$. The Hamiltonian is

$$H = -i\frac{\partial}{\partial x} + \sum_{n=1}^{\infty} W(x-n)\,\sigma_x^{(n)}(1-\sigma_z^{(U)})$$

where $W(x) = 0$ for $|x| > r$ and $\int_{-\infty}^{\infty} W(x)\,dx = \pi/4$

(a) Show that

$$\psi_\pm(t, x) = f(x-t)\,u_\pm^{(U)} v_\pm^{(1)}(x-1)\ldots v_\pm^{(n)}(x-n)\ldots$$

are solutions of the Schrödinger equation, where

$$v_+^{(n)}(x-n) = u_+^{(n)}; \quad v_-^{(n)}(x-n) = \exp\left[-2i\,G(x-n)\,\sigma_x^{(n)}\right]u_+^{(n)};$$

$$G(x) = \int_{-\infty}^{x} W(y)\,dy; \quad \sigma_z u_+ = u_+$$

and that $f(x) = 0$ for $|x| > w$ implies $v_-^{(n)}(x-n) = u_+^{(n)}(u_-^{(n)})$ for $n > t + r + w (n < t - r - w)$.

(b) Show that any observable Q that involves only $\sigma^{(1)}, \ldots, \sigma^{(N)}$ with N finite is such that $(\psi_+, Q\psi_-)$ goes to zero when $t \to \infty$. Apply to measurement problem, starting with all lattice spins up.

(c) However, show that if $M < t - r - w < M + 1$ and if $R = \sigma_x^{(U)} \sigma_y^{(1)} \ldots \sigma_y^{(M)}$ is measurable the problem of measurement is *not* solved.

(Hepp and Bell)

PART FIVE

KNOWLEDGE AND THE PHYSICAL WORLD

What is the exact scope of the results accumulated in the preceding chapters? Do they really forbid us to think in terms of finite separable *things*, as common sense tells us to do? If not, what significance do they still allow us to give to such a concept; in other words, to what extent are we still permitted to think of finite things as existing independently of our own existence? Even if the things of our experience are relative to *our* abilities to feel and act, can we think of some more fundamental *physical reality* that is not? Then, again, can this more fundamental reality be regarded as composed of a multitude of objects, in the spirit of ancient and of classical atomism, or should we consider it as an essentially unseparable whole? In the latter case, to what extent is it accessible to experience, and, more fundamentally, to what extent does the notion of such a reality preserve a *meaning*? In spite of their partly philosophical character, these questions are undoubtedly raised by modern science, as the existence of the seemingly paradoxical results reviewed in Part Four demonstrates.

Under these conditions, it is clear that the best method for dealing with these questions is to test the appropriateness of the various conceivable answers against our experience, as embodied in *the* existing successful and nearly universal physical theory, namely, quantum mechanics. Now, anyone who tries to discover what general frames of ideas underlie the various investigations currently made into the foundations of quantum physics easily discovers the existence of two main tendencies. These correspond to two different—and, unfortunately, hardly reconcilable—conceptions of the notions of scientific truth. One of them is represented by logical positivism and the related trends of thought. Its description and discussion constitute the main part of Chapter 20. The other is realism, but not at all in the medieval sense. Indeed, many investigations into the foundations of quantum physics are based on a kind of modern realism that remains in them as an *implicit* assumption. One of our tasks should be to analyze this tendency. This is the main subject of Chapter 19. Chapters 21 and 22 are more specifically devoted to modern epistemologies of physics, namely, the Copenhagen interpretation and the possible alternatives to it.

REALITY AND OBJECTS

The first section is an attempt to give a moderately precise shape to the idea that physical reality is independent of ourselves, an idea deeply rooted in the minds of scientists and laymen alike. In the following sections this idea is particularized through the introduction of supplementary assumptions that are inspired by common sense. The versions of realism thus obtained are discussed in the light of quantum physics. They are found to be unacceptable.

19.1 REALISM

In view of the remarks stated in the introduction above, it is appropriate to begin by characterizing realism rapidly through its difference from logical positivism. Logical positivism is studied in the next chapter, but its main idea is well known. It is that a statement has a meaning only if it can be verified by human beings by means of some sequence of operations and observations. As a consequence, logical positivism asserts that no meaning can be attached to this statement: *physical reality would exist even if no observer existed.* In contradistinction we designate as realism the view that somehow the underlined statement is meaningful and that, moreover, it is true.

This tentative definition is open to the criticism that it attributes a meaning to a statement whose validity cannot—for quite obvious reasons!—be verified. To the logical positivists it is not even understandable. On the other hand, to the modern layman the idea that a physical reality would exist, in an absolute sense, also in the absence of human beings, and that indeed it already existed, in such a sense, before the advent of man or other life, appears as quite an obvious one. In other words, the most our layman might be willing to concede to the logical positivist is that perhaps statements of such kinds have no *scientific* meaning. But he would consider that this is merely a restrictive definition of *scientific* knowledge and that the statement in question *does* have a respectable, assertive meaning even if that meaning is not of the kind in which scientists are normally interested.

In view of the gulf that separates the two conceptions, and all the possible misunderstandings that this situation implies, we should not hasten to ban the realistic standpoint. At any rate, we should refrain from doing so until we have examined in detail the weaknesses of the views of positivism. Meanwhile, it is necessary that we state precisely the fundamental assumptions of a philosophy which will be conveniently referred to as the "realist philosophy." This we choose to do as follows.

Assumption 1. It is meaningful to define reality as everything that does exist.

Assumption 2. Although we are ourselves embedded in reality, reality is independent of us in the senses that it existed before the advent of life and consciousness and that it would exist even if no human being existed.

Assumption 3. Some features of reality are accessible to our knowledge.

According to Assumption 1, the concept of existence is a primeval one that need not be defined (on the other hand, in some specific cases it is of course difficult to determine whether there "really exists" some entity corresponding to a concept that has been defined.) According to Assumption 2, reality—the set of everything that exists—is by no means a mere creation of the human mind, although it may be difficult and even partly impossible to know. Assumption 2 thus offers an alternative to the tacit anthropocentrism that underlies Logical positivism and the related philosophies.

In order for realism to be something more than an arbitrary metaphysical construction, we should have some grip on reality through our experience. This is the reason for Assumption 3. This postulate is purposely kept extremely vague at this stage so as not to decrease the scope of the realistic philosophy.

For the same reason, we call "reality" rather than "physical reality" the totality of what exists. The word "physical" can only mean "obeying the laws of physics" and is therefore ill defined as long as we have not specified exactly what is the scope of physics. In spite of that fact, it is a common practice to consider that *some* conceptions of reality fall outside what is referred to as "physical reality." This is, for instance, the case in regard to Platonic realism. Now there would be no point in arbitrarily rejecting such philosophies outside the category of realism, especially since we want the latter to be as widely open as possible. On the other hand, our definition of realism clearly does not *commit* us to Platonic realism. Indeed, it also incorporates, for instance, most of the definitions that have been given of materialism. All these more specific philosophies can be obtained from realism by the adjunction of one supplementary assumption or several such assumptions.

19.2 MICRO-OBJECTIVISM

We are accustomed to the idea that some parts of reality, called "physical objects," can be considered separately and are endowed with properties which are specific, that is, which belong to one particular object. Since matter is composed of atoms, elementary particles, and so forth, this view is naturally extended to such entities, commonly called "micro-objects." Of course, micro-objects must satisfy the uncertainty relations that—in a theory without hidden variables—prevent them from possessing too many such properties at the same time. But a hypothesis quite often made by beginners is that micro-objects *possess the maximum number of specific physical properties that is compatible with these relations.* This assumption is supplemented by the idea that, although these properties can be modified by an interaction with some other system such as an instrument of measurement, at any rate they exist quite independently of the mere presence or absence of such instruments. Let this theory be named "micro-objectivism."

In conventional quantum mechanics, the hypothesis stated above implies that at any time in its history a microscopic particle always possesses a complete set of compatible observables endowed with definite values. An ensemble of such particles would thus always be describable by means of a definite state vector pertaining to the Hilbert space of the system, even after an interaction with other systems had taken place. This, as we know from Chapters 7 and 8, is not possible. Hence micro-objectivism is not compatible with the assumption that hidden variables do not exist.

Elementary as it is, this result entails far-reaching consequences in regard to deeply engrained ways of thought. For instance, the approach toward a general description of the world initiated by Leucippus and Democritus under the name of "atomism" is just a version of micro-objectivism. It is therefore condemned by the foregoing considerations. Similarly, it seems quite difficult, to say the least, to reconcile the above result with the philosophies that we may call "multitudinist." By this expression we designate general descriptions of the world which were mainly in favor toward the end of the nineteenth century, partly because of the influence of such scientific advances as the discovery of molecules and the elaboration of the classical kinetic theory of gases.

Many types of multitudinist philosophies have been proposed. Their common and characteristic feature is that they seek to describe natural phenomena as resulting from the (coordinated or incoherent) interactions between extremely many simple elements, each of which has its own properties. On the contrary, what happens according to quantum mechanics is that, even if it were true that at a certain time a large number of microscopic systems each possessed its *own* independent maximum set of properties, this could not remain true at later times, subsequent to the time at which the first interaction took place between the systems. Indeed, when an interaction takes place (it is usually a two-body phenomenon) the elementary systems participating in it lose, so to speak, some properties that they possessed. New properties appear in their place, but these are now common to both systems, that is, they really are properties of the composite system constituted by the original two (see chapter 8). The new system should thus be considered conceptually as an unseparable whole in spite of the fact that, at least in any conceivable classical analogy, its two "constituents" generally cease to interact after some time. As we have seen, such a nonlocality— or nonseparability, as we called it— is something more than a mere statistical correlation. When, because of some previous interaction two separate distant systems (or rather ensembles thereof) are just statistically correlated, in classical physics we do not modify either one of them by acting on the other one (we only modify our *knowledge* concerning it, if the action in question is a measurement). In the case of nonseparability, we are not even allowed to think of either of these two systems as remaining unchanged when something is done with the other one.

Thus an "atomistic" description of Nature that, in particular, would attach definite, individual, specific properties to noninteracting or practically noninteracting microsystems would not be stable against evolution in time. After a sufficient time, we would have to replace it by a more "monist" description, that is, by one in which we could correctly attribute objective properties only to some indivisible whole. If, for instance, some galactic gases or parts of stars are made of diffuse plasma, the foregoing considerations show that a description of these objects as composed of an immense number of small particles moving at random and occasionally colliding should not be taken as expressing a rigorous truth. Although it is a very useful description, others of a similar kind could in principle be disproved by sufficiently accurate experiments. Even for some positivists, and a fortiori for realists, this is indeed quite different from what can be meant by the expression "an objectively true description."

19.3 MACRO-OBJECTIVISM

Our experience of atoms and of elementary particles is considerably more indirect than our experience of macroscopic objects. In regard to the latter, we have a very strong tendency to attribute an absolute validity to the testimony of our senses that endows them with some well-defined properties known as their macroproperties. For instance, we are strongly inclined to believe that, quite independently of our own existence, any such object *is* in some given region of space and *is not* in other regions. This is a particular variety of realism. Let us call it "macro-objectivism."

As we define it here, macro-objectivism, in spite of our reference to experience, is distinctly a realistic philosophy, since it assumes that physical objects would exist and have specific properties even if no human beings existed. Accordingly, macro-objectivism must consider that the fact of a macro-object possessing macroproperties is a general one and, in particular, cannot depend on whether or not we intend to use this macro-object as an instrument to measure. But then the analysis carried out in Chapter 18 shows that macro-objectivism cannot be reconciled with ordinary quantum mechanics. More precisely, macro-objectivism cannot be reconciled with the assumption that macro-objects are composite systems ultimately describable, at least in principle, by means of the usual quantal algorithms, that is, by state vectors and density matrices. Logically this conclusion cannot be altered simply by defining new concepts, such as the notions of classical systems or of irreversible processes, unless these notions are given a role in the formulation of the axioms. But, as shown in Chapters 15 and 16, the notions just mentioned are partly subjective, in the sense that they cannot be defined without reference to some inabilities of human beings. Consequently, they cannot be introduced as primary concepts in a realistic description of physical reality.

We could not reach such a definite conclusion in the case in which the macro-objects are assumed to be describable merely by means of mathematical algorithms more general than those of elementary quantum mechanics. If the microobjects are assumed to obey quantum mechanics, the case just mentioned above can happen only when the number of degrees of freedom of the considered macro-objects is infinite. However, the generalized formalisms that should be used in this case have not yet been applied successfully to our present problem.

Remark

An approach along these lines was discussed in Section 16.3. It was found to be unconvincing. On the other hand, at least one of the generalized formalisms alluded to, namely, the algebra of observables, has a feature that at first sight looks somewhat promising. A state $|\psi\rangle$ that describes a system of particles inside a finite volume V is considered, and the functional

$$\Phi(A) = \langle \psi | A | \psi \rangle$$

is written. As mentioned above (see Chapter 6), the knowledge of the functional $\Phi(A)$ is another way to describe the ensemble. In the present case, this ensemble is quite obviously a pure state. Next, the limit $N \to \infty$, N/V finite, is taken, where N is the number of degrees of freedom, and a surprising result then emerges: the functional thus obtained is no longer one of those that are formally associated with a pure state. Rather, it is a sum over a collection of pure states [1]. Thus the limiting process considered here entails a transition from a pure case to what is formally a mixture. This reminds us of a condition often required in measurement theories, namely, that it be possible to describe the final state of instrument plus system as a mixture even if the corresponding initial state is a pure case.

It is conceivable that a quantum measurement theory compatible with macro-objectivism can eventually be constructed on that basis, but we should not be too optimistic in this respect. The fact that a mixture can be found that describes the final ensemble does not imply that an adequate mixture can be found. Indeed, in the elementary treatment described in Chapters 17 and 18, it is not even required that the ensemble be described at any time as a pure case. There, however, no mixture of states corresponding to *definite pointer positions* could be found that would obey the quantum rules.

In answer to such an argument it can admittedly be pointed out that the incompatibility theorem proved in Section 17.1 does not apply to the present case, since it postulates that the density matrix is measurable, and this means that no superselection rule is present. In the case considered here, however, *some* superselection rules obviously operate, and therefore the density matrix is not measurable. On the other hand, the proof given in Chapter 18 is quite independent of whether or not the density matrix is a measurable entity. Under these conditions, it is clear that the remarkable appearance of a set of selection rules in the case here discussed does not by itself solve our problem. In other words, there are no convincing arguments in favor of the thesis that the introduction of infinite systems—and of the new superselection rules that are associated with them—would alleviate in any way the difficulties of the quantum measurement theory, which have their origin in the linearity of quantum mechanics. Hence it seems fair to state that the burden of the proof should rest on the physicists who nevertheless believe that the thesis in question is correct.

Another point worth mentioning is that the condition that the limit $V \to \infty$ be taken is somewhat unattractive. In such a theory it would give an essential role to the mutual interactions of objects that lie in arbitrarily distant regions of space, thus weakening the assumptions of macro-objectivism in regard to the essential finiteness of macro-objects.

19.4   MACRO-OBJECTIVISM AND LONG-RANGE CORRELATIONS

The Einstein-Rosen-Podolsky effect discussed in Chapter 8 is typical of what we understand here by long-range correlation effects. The results of the measurements of two quantities ($S_z^{(U)}$ and $S_z^{(V)}$ in the example) are correlated in spite of the fact that the measurements take place in two regions of space that are arbitrarily far apart. Indeed, the general rules of quantum mechanics predict that these results are correlated even in the case of a spacelike interval.

Quite generally, in the spirit of macro-objectivist realism, two distinct ways of accounting for long-range correlation effects can be thought of, both of which look quite natural because we are accustomed to using them in familiar cases. The first way is to assume that somehow the two correlated effects are determined by one and the same cause. This type of correlation has the particularity that we expect it to take place also between events that are separated by a spacelike interval. The second way of accounting for long-range correlation is to consider that one effect is the "cause" of the other one. In this case the principle of causality (in the technical sense used, e.g., in quantum scattering theory) demands that no correlation take place if the space-time interval between the two events is spacelike.

The difficulty with which we are confronted is that in the case of long-range quantum correlations neigher of these two trivial ways of accounting for the effect is acceptable. The first one must be excluded, since conventional quantum mechanics assumes the nonexistence of hidden variables. Moreover, even if we did not exclude a priori their existence, we would know from Bell's theorem (see Chapter 11) that it is impossible to introduce such variables without violating either some generally accepted consequences of relativity or some verifiable predictions of quantum mechanics. As for the second way, it cannot be reconciled with all the predictions of quantum mechanics, since that theory *does* predict the existence of such correlations in many instances in which the two measurements are separated by a spacelike interval.

At first sight, the strength of the latter argument could be questioned on the ground that relativistic quantum mechanics is still far from being a reliable theory. On the other hand, the fact that correlations are observed between the two photons that emerge from π^0 decay seems to show, quite independently of any theoretical argument, that we cannot generally assume one measurement event to be the cause of the other in the technical sense alluded to above. Moreover, in the present context we must deny any relevance to the argument which states that the mechanism constituted by such correlations can transmit no *instruction* and that therefore causality is not really violated in the phenomena under consideration. To ground causality on the notions of *instruction* or of *information* is to consider these notions (or equivalent ones) as primary notions in our description of nature. But this conception is borrowed from Positivism, and in fact it contradicts the whole spirit of scientific realism.

Under these conditions, macro-objectivism could be reconciled with the existence of long range correlations only if we were able to find some way of accounting for these correlations that would be different from the two considered above. This is far from easy. The difficulty is connected with the fact, stressed in Chapter 8, that the reduction of the wave function is a noninvariant process. It constitutes a strong supplementary argument for rejecting macro-objectivism.

19.5   OTHER VERSIONS OF SCIENTIFIC REALISM

In the spirit of realism, fundamental science can have some value only if it bears on aspects of reality. This does not imply that it can give a fully accurate description of reality as it is, but it *does* mean that the descriptions it gives should have *some* connection with reality. A scientific realist thus necessarily believes that the world possesses some regularities (the "laws of nature") which are independent of men but are knowable by them, at least in some approximate manner. Hence the scientific realist is bound to consider that one purpose of the fundamental principles of physics should be to give a description of these regularities that is as accurate as possible. Under these conditions, the minimal achievement that we are entitled to demand from this scientist is that he describe the said principles without referring—even implicitly—to the knowledge, abilities, or inabilities of the experimenters, since these qualities and defects obviously play no part in the regularities of the world.

Unfortunately, these requirements are not met by present-day theoretical physics, as we saw in preceding chapters. The question thus arises as to whether or not it is possible to translate the fundamental principles of this science into a language that would not refer in any way to the abilities of human beings. The fact that neither micro-objectivism nor even macro-objectivism could be reconciled with elementary quantum mechanics shows that this problem does not admit of any easy solution. Indeed, it shows that the kind of physical reality whose regularities the principles of physics are expected to map must presumably be a nonseparable whole, with properties quite different from those we are accustomed to attribute to any kind of real entity.

Under these conditions the distinction between realism in a "materialistic" sense and realism in the Platonic or medieval sense becomes quite difficult to make. It could be claimed that a realist philosophy is of the first type whenever it assumes that *physical* reality ultimately constitutes the totality of reality. But the word "physical" should then be precisely defined. Since nowadays the general principles of quantum mechanics underlie the whole of physics, a natural definition would be to call "physical" what is subjected to these principles, that is, the set of what can be ascribed either a state vector or a density matrix or some more general algorithm of a similar nature. We may call *unirealism* [2] the philosophy according to which *that* reality is the whole of reality.

It is very difficult to reconcile unirealism with quantum physics. The theory reviewed in Chapter 23 represents what is probably the most serious attempt along these lines. However, we shall finally come to the conclusion that even that theory does not satisfy the unirealistic postulates.

REFERENCES

[1] H. Araki and E. J. Woods, *J. Math. Phys. 4*, 637 (1963).
[2] B. d'Espagnat, *Conceptions de la Physique Contemporaine*, Hermann, Paris, 1965.

CHAPTER 20

POSITIVISM

The main result of Chapter 19 is that any attempt to build a description of the world that would accord with the general postulates of realism and at the same time would not merge—in a way—into Platonism meets with considerable difficulties if the quantum rules are correct. Still, a number of physicists who do not consider themselves Platonists believe in quantum physics. How can this be?

20.1 STRUCTURES; WEAK AND STRONG OBJECTIVITY

One element of the answer lies in the fact that different people attach different meanings to the word "scientific." The existence of at least two very different acceptations of this word among scientists themselves can be illustrated by considering the two quite different meanings that are commonly ascribed to the word "structure" even in scientific circles. It is true that a general agreement seems nowadays to exist among physicists that the aim of their scientific investigations is to discover structural relationships between individual "happenings." Indeed, it was stressed by Roman [1], for instance, that this observation is sufficient to determine both the language and the form of physics. Since the science of structures as such is mathematics, mathematics must of course necessarily be the language of physics. And since the fundamental mathematical structures are either analytical or algebraic, it is not surprising that analytical properties and algebras should be the backbone of fundamental theoretical physics.

Do we find here, at least, a unanimity of views among physicists? Yes, of course, but only in parts. Indeed, the expression "structural relationships between individual happenings" used above is obviously ambiguous. Is the word "happening" merely a substitute for "phenomenon" (in the original sense of "communicable observation"), or does it mean something that really happens in the sense that it would happen also if there were no observer? Both opinions are held.

The latter view is the one whose consequences have been extensively discussed in preceding chapters. The physicists who adopt it motivate their choice by arguing that the observer is just a physical system—and a very complex and delicate one, moreover—and that it would be ludicrous indeed to base physics on some of the most obscure and most accidental properties of such systems, especially since these systems are not present in most parts of the universe. For these physicists the ultimate purpose of science is dictated by their *realist* convictions: it is to discover "how things really are." They therefore insist that a description of a physical law or principle should not refer, even implicitly, to any specific ability or inability of the observers. This is their idea of objectivity. Let us call it "strong objectivity." And let us note parenthetically that it is possible to aspire to strong objectivity without being *ipso facto* an objectivist; it suffices to be a realist.

The physicists who take the radically opposite view stress the fact that we should always be careful not to construct sentences that are void of any meaning. They claim, moreover, that the only method for ascertaining that a sentence or an expression has a meaning is to define the concepts it involves in terms of possible operations and of subsequent observations. (Such *operational* definitions are discussed in Section 20.3.) Since this definition of a concept explicitly refers to the *observations* (nay, even to the acts) of some observer, it is clear that these physicists cannot consistently define the objectivity of a statement independently of the abilities (or inabilities) of human observers. They therefore base their definition on communicability. For them, the objectivity of a statement is nothing else than the validity of this statement *for any human observer*. Let us call "weak objectivity" such invariance under any change of observer (an equally valid name is "intersubjectivity"). The difference between strong and weak objectivity is that a statement of physics can be objective in the weak sense and still refer in a decisive way to the abilities of an observer, if these abilities are common to all observers.

20.2 THE POSITIVISTIC STANDPOINT

The differences in the two definitions of objectivity (weak versus strong) can, in a sense, be considered as merely underlining a difference of emphasis between two methodological approaches. In other words, even a realist—nay, even an objectivist,—can make use of the concept of weak objectivity. As a matter of fact, science is, by definition, exclusively constituted of communicable experience, so that every scientist *must* use the criterion of weak objectivity in order to ascertain that his assertions make sense. But the realist is also concerned with verifying that the statements he makes, at least do not flatly contradict the requirements of strong objectivity. At any rate, he should do so if he wants to be consistent.

Since we have already studied at length the consequences of the intuitively most appealing version of realism, namely, objectivism (micro- and macro-), let us now explore the other end of the philosophical spectrum. In other words, let us for a while study conceptions that depart radically from objectivism and even, to begin with, from realism. The best-known and important of these is scientific positivism.

In the form that concerns us here, scientific positivism (also called "scientific empiricism") should of course not be confused with the practical empiricism of which we make daily use as scientists. Every scientist either develops or applies phenomenological models that reproduce the most important observations, at least in some domain of experience, but are not believed to describe the real nature of the underlying processes. Clearly, a scientist who does this is not *ipso facto* a positivist. He can, for instance, be a realist. As such, his hope is that the knowledge of these regularities will pave the way for discoveries bearing more closely on the realities of things. But the scientist for whom empiricism is merely a practical method is of course not helped by empiricism when he encounters the difficulties reviewed in the foregoing chapters, which are difficulties of principle. Indeed, the only form of empiricism that can—and does—help in this connection is one that changes the basis of the whole argumentation by making *observation* and *meaning* the really primeval concept, from which every other concept should be derived. Reality, in particular, thus reduces to phenomena; and the latter word, moreover, can here be given no other meaning than its etymological one: *what is perceived.*

We should not claim, of course, that such radical views were (or are) those of every empiricist or positivist. The words "empiricism," "pragmatism," "operationalism," "logical positivism," and so forth, designate various related epistemologies and methodologies that made their apperance before the advent of quantum physics. Hence, on the question of what are the acceptable meanings of the word "reality," their proponents enjoyed more freedom than we have now. For instance, Locke and perhaps even Hume accepted to some extent the commonsense views in regard to the existence of the external world and even of the material objects. Hume stressed that the question of whether the laws of nature are subject to change is one that is open to doubt, and this seems to imply that, at any rate, some external "nature" exists. At least the empiricists of these times do not seem to have adopted unreservedly the extreme position of some idealists *against* an idea of this sort. The same attitude in revealed in the writings of the nineteenth century pragmatists, and again, considering, the scientific knowledge of the time, it was a reasonable one.

It was only with the advent of logical positivism that this situation began to change. The philosophers of the Vienna Circle placed the emphasis quite strongly on the necessity of a logical analysis of our language that, in particular, would eliminate meaningless—though apparently meaningful—statements They postulated (verifiability requirement) that a statement has a meaning only if it can be verified or falsified by means of some definite experimental procedure. All statements that do not fall in this category they called "pseudostatements," and considered them to be void of any cognitive content. This led their most eminent representatives, such as Schlick and Carnap [2], to explicit rejection both of the thesis of the reality of the external world and of the thesis of its irreality, on the ground that such theses are pseudostatements (what we can verify is always the existence of some sense data, not of an external object; and it is obvious that the negation of a pseudostatement is also a pseudo-statement).

Logical positivism asserted that statements should be verifiable. Operationalism extended these views also to the concepts used in formulating such statements. It insisted that the definitions of scientific terms should be stated as rules "to the effect that the term is to apply to a specific case if the performance of specified operations in that case yields a certain characteristic result" (Hempel [3]). Now the concept of a reality that will exist even in the absence of *any* observer can obviously not be given an operational definition. Thus, as we pointed out (in Chapter 19), this concept should, in the operationalist conception, be considered as meaningless. Indeed, the only use of the word "reality" that operationalism can tolerate is the one in which it is identified with the set of all intersubjective appearances, that, is, subordinated to men's abilities. This is essentially the same conclusion as the one obtained above.

It must be granted that, as a result of further investigations (based, in particular, on the observation that science is steadily becoming more formal), contemporary positivists have reached the conclusion that both the verifiability postulate and the requirement that all definitions be strictly operational have to be somewhat alleviated. In particular, they are now willing to accept the view that many symbols and words referring to nonobservable quantities may creep into the scientist's formulas and sentences. However, they require that all these symbols and words serve only as *intermediate links* in the formulation of rules exclusively bearing on what can be observed. In Carnap's terminology, these terms constitute a *linguistic framework*. According to this author, the question of whether the set of all the entities constituting a given framework "really exists" cannot possibly be a theoretical one, bearing on the absolute existence of such entities, independently of ourselves. Rather, it is a *practical* question, that is, it can only mean, "Shall we decide to use the considered framework?" Of course, our decision on this point is influenced by questions of efficiency: in the case of the successful theoretical concepts we choose to accept them because they are useful *to us*. But Carnap [2] stresses very strongly that, since our decision is essentially a *practical* one, it should, above all, not be interpreted as a belief in the "fundamental reality of entities," and that such an expression is and remains void of any cognitive content.

Among scientists, a not infrequent opinion is that positivism is essentially a *method*, and that therefore it cannot alter our conviction that an independent physical reality "obviously" exists. As already stressed, this opinion would make positivism entirely irrelevant to the problem of solving the difficulties described earlier in this book. But it does not, in fact, correspond to the truth. Positivism *is* indeed a powerful method of solving these difficulties *because* it suppresses what is at the root of all of them, namely, the "dogma" that a reality somehow exists that is entirely independent of us and whose features science should progressively approach.

In order to keep this fundamental point in mind, we shall in our conclusion use the expression "purely linguistic standpoint" for designating a general philosophy that incorporates, in particular, the kind of positivism which we have defined above. In conformity with the views of Carnap and of other philosophers of the same school, the purely linguistic standpoint thus rejects as meaningless the thesis of the reality of the external world (and the thesis of its irreality). At that price it does indeed solve such problems as those raised by the linearity of the quantum law of evolution. But it can hardly claim not to be anthropocentrist.

20.3 OBJECTIONS TO POSITIVISM VIEWED AS A BASIC PHILOSOPHY

Just as the content of the preceding section is not sufficient for giving a detailed account of positivism and of its success, similarly the limited extension we want to give to investigations bearing on these matters unavoidably prevents us from describing adequately several difficulties that this philosop. y encounters. Nevertheless, we cannot ignore a few classical objections that have been raised against it.

Most of these objections bear on the verifiability postulate—first of all, on its consistency. How do we *know* that the statement through which this postulate is expressed is correct, since it is not verifiable? Moreover, we cannot have any direct experience of the past or of other minds. If it is true that a statement has a cognitive meaning only if it can be verified or falsified by some experimental procedure that ultimately refers to our sense data, should we consider as meaningless all statements about the past or about the experience of other men?

In regard to the objection concerning the meaning of statements on past events, the standard positivist answer is that they are really statements about the future. They are statements about the possible future pieces of evidence that will eventually verify them. To borrow an example from Feynman [4], a historian who makes a statement about Napoleon simply means that, if he goes to a library and opens some books and other documents about Napoleon he will find some written statements that are substantially similar to his own. A difficulty with this view is that, if we make such a statement in 1971 and reiterate it later, its content will have changed: For example, the conception studied here renders meaningless a statement such as, "The man who painted the first wild ox in the cave of Lascaux had a wart on his nose." However, there was a time when this statement was either right or wrong, that is, unquestionably meaningful.

To cope with such difficulties, it is often suggested that we should consider a statement as meaningful as soon as we can *imagine* the performances of some human being engaged in the process of carrying out the appropriate verifications. This view seems quite reasonable. It is gratifying to be able to think that neutrinos existed in the past in precisely the same sense as the one in which we say they exist *now*. However, it is difficult to set a priori limits on men's technical ingenuity. Considering the technical advances that took place during the last two centuries, we are entitled to believe that in a few thousand years men will be able to make many measurements that seem *to us* practically quite impossible to perform. Hence, in the view presently discussed, we should absolutely refrain from basing any argument on the fact that some experiments seem to us to be fantastically difficult, although no principle of physics actually makes them impossible. Unfortunately this is just what both the thermodynamician and the quantum physicist customarily do when they discuss irreversibility and quantum measurement theory, so that the proposed extension of the verifiability postulate considerably lowers its efficiency as a tool for the physicist. To this, it should be added that, generally speaking, as soon as we allow ourselves to *imagine* observations that could conceivably be done in some unspecified future and by human—or superhuman—beings who might be very different from our experimentalist colleagues, we open the door to rather wild thinking and can hardly exclude metaphysics, which was one of the main purposes of the positivist movement.

As previously mentioned, another well-known objection to the verifiability principle is that none of us has any direct experience of the sense data of others. To this objection the answer usually given is that our statements really bear not on the sense data of other human beings but exclusively on the observable behavior of the latter, and that the language of ideas and feelings, when we apply it to persons other than ourselves, is for us just a useful tool. Ultimately our alleged statements about other men's sense experience are therefore just statements about our *own* personal experience. The positivists who take this position vehemently deny being solipsists. Nevertheless their defence is weak. For instance, they point out quite correctly that the evidence we have of the existence of our own sense data is no more an evidence for the persistence of our own personality than it is for the existence of the external world. But whereas this is an acceptable argument against some versions of idealism, it does not abolish the unpleasant disparity introduced above between our own sense experience and the one of other men. Since these sense data are considered as being the ultimate foundations of our scientific knowledge, it would seem that a philosopher who argues along such lines is driven to the conception that, to quote Passmore [5], "the ultimate meaning of scientific statements is, after all, ineffable; a strange conclusion for a positivist!"

It would be unfair to suggest that the positivists have ignored these objections. On the contrary, they were aware of them and did their best to cope with them, but these attempts show too great a diversity to be reviewed in a few pages. Moreover, the present writer is not qualified for this task, since he cannot claim to have fully understood them.

Along with the difficulties hitherto mentioned, which bear essentially on the empiricist or positivist approach, another quite general one must be mentioned, which concerns both the realistic and the positivistic standpoints but which—in our opinion—is more serious in regard to the latter. This is the problem of how to justify inductive inference.

The procedure known as inductive inference is the one through which we conclude from an observed regularity to a law. As such, it is fundamental to science. The problem of how to justify it was first raised by Hume [6], and although Hume's criticism is more than two centuries old it has never been successfully refuted. Its essence is well known: in spite of the fact that, say, white crows have never been observed as yet, and no matter how many crows have been found until now to be black, we cannot *derive* from this experience the conclusion that the next crow we see will be black. Indeed, we should not even say that this event is most probable in the sense that we expect it to happen, for we can base this expectation only on the observation that similar recipes have already proved correct in many different instances. But when we argue that, since similar deductions have happened to be correct, the present one should also presumably be correct, we are simply advancing an argument of the crow type: we are already applying (this time, to a kind of recipe that is quite general) the principle of inductive inference that precisely we are trying to justify. We are thus turning in a circle.

The consequence of Hume's objection is that, in order to accept the validity of inductive inference, both the empiricist and the realist philosophies require something which is very much in the nature of an act of faith. The act of faith of the empiricist is simply that inductive inference can be applied to his various observations. Outwardly, that of the realist is more complex: It consists in believing in the existence of an external reality that, to some extent, is knowable *and* whose regularities do not change in time.

We would like to sketch here the main lines of an argument showing that, in a way, Hume's argument speaks nevertheless in favor of the realist. First of all, it must be granted that, when the positivist blames the realist for believing in the existence of a few nonoperationally defined physical entities, the latter can answer that a comparable act of faith (in the value of the recipes based on inductive inference) is also made by his critic. If the positivist then argues that his own act of faith is at any rate less complex than that of the realist, the latter can answer that this is merely an appearance. The positivist has somehow to make a *renewal* of his act of faith for every new phenomenon that he wants to investigate (since he does not believe in universal entities, he cannot make it for "inductive inference in general"); the realist, one the other hand, can make his own act of faith *only once*, in a reality which is subject to simple (i. e., easily expressible) laws. Then he can justify inductive inference, at any rate in the important case in which only *one* simple law can be stated that correctly describes the facts (of course, equivalent formulations here count as one). Moreover, he can understand that "simple" and "easily expressible by men" should be synonymous also under the realist viewpoint, since men are themselves a part of reality and therefore structures of their minds may well reflect rather fundamental structures of reality itself.* It should be noticed that within these views fundamental reality is not wholly unattainable by experience, since its formal *structures*, at least, should be knowable in principle.

Admittedly, the positivist can still object that in his view the notion of an absolute reality can have no meaning. But we may wonder at this stage whether this opinion could not be simply due to the fact that the positivist is a captive of too technical a definition. To consider again Feynman's example about Napoleon, there is no doubt that any statement about the past has the kind of *implications* that Feynman mentions. There is also no doubt that implications falling into this class are the only *criteria* at our disposal for verifying that the statement is true. The statement about Napoleon, therefore, certainly means what it has been asserted to mean, and we have no way to verify any other possible meaning that we might attribute to it. However, the question is this: "Does this really *imply* that no other meaning can legitimately be attached to the considered statement?" To this question the positivist philosopher is forced by his very principles to give an affirmative answer. Innumerable, however, are the scientists and the laymen who—even after thorough reflection—would not answer along these lines. Indeed, a statement about Napoleon—or about the mountains that existed during the Mesozoic age—is almost universally understood as having an intrinsic content of its own, a content which indeed is more than that of a mere summary of the impressions *we shall* have when we read books or dig the ground. If we really thought otherwise, most of us would very soon lose interest in reading about history, and most paleontologists would put an end to their researches!

In conclusion, it seems that the positivist scientist must really face a dilem-

---

*Obviously, these arguments should be developed in order to gain convincing power. It is the author's opinion that such developments would indeed be possible, although the matter must, by its very nature, always remain controversial.

Concerning the general problem of the role of induction in physics, see Ref. [7], where a bibliography is given.

ma. Either he describes his standpoint as being merely a *method*, or he claims that it brings about a real elucidation of the subject-object relationship. If his standpoint is a method, it is one that has proved its excellence in several outstanding cases. But, being simply a method, it does not yield a solution to the fundamental problems discussed in the foregoing chapters, because a valid argument cannot be overthrown just by advancing a maxim: if we believe in realism, we must face all the difficulties it entails, whatever our methods may be. If the positivist standpoint is a philosophy, it really dissipates most of the difficulties we have found. However, it succeeds in doing so only by resorting —be it in a sort of hidden and roundabout way—to the extreme view that, after all, man is the center—nay, the source,—of all conceivable realities. Such opinions can be upheld (see, e. g., Chapter 22), although they have, like all others, a few weaknesses. Yet they should preferably be upheld openly, in full awareness of their real character. It is perhaps to be regretted that scientists who borrow their conceptions from positivism often do not recognize, under the mask of sound *methods*, the true nature of their option.

REFERENCES

[1] P. Roman, Symmetries in Physics, address to the Boston Colloquium for the Philosophy of Science; *Boston Studies in the Philosophy of Science*, 1966.
[2] R. Carnap, Empiricism, Semantics, and Ontology, *Rev. Int. Philos. 4*, 20 (1950); reproduced in *Meaning and Necessity*, University of Chicago Press, Chicago, Ill., 1956.
[3] C. G. Hempel, A Logical Appraisal of Operationalism, in *The Validation of Scientific Theories*, P. Frank, Ed., Beacon Press, Boston, 1956.
[4] R. P. Feyman, *The Character of Physical Laws*, The M.I.T. Press, Cambridge, Mass., 1967.
[5] G. Passmore, *Hundred Years of Philosophy*, Duckworth, London.
[6] D. Hume, *An Essay on Human Understanding*.
[7] G. Toraldo di Francia, *Riv. Nuovo Cimento 4*, 144 (1974).

Further Reading

H. Reichenbach, *The Rise of Scientific Philosophy*, University of California Press, Berkeley, 1964.

W. Heisenberg, *Physics and Philosophy*, Harpers, New York, 1958.

A. C. Ewing, *The Fundamental Questions of Philosophy*, Colliers Books, New York, 1962.

CHAPTER 21

BOHR AND HEISENBERG

One of the various interpretations of quantum mechanics occupies a privileged position in the opinion of the majority of physicists. This is the Copenhagen view, which owes its fame and popularity not only to historical reasons but also to its simplicity and practical efficiency. We must examine it here with very special attention.

As is well known, the Copenhagen outlook was elaborated collectively. Nevertheless, Bohr's and Heisenberg's contributions should be singled out as particularly outstanding. Bohr's articles [1, 2, 3, 4] are especially important for defining what has come to be known as *the* Copenhagen viewpoint, although a contribution from Rosenfeld [5] is also quite valuable in this respect. As for Heisenberg, his views never departed from those that he developed in Copenhagen in close association with Bohr, but later he elaborated on that canvas and reached some conclusions which can be attributed only to him.*

21.1 SUMMARY OF BOHR'S THESIS*

Bohr takes as his starting point the very simple observation that "by the word *experiment* we can only mean a procedure regarding which we are able to communicate to others what we have done and what we have learnt" [4]. According to him, this implies that the description of the experimental arrangement and of the experimental results must be given "in plain language, suitably refined by the usual physical terminology" [4]. By this he means that, when we speak of the intensity of an electric current in a coil, or of the position of a needle on a graduated scale, we understand one another exactly, which is not necessarily the case when we use other, more elaborate concepts that are less directly related to experience. Since this language is precisely the

*A recent article by Stapp [11] is particularly illuminating in regard to the philosophy of the Copenhagen interpretation (see especially the appendix: Correspondence with Heisenberg).

*Note added to the special edition: Not all the experts agree that the content of this section correctly reproduces all the subtle facets of Bohr's standpoint. Perhaps it should be described only as the way the present author manages to understand in a consistent way the essential message of Bohr's main writings.

language of classical physics, what Bohr asserts is that ultimately it is in the language of classical physics that our statements about physical phenomena of any kind receive unambiguous meanings.

Does this imply that according to Bohr we should describe atoms, electrons, and so forth in the language of classical physics? Obviously not. What Bohr means is that, when we want to speak of entities not subject to the laws of classical physics, we should not strive to formulate statements about these entities alone, thus isolating them from their environment. On the contrary, he claims that in most cases the only way in which we can ensure unambiguous communication of what we have done and observed about quantum systems is to make statements exclusively about the instruments by which these systems are prepared and by use of which they are observed. These latter statements we *can* formulate in a classical language, and from the foregoing argument we know that we *should* indeed formulate them in such language.

From this analysis we can infer two principles, which are in fact the two pillars on which Bohr's doctrine is built. These are as follows.

Principle (a)
The working of measuring instruments must be accounted for in *purely classical terms.*
Principle (b)
In general, quantum systems should not even be thought of as possessing individual properties* independently of the experimental arrangement.

Principle (*b*) warns us that from observations carried out on the instruments we should not try to infer any unwarranted conclusion bearing on the systems themselves. By "unwarranted conclusions" we mean, in particular, those that could emerge only from applying, to the systems themselves, concepts whose validity has been ascertained only for other types of systems. This does not imply that we should in no case issue any statement bearing on the microsystems we observe, but it does imply that these statements should as a rule refer explicitly *or implicitly* to the experimental arrangements used for the observation. The idea embodied in Principle (*b*) can also be expressed by stating that the instrument and the observed quantum system constitute together an inseparable whole [5]. This means that the observed system shares, so to speak, its properties with the instrument.

Concerning Principle (*b*), Bohr stresses what he considers to be the most important difference between classical and quantum physics. To determine the properties of objects we must, *both* in classical *and* in quantum physics, make use of experiments, so that ultimately these properties are *defined* by the experiments in question. But in classical physics we can, if we like, develop an

---

*Even if it is not absolute, the distinction between *specific* properties (e.g., mass and spin) and *individual* properties (e.g., position, velocity, spin, and components) is often a convenient one.

apparatus so complicated that in one operation it gives us the totality of the conceivable information about the object. This is not the case, however, in quantum physics. In the domain of experience that the latter describes, we can set up an instrument that measures the momentum of a particle *or* an instrument that measures its position, but we cannot construct an instrument that measures both at the same time. Although, strictly speaking, the particle has no properties that it does not share with the instrument, it is in both cases a harmless misuse of terms to attribute conventionally the property in question to the system alone. More precisely, such a convention is indeed harmless (and even useful) as long as we are aware of the limitations imposed by the very fact that it is nothing else than a convention. Although these limitations were not felt in classical physics, they are extremely important in quantum physics; they account very naturally for the fact that in quantum physics it would be meaningless to introduce into our descriptions the concept of simultaneous, precise values of the position and the momentum of a particle.

We have already noticed in this connection (see Chapter 18) that we can in some cases artificially define simultaneous values of position and momentum. For instance, we can create an almost perfectly monochromatic beam of particles, measure the position of one particular particle at a given time, and calculate the position that a classical particle having the same position and velocity at the considered time would have had at an earlier time. We can then, if we like, *define* the *position of the quantum particle at the earlier time* as being determined by the number thus obtained. But the knowledge of this number does not help us in any way when we try to predict the results of future observations on the particles. It is therefore generally asserted that such a definition is meaningless.

Clearly, the considerations about the impossibility of attributing simultaneously—and by convention—a precise position and a precise momentum to a particle should be generalized to other observables. The result constitutes what is known as *the principle of complementarity*, which can be formulated as follows.

Principle of Complementarity

The nonseparable whole constituted by the quantum system and a definite instrument can be described by using a simplification of our language, according to which some of the properties that the system and the instrument share with one another are conventionally attributed to the system. However, other properties, which our classical experience leads us to think of, cannot then be attributed to the system. They are said to be complementary to the first ones. They can also be attributed to a quantum system similar in type to the one considered so far, but this is possible only if that system builds up an indivisible whole with some *new* instrument, which is appropriate for a measurement of the new quantities.

In the particular case of the pair of observables "position" and "momentum," the Heisenberg uncertainty relations can be considered as a generalization of the complementarity principle. It can happen that a definite experimental setup corresponds *neither* to a precise definition of the position of a particle *nor* to a precise definition of its momentum. In that case, neither a precise position nor a precise momentum can be attributed to the particle. Nevertheless, smeared values of *both* these variables can conventionally be simultaneously attributed to the particle, provided that the product of the root mean squares of the two smearing functions is larger than Planck's constant.

Bohr's view that microscopic systems have essentially no intrinsic properties (i.e., no properties that would not depend on the instruments) explains why he did not consider the state vector (or the density matrix) as being objective, that is, as reflecting some independently existing state of a system or of an ensemble of systems. As we saw in Chapter 9, it also explains why he could simultaneously consider the state vector as a nonobjective quantity and yet claim that quantum mechanics is essentially complete and that hidden variables do not exist.

21.2   DISCUSSION

Schematic and elementary as it is, the foregoing summary of Bohr's standpoint suffices to show the outstanding role that it imparts to macroscopic objects or, at any rate, to a certain category of these, namely, the instruments used for measurement. If we focused our attention exclusively on that remark, we could be tempted to make Bohr's views more definite by identifying them with the philosophy of nature that we called macro-objectivism in preceding chapters. However, two important reasons can be given for refraining from doing so. One of them is, of course, the fact that macro-objectivism meets with difficulties that have been found to be considerable. The other one is that such an identification would sharply contradict many statements which have been made by Bohr himself. For instance, this author has stressed that some of the diaphragms that are parts of the experimental setup used in several thought experiments are to be considered as quantum systems in many discussions bearing on these experiments. He then contrasts these diaphragms with what he calls the "proper measuring instruments." According to him, the latter "serve to define, in classical terms, the conditions under which the phenomena appear" [3]. And he stresses the fact that the extreme difficulty, or practical impossibility, of proving experimentally the quantum nature of such macroscopic bodies is *not relevant* to his interpretation of the facts.

Obviously, such statements show that Bohr simply does not accept macro-objectivism. Indeed, they show that Bohr's fundamental distinction between instruments and quantum systems is based not at all on the macroscopic nature of the former, but entirely on the use that *we* make of them for defining the conditions under which we observe. This puts in the forefront the notion of the observing subject. And, indeed, Bohr went as far as to mention the "fundamental limitations, met with in atomic physics, of the objective existence of phenomena independent of their means of observation" (Ref. [3], p. 7). As a matter of fact, he even emphasized repeatedly the necessity of paying proper attention to the object-subject separation, and he also stressed that such a separation could be made in an infinite variety of ways.

Of course, the latter statements are straightforward interpretations of the elementary formalism of measurement theory described here in Chapter 14. In this respect, they are common to Bohr and Heisenberg, on the one hand, and to von Neumann, London, and Bauer and Wigner, on the other hand. However, a considerable difference exists in the manners in which these two groups really understand the role of the subject and the object-subject separation. The views of the latter group, which are summarized in the next chapter, are very roughly described by the assertion that two fundamental laws of evolution in time exist in Nature, one of them—the wave packet reduction—operating only when some perceiving subject makes an act of observation, and the other—the causal time evolution—operating in every other circumstance (the arbitrariness as to where the object-subject separation occurs in practical calculations is then considered, as in Section 14.2, as a technical point rather than as a fundamental fact).

Bohr's views are entirely different in this respect. To see this better, let us bring together two statements made by this author. The first is the one quoted above, according to which the objective existence of phenomena is not independent of their means of observation. In the other one, Bohr mentions the "conditions which define the possible types of predictions regarding the future behavior of the system" and asserts that "these conditions constitute an inherent element of the description of *any phenomenon to which the term* physical reality *can be properly attached*" [2] (the italic is ours). Taken together, the two statements show that, in Bohr's view, the conditions of observation influence in a fundamental way physical reality itself, that is, anything that we can call "physical reality." In other words, Bohr does not consider the arbitrariness of the object-subject separation as merely reflecting a set of technical possibilities which enables us to simplify the calculations whenever we are not interested in complex correlations between object and instrument. On the contrary he attributes to that arbitrariness in the separation—and, in particular, to the possibility of putting the instrument on the same side as the subject —a fundamental significance. Following Shimony [6], we can state that Bohr introduces a complementarity between object and subject and that this generalized complementarity, far from being a mere appendix to the physical theory, is, on the contrary, the most essential part of it.

An implication of these views has already been analyzed in the foregoing

chapters, particularly in chapter 9: the state vector is not objective, and therefore the wave packet reduction is not a physical process which would take
place when an instrument comes into play or when a subject has a perception.
Rather, it has already taken place at the time when we consider any definite
experimental setup. In other words, it is ultimately determined by the actions
we take in order to acquire knowledge, as well as by the consciousness that we
then have of purposefully performing such actions.

In a sense, this leads to a violation of the causality principle (see Chapter 9).
As for the Lorentz invariance of the measurement process, it raises less acute
problems than occur in the theory in which the wave packet reduction is an
objective phenomenon. However, in the correlation-at-a-distance phenomena,
the fact that an instrument at one given place should instantaneously modify
the list of observables that have sharp values (i.e., that can conventionally be
attributed to the quantum system alone) at some other, distant place remains
disturbing. At least, this is the case as long as it is considered that a *realistic*
conception of the relativity requirements, such as the one discussed in Section
19.4, should be kept.

The objection is often formulated against Bohr that the descriptions of the
Copenhagen school are "not objective." Sometimes they are even criticized as
not being "materialistic." The first objection is largely a question of semantics. If we define "objectivity" as being identical to *strong* objectivity in the
sense set forth in Section 20.1, it is quite true that Bohr's views are not objective. On the other hand, if what we mean is simply *weak* objectivity, then these
same views are unquestionably objective, since they are constructed precisely
in order to secure an unambiguous communicability of all pieces of information. In regard to the second objection, it is similarly, to a very large extent,
just a question of definition. If materialism is essentially identified with the
philosophy called "unirealism" in Section 19.5 (and this is the most frequent
case), it is quite true that Bohr's views cannot be reconciled with materialism.
On the other hand, dialectical materialism has been defined, at least by some
authors (see, e.g., Ref. [5]) as "the ever more refined expression of the relationships that exist between the external world and its representation in our
mind." What is remarkable about this definition is that it uses the word
"mind" in order to define materialism. It thus makes the concept of *mind*, as
far as the reader can judge, at least as fundamental as the concept of matter—
and perhaps even more so.

If this is what is really meant, then the question of whether or not Bohr's
doctrine can be reconciled with "materialism" as thus defined reaches some
degree of subtlety. To a noncommitted reader it may appear that Bohr's
conception of reality describes at most an *empirical reality*, that is, a reality
entirely relative to accidents in our knowledge. The view could therefore be
expressed that some fundamental reality should underlie Bohr's reality.
Although we could read between the lines of Heisenberg's writings an opinion

resembling this one (see Section 21.3), it seems that Bohr always rejected such an idea. Most certainly, this testifies to Bohr's reluctance to formulate anything that could appear as an ontology. However, in the present context this refusal of Bohr makes his standpoint very obscure to many people. Even if no ontology is free from arbitrariness, we would like to know whether any one —or perhaps several—of them can be accommodated to the facts. And it is certainly not clear for what reason such a question should be considered as meaningless. However, we must at the same time carefully keep in mind Bohr's pronouncement that a "renunciation of accustomed demands on explanation" may well be necessary as a counterpart to the acquisition of logical means of comprehending much wider fields of experience ([3], p. 78).

21.3 SOME ASPECTS OF HEISENBERG'S PHILOSOPHY

It is an open question whether or not Heisenberg would unreservedly subscribe to a detailed account of Bohr's standpoint such as the one given above, but there is no doubt that he agrees in general with the views expressed by Bohr. On the other hand, it is also quite clear that he has added to these views. Our present task is to summarize some of these additions. Heisenberg's most important contribution in that respect might well be his redefinition of the Kantian *a-priori*. As everybody knows Kant felt very much concerned about Hume's paradoxical result that inductive inference cannot be justified. As a first step toward the solution of this problem, he suggested that some of our concepts are *not* derived from experience; rather they constitute an a priori framework which we cannot help using as soon as we want to organize our sense data.

Kant considered absolute time, Euclidian space, causality, and substance as being the main a priori concepts, and he claimed that such concepts had absolute validity. In this he was unfortunate. What Heisenberg asserts is that the use of these concepts—and, more generally, of the *classical* concepts—is indeed necessary for investigating atomic phenomena and is therefore an a priori in that sense. But this, he claims, does *not* imply that the concepts under discussion have an *absolute* value. What Kant did not foresee is that some notions could be the conditions of science and *at the same time* have only a limited range of validity; and this, according to Heisenberg, is precisely what happens with the classical concepts. Hence this author concludes that modern physics has modified rather than overthrown the status of Kant's assumption regarding the existence of what he called "a priori synthetic judgments." It has changed it from a metaphysical to a practical postulate. Thus a priori synthetic judgments have the character of relative truths.

It might seem at first sight that such a view contains an internal contradiction, but this is due only to the fact that the expression "a priori" can be understood in two different ways. In Kant's language "a priori" has the same connotation as the words "unalterable," "transcendental," and so forth. This is the sense that Heisenberg rejects. But it can also be used to qualify simply the set of concepts which we, presently living human beings, *must* use in order to organize our experience, or, in other words (in a realist conception), the colored spectacles that we cannot remove when we look at reality. In this sense a priori judgments and concepts can indeed be relative. They could even be intuited from some ancestral experience, for example, through education.

We just mentioned realism. The foregoing discussion is purely epistemological and does not compel us, therefore, to a definite choice either in favor of realism or against it. Although Heisenberg does not commit himself unreservedly to such a choice, he does not exclude the thesis of a realism similar to Kant's realism. Only, the "things-in-themselves" of Kant are for him the mathematical structure we disclose by our theories. In contrast to Kant's view, they are therefore indirectly attainable by experience.

These aspects of Heisenberg's philosophy can be easily reconciled with the considerations developed in the more technical sections of this book and do indeed help in making understandable the results gathered therein. But Heisenberg's outlook also has other, and perhaps better known, feature which do not fit so well with what we found. One of them is the introduction of *potentia* and the use made of this concept. The potentia are the possible as contrasted to the actual, and Heisenberg considers the reduction of the wave function in a measurement act as a transition from the possible to the actual. He states explicitly, "If we want to describe what happens in an atomic event, we have to realize that the word *happens* can apply only to the observation, not to the state of affairs between two observations. It applies to the physical not to the psychical, act of observation, and we may say that the transition from the *possible* to the *actual* takes place as soon as the interaction with the measuring device, and thereby with the rest of the world, has come into play" [7].

Moreover, according to him the explanation of this reductive process is that "the equation of motion for the probability function does now contain the influence of the interaction with the measuring device. This influence introduces a new element of uncertainty since the measuring device is necessarily described in terms of classical physics; such a description contains all the uncertainties concerning the microscopic structure of the device which we know from thermodynamics and since the device is connected with the rest of the world it contains in fact the uncertainties of the microscopic structure of the whole world" [7].

As Earman and Shimony [8] have pointed out, the obvious interpretation of these passages is that the *pointer reading* is definite though unknown when the final state of object plus apparatus is reached (but before registration upon the consciousness of the observer), and that this comes about in spite of the initial indefiniteness of the measured quantity, because of the initial uncertainties concerning the state of the apparatus. But if instruments are finite, the fallaciousness of this analysis is essentially the content of the results reported in Chapters 17 and 18.

In the foregoing statements, Heisenberg mentions the fact that the macroscopic device interacts with the rest of the world. Indeed, arguments have been given [9] to the effect that macroscopic systems can never be considered as practically isolated when their microscopic structure is taken into consideration. However, the considerations reviewed in Chapters 17 and 18 can be applied to a system which, along with the measurement device, incorporates arbitrarily large portions of the external world, as long as the quantum equations to which this system is subject do not violate linearity. Moreover, it is difficult to see how the rest of the world can be incorporated into the quantum system without the observer's consciousness also being incorporated in this system—a view which is discussed here in Chapter 23, and which is alien to the Copenhagen interpretation. Under these conditions, the consideration of the rest of the world is not, when taken by itself, a very convincing argument.

All the same, it is possible to reconcile Heisenberg's first-quoted passage with the developments of Chapters 17 and 18. But for that purpose this text should not be considered in the light of macro-objectivism (see Chapter 19). On the contrary, it should be associated with Heisenberg's general philosophy, which is summarized above. Then it becomes clear that for this author *the actual*—the position of the pointer—has no *absolute* reality. It is relative to us, that is, to the conditions which make man's experience possible and which in particular make possible the special set of observations considered in the experiment under study. This obviously wipes out the difficulty. On the other hand, the potentia of Heisenberg are something more than what mere possibilities would be within the framework of a philosophy such as macro-objectivism. If an independent reality is said to exist (and, as we have seen, Heisenberg does not object to this idea), then the potentia are somehow a part of that reality. This reflects the fact that Heisenberg considers these potentia as either totally or at least partly objective.

Even if one does not subscribe to the totality of Heisenberg's statements in the philosophy of science, he is bound to recognize in the general philosophy of this author a remarkable balance of empiricism and realism (in the Platonic sense) that goes very far toward elucidating the apparent paradoxes of modern theoretical physics.

21.4 CONCLUSION

The Copenhagen interpretation of quantum mechanics is a remarkably efficient practical tool. In particular, insistence that we should always ask our questions in terms of conceptually possible experimental arrangements is of considerable help in the task of formulating the right questions and of avoiding spurious paradoxes and ambiguities.

The Copenhagen interpretation avoids the difficulties posed by more naive theories such as micro- or macro-objectivism. More precisely, it avoids their most blatant inconsistencies. For example, in the long-range correlation phenomena that involve correlations at a distance, it short-circuits the difficulty that the wave packet reduction is a noncovariant process by considering that the state vector is not objective. However, it cannot avoid a kind of violation of the causality principle, which is embodied in the relationships between the instruments of measurement and the measured systems (see Chapter 9). Similarly, it can avoid a violation of the relativistic principles in the long-range correlation effects only if it states explicitly that relativistic covariance should refer to the exchange of *information*, rather than to the interaction of events occurring independently of observers (see Section 19.4).

Unavoidably we are thus led to the ontological problem. Bohr seems to have carefully avoided committing himself to any particular ontology [10]. Hence some authors have erroneously believed that the Copenhagen interpretation could be reconciled with the assumption that an *independent* physical reality exists, individual aspects of which (such as *events*) are knowable. The analysis made in this book shows the fallacy of this opinion.

REFERENCES

[1] N. Bohr, *Atomic Theory and the Description of Nature*, Cambridge University Press, 1934.

[2] N. Bohr, *Phys. Rev.* **48**, 696 (1935).

[3] N. Bohr, *Atomic Physics and Human Knowledge*, Science Editions, New York, 1961.

[4] N. Bohr, Quantum Physics and Philosophy—Causality and Complementarity, in *Philosophy in the Mid-Century*, R. Kilbansky, Ed., La Nuova Italia Editrice, Florence, 1958.

[5] L. Rosenfeld in *Louis de Broglie, Physicien et Penseur*, André George, ed., Albin Michel, Paris, 1953.

[6] A Shimony, *Am. J. Phys.* **31**, 755 (1963).

[7] W. Heisenberg, *Physics and Philosophy*, Harper, New York, 1958.

[8] J. Earman and A. Shimony, *Nuovo Cimento* **54B**, 332 (1968).

[9] H. D. Zeh, *Found. Phys.* **1**, 67 (1970).

[10] A. Petersen, *Quantum Physics and the Philosophical Tradition*, The M.I.T. Press, Cambridge, Mass., 1969.

[11] H. P. Stapp. *Am. J. Phys.* **40**, 1098 (1972).

Further Reading

P. Heelan, *Quantum Mechanics and Objectivity*, Martinus Nijhoff, The Hague, 1965.

W. Heisenberg, *The Physicist's Conception of Nature*, Hutchinson, London, 1958.

E. Scheibe, *The Logical Analysis of Quantum Mechanics*, Pergamon Press, Oxford, 1973.

CHAPTER 22

WIGNER'S FRIEND

In view the persistence of our interpretation problems, it is legitimate for us to explore once more the spectrum of well-known general truths, in order to find out whether, by any chance, one of them will open some new alley for our research. In that spirit, we may, for instance, remember that science always had two purposes. The ambitious one is to know "what exists." The more modest one is to systematize our experience. Although at times these two purposes can harmoniously coexist, at other times, such as ours, this duality raises serious difficulties.

In such cases, the normal position of a scientist is always to favor the more modest of the two goals; indeed, the positivist attitude we have already described is but a systematization of this natural behavior. However, we should then acknowledge a simple truth which is somewhat blurred in the standard expatiations of that doctrine, namely, that science aims primarily at describing men's perceptions. That this is a correct standpoint is confirmed by the fact that the other, more ambitious, choice often leads to ambiguities. As a matter of fact, as soon as we leave the domain of our familiar, everyday experience, we do not really know in most cases what meaning should be given to the words "to exist." For instance, we *do* feel some uneasiness about deciding whether—in a model of an expanding universe in which some objects would increase their velocity beyond the limits of observability—such objects should still be considered as existing. In any case, it becomes increasingly clear that our alleged descriptions of the *world as it is* are, up to now, at the best defective images. On the other hand, the mathematical formalism of modern physics, which is the alternative choice, increasingly refers to our observations. Indeed, nowadays it is so much dominated by this concept, treated as a primeval one, that it is best described—as in Chapter 3—by a mere set of *rules* interconnecting past and future observations. All these considerations, then, support the view that observations and, more generally, perceptions truly constitute the backbone of physics. Let us therefore try to separate this conclusion from the methodological developments it is often associated with and to consider it for its own sake.

As is well known, perceptions are complex events. They involve our immediate sensations, and they also incorporate parts of our already existing theories about things. However, the latter—inasmuch as they are theories about *what really is*—partake of the general ambiguity and uncertainty that we just criticized, so that we are left with our immediate sensations (and with our logical a priori if there is such an entity) as the only elements of which we are absolutely sure.

Arguing along these lines, we are easily and rather convincingly brought to a view rather similar to one that was formulated long ago by Descartes. It is that our immediate impressions are the only things the existence of which we cannot and should not question, and that anything else, as Margenau [1], for instance, has particularly emphasized, is a construct.* It can of course happen that according to our prevailing theories some of our impressions should be considered as illusions. Dreams, for instance, are (rightly) considered as illusions. Yet the very fact that we *do* dream—or at least that on these moments we do have impressions—is one of which we can be sure. Similarly, when we see a signal, it can happen that according to the prevailing theories this is but an error made by our senses: in the language in which we usually describe the world to one another, there *really* is no such signal. However, the fact that we think we see a signal is a reality in its own sake; and it falls into the category of immediate realities about which—contrary to those that are interpretations of data—we cannot entertain even the slightest shadow of a doubt.

Science aims at certitudes, and the only certitudes each of us has concern his immediate impressions. Hence, if there is one datum that science may not contradict, it is precisely the fact that each of us has definite impressions. Now let us again consider a *generalized* measurement process; that is, as in Chapter 14, we consider a process in which, initially, the measured system S is not an eigenstate of the measured observable. Let us, moreover, assume that the observation consists in seeing or in not seeing a signal. Then, if we describe the observer A (plus his instrument, of course) by a density matrix, and if the quantum rules are strictly valid, we know from the analyses of Part Four that after the interaction A cannot be considered as being *either* in a state in which he saw a signal *or* in a state in which he did not [because (i) on an ensemble of $S + A$ systems correlation experiments could *in principle* be performed that would invalidate any such description, unless perhaps A is infinitely extended; and (ii) the transition that would lead to such a situation would not be a relativistically invariant process]. However, we just asserted that the fact that an observer has the impression of seeing a signal is precisely of the

---

*Descartes actually claimed more, but his principal merit is to have noticed the preceding truth.

kind that science may not contradict. In the present case the observer A (i) has the definite impression that he saw a signal (or, as the case may be, the definite impression that he did not), and (ii) incidentally also has the definite impression that he does not extend over the entire universe. Should we then conclude that the first of these impressions is real but that it is merely an *impression*, in no strong correlation with the facts? If—as many biologists would have it— consciousness is just a physical property of a few special physical systems, this conclusion again implies that quantum mechanics does not apply to all physical properties.

Wigner [2] has forcefully stressed this point (whence the name of this chapter) by means of a variant of the cat paradox (see Chapter 17), in which an exchange of information is introduced. He considers three systems, which we call here S, A, and B. S is the quantum system on which a quantity L (with eigenvalues l_n and eigenvectors $|\psi_n\rangle$) is measured. B is the observer. A is either an instrument that is fitted for measuring L or one of the observer's friends, whom the observer B has asked to perform in his place the measurement of L. Again, for simplicity, let us consider a quantity L with two eigenvalues l_1 and l_2 only. A then ascertains the value L has on S by registering or not registering a signal.

For the sake of simplicity, let us describe both S and A by state vectors (for describing A we could also use density matrices; however, this, as we know, would make no essential difference). If the initial state of the system S is any $|\psi_n\rangle$, the final state of the $S + A$ system is $|\psi_n\rangle|n, s_n\rangle$, where, as in Chapter 14, $|n, s_n\rangle$ describes the final state of A. Therefore A can be said to have registered the value l_n (i.e., to have seen, or not seen, the signal). When B looks at A or, as the case may be, listens to A's report, he is thus merely informed of a fact. However, when the initial state of S is

$$\sum_n c_n |\psi_n\rangle \qquad\qquad (22\text{–}1)$$

the final state of the $S + A$ system is, as we saw (because of the linearity of the equation of motion),

$$\sum_n c_n |\psi_n\rangle \, |n, s_n\rangle \qquad\qquad (22\text{–}2)$$

If the observer B repeats his experiment a large number of times, he can ascertain, by studying $S - A$ correlations on some subensemble E' of the ensemble E he has constructed, that the $S + A$ systems are indeed described by (22–2) and *not* by any mixture corresponding to definite values of n. Thus he cannot consistently describe any other element of E (i.e., one of those he neither looked at nor listened to) by saying that its A either *has* or *has not* registered a signal. It is only after B has looked at (or listened to) A that he can use that description. However, in the case in which A is not an instrument, but a friend of B, B can then ask him what he (A) felt when he interacted with S. If A is of the same nature as B, he must necessarily have already had definite impressions at that time. Either he felt he saw a signal, or he felt otherwise. The conclusion is, therefore, that if observer B believes in the strict validity of the quantum rules, he cannot assume that his friend A receives, like himself, definite impressions from the outside world. B is the only being of his species.

This conclusion—a variant of solipsism—is not the final conclusion of Wigner's [2] article. However, in order to escape from it, this author points out that, according to the considerations just developed, it is necessary to modify quantum mechanics. More specifically, he stresses the fact that, since the linearity of the laws of evolution is at the origin of the difficulties, it is linearity that should be given up.

Specifically, what, then, are the phenomena that violate the linear laws of quantum mechanics? This is a question that of course cannot be answered in an a priori manner. It could be, for instance, that even the phenomena we are accustomed to consider as purely physical ones violate linearity. However, in view of the absence of any material evidence in favor of the existence of phenomena of this kind, Wigner expresses a preference for a theory that could be called one of "minimal violation." Since the only phenomena that exhibit such a violation are those involving consciousness, he assumes that it is consciousness that violates linearity (a similar view had already been expressed, but less explicitly, by London and Bauer [3] and von Neumann [4]).

Of course, such a view can only appear as being inherently absurd to all of us who start from the a-priori belief that "consciousness does not exist," at least not in the sense that "material objects" or properties thereof exist. However, we have already mentioned that the very concept of physical existence is far from being a priori given, or even clear. Indeed, when—as scientists—we say that some physical entity "exists," we always mean that the corresponding concept is useful in describing *our* experience in a reasonably concise way. In regards to the "material objects" of our daily life, a little reflection shows that exactly the same holds. Moreover, we ascertained in the foregoing chapters that the extent to which that concept is useful is in fact severely limited: if we raise the idea of the existence of finite, separate objects to the heights of a transcendental (or a priori) principle, we arrive at insuperable difficulties. On the other hand, consciousness—let this fundamental point be repeated once more—is the basis of the whole construction of science. Thus the a priori conviction that consciousness should necessarily be considered as being "less real" than, or a "mere property of," some physical objects (implicitly viewed as having an absolute existence) is obviously too naive.

Having discarded the *dogmatic* objection, it is of course important that we should examine any real (i.e., a posteriori) difficulties that the theory may present. Let us now survey some of these.

For this investigation, it is convenient to agree to call "physical reality" everything that strictly obeys the linear laws of quantum mechanics as long as no consciousness comes into play. In the theory presently analyzed, consciousness is therefore, by definition, a nonphysical entity. However, this is not an outstanding peculiarity of this theory, since in most of the current philosophical views on the subject consciousness *is* considered already as a property of reality that cannot be fully reduced to the "ordinary" physical properties. Indeed, these views radically distinguish consciousness from any other phenomenon by asserting that it violates the otherwise universal law that there is no action without reaction. It can be acted upon but it never acts (the fact of being aware of some physical situation does not affect that situation). In this respect, Wigner is even more respectful of the universality of great principles, for in his description, a reaction of consciousness on the rest of the universe—that is, on what we agreed to call "physical reality"—indeed exists.

Is this reaction practically observable? In attempting to answer this question, let us first remark that, as we formulate it here, it is somewhat ambiguous. If, in the spirit of Wigner's work, we assume that consciousness is the agent that reduces the wave packet, then, if the wave packet is objective, the effects of consciousness on physical reality are of course observed in extremely many cases and are sometimes quite enormous. Let us, however, postpone the discussion of this first point. In another, more conventional sense, the observability of the consciousness-induced processes would be tantamount to the nonexistence of some quite delicate correlations that systems should exhibit, according to the conventional theory, once they have interacted with one another. Such differences are, however, presumably inobservable in practice. Unless, therefore, we find good reasons to attribute consciousness to extremely minute systems, we shall probably remain at a loss as to how to observe the reactions of consciousness.

Another possible objection to the theory discussed here concerns the occurrence of two successive observations, made by two different observers, A and B. In such a case, in order to preserve the observed correlations, we need simply consider that the wave function is reduced twice: once when A becomes conscious of the event he measures, and a second time when, in his turn, B becomes similarly conscious. However, the following difficulty is then quite often pointed out. Let a system S be photographed successively by two observers, A and B. The pictures are looked at in the order which is the reverse of the one in which they were taken, that is, B looks at his own picture first. It is then pointed out that in the present theory the first wave reduction occurs when B looks at his picture. By so doing, B somehow selects a specific image on A's film among the whole set of images that are compatible with the original superposition. This selection is a kind of causation, even though there need be no physical interaction between A and B.

Even if we observe that the A's film really does not have independent existence before *B* looks at his picture (since we then have to do with just *one* system—see Chapter 8), still this whole description seems strange. It looks of course even more so if the two observational acts are events that lie outside each other's light cones. However, this is but one special instance of a phenomenon that we acknowledged as a general one in Chapter 8: the wave packet reduction is *not* a relativistically invariant process.

Now, the fact that the wave reduction is not an invariant process is certainly a very serious objection to any theory that would interpret it as a real physical event occurring between two purely physical systems (as macro-objectivism, for example, would have it). It is perhaps not a really drastic objection to the theory presently discussed, however, simply because the nonlinear phenomena that involve consciousness are so different from the usual ones to which we are accustomed in physics—and so completely unknown physically—that we may as well postulate that they do not satisfy Lorentz invariance.

Thus, is regard to possible criticisms of the theory under discussion, we are essentially left with the rather weak objection that it is unpleasant to imagine that consciousness can have very large effects on reality, even though we know that the specificity of these effects—in the present status of our techniques— cannot be observed in practice. In an essay that is different from the one analyzed here [5], Wigner has advocated a kind of solipsism that could circumvent this last objection. This view classifies all the realities external to a given ego as realities of a second type. Such realities are considered as having degrees of existence that depend essentially on the usefulness of the corresponding concepts. The reader is referred to the article cited for a discussion of the advantages and difficulties of this latter standpoint.

REFERENCES

[1] H. Margenau, *The Nature of Physical Reality*, McGraw-Hill, New York, 1950.
[2] E. P. Wigner, Remarks on the Mind-Body Question, in *The Scientist Speculates*, I. J. Good, Ed., W. Heinemann, London, 1961.
[3] F. London and E. Bauer, *La Théorie de l'Observation en Mécanique Quantique*, Hermann, Paris, 1939.
[4] J. von Neumann, *Mathematical Foundations of Quantum Mechanics*, Princeton University Press, Princeton, N.J., 1955.
[5] E. P. Wigner, Two Kinds of Reality, *Monist*, Vol. 48 (2) (April 1964).

Further Reading

E. P. Wigner, *Symmetries and Reflections*, Indiana University Press, Bloomington and London, 1967.

E. P. Wigner in *Foundations of Quantum Mechanics, Proceedings of the Enrico Fermi International Summer School*, Course IL, Academic Press, New York, 1971.

CHAPTER 23

THE RELATIVITY OF STATES

As shown in Part Four, the particular version of realism that is called there "macro-objectivism" seems hardly compatible with quantum mechanics. On the other hand, we also observed in the foregoing chapters that it is difficult to abandon realism altogether, and that attempts to steer a midcourse between realism and positivism (we think of the Copenhagen viewpoint) also present nonnegligible ambiguities and difficulties of their own. The theory considered in the present chapter can be described as a variety of realism. However, it is a kind of realism that is at the antipodes of micro- or macro-objectivism. In a sense, its first promoter was Schrödinger [1], but it was systematized and carried to its logical extreme 10 years ago by Everett [2]. This theory received support from Wheeler [3] and, more recently, from De Witt [4].* Here we follow the latter's description for the most part.

23.1 OUTLINE OF THE THEORY

The general idea of the approach under discussion is, in a sense, quite elementary: a one-to-one correspondence exists between the reality of the universe and a state vector that obeys the first law of evolution in time of quantum mechanics (the Schrödinger equation or its generalizations). In other words, no wave packet reduction ever occurs.

In view of all our preceding discussions, particularly those of Chapter 14, such a statement seems at first very strange. Even in a simplified universe that consisted merely of a system S and an instrument A, the evolution would, in general, produce a state in which the pointer of A would be spread over several of its possible macroscopically distinct positions. Similarly, if A involved an observer, this observer would, in general, evolve into a superposition of distinct states of consciousness. The originality of the theory under discussion

---

*A very similar theory has been advanced by Cooper and van Vechten [5].

266

is that it serenely accepts these apparently absurd conclusions. Yes, it asserts, measurement-like interactions occur all the time between the various components of the universe. Yes, as a result, this universe of ours is continously splitting into a stupendous number of branches, each of them as real as any other one. Yes, we, as parts of that universe, are ourselves continuously splitting.

To those of us who object that we do not feel ourselves split, the proponents of the theory answer by proving a proposition: To the extent that we can be regarded as mere automata, *the laws of quantum mechanics do not allow us to feel the split.* Since this is obviously the key point, let us first show that the proposition does indeed follow from the principles of the theory.

Let us again consider a simplified world in which there exists only one system S, whose state vector is designated by the symbol $|\psi\rangle$, and one instrument A, whose state vector is $|n, r\rangle$. To the extent that observers are simply automata, A can also, of course, be an observer. Let us assume, moreover, that the evolution in time of systems S and A is described as in Chapter 14 by the transition

$$(\sum_n a_n|\psi_n\rangle)|0, r\rangle \to \sum_n a_n|\psi_n\rangle |n, s_{n,r}\rangle \qquad (23\text{--}1)$$

It must be stressed that, contrary to the case in Chapter 14, the state vectors in (23–1) describe just *one* system $S + A$, namely, the simplified universe under consideration. It also must be stressed that, correlatively, in a simple process like the one described in (23–1) there is as yet no place for a statistical interpretation.

Following Everett, let us rewrite the right-hand side of (23–1) as

$$\sum_n |\psi_n\rangle |\Phi_n\rangle \qquad (23\text{--}2)$$

where

$$|\Phi_n\rangle = a_n|n, s_{n,r}\rangle \qquad (23\text{--}3)$$

is what this author calls a "relative state" of the instrument. Obviously, (23–2) closely associates any $|\Phi_n\rangle$ with a given value of n to the $|\psi_n\rangle$ that has the same value of n. This is the reason why $|\Phi_n\rangle$ is said to be *relative* to $|\psi_n\rangle$, and similarly for $|\psi_n\rangle$. When all the a_n are zero except one, (23–2) has, of course, the simple and straightforward interpretation that S *is* in the state $|\psi_n\rangle$ and that, correlatively, A *is* in the state $|\Phi_n\rangle$. When several of the a_n are different from zero, on the other hand, the association between $|\psi_n\rangle$ and $|\Phi_n\rangle$ is not the one that can be described in terms of the familiar notion of a *correlation*, since we lack here the statistical element. However, (23–2) *does* formally imply an association between state vectors of S and state vectors of A, and this is sufficient to allow us to proceed.

The next step is to consider iterated measurement processes, that is, processes that can be described by relation (14–11). This category of processes encompasses the measurements by a second instrument B of the value found by a first instrument A. It also includes all the processes of repeated measurements, when B measures again, on S, the observable that A has already measured. B and A can then, of course, also be considered as two distinct parts of the same instrument, so that (14–11) also describes a mere repetition of a measurement already made once. Then the states $c_m|a_m\rangle|b_m\rangle$ appearing in (14–11) are just the relative states of the instrument. In spite of the fact that no reduction of the wave packet ever takes place, it is of course apparent from (14–11) that the formal association mentioned before also holds between the results of the two separate instruments. In other words, with a given $|\psi_n\rangle$ (14–11) associates unambiguously relative states of A and B that are labeled by the same index n. When this is the case, we say that *by definition* "the results of the two measurements agree with one another." As we begin to see, an attractive feature of the theory under discussion is that the formalism itself (as emphasized by De Witt) seems capable of yielding its own interpretation.

More complex questions can be asked. For instance, let us assume that a quantum system S is split into two parts, U and V, as in the examples of Chapter 8; that U and V fly far apart; that a quantity $L^{(U)}$ is measured on U by an instrument A (which registers the result on punched cards); that a quantity $L^{(V)}$—strongly correlated with $L^{(U)}$—is similarly measured on V by an instrument B; and that, in the end, instrument A measures the value registered in the memory of B and compares it with the value it has already registered in its own memory upon measurement of $L^{(U)}$. Because of the one-to-one correlation which we assumed between $L^{(U)}$ and $L^{(V)}$, the two results obtained by A should agree with one another if the theory is correct. To verify that this is indeed the case, let us write the system $U + V$ as (with obvious notations)

$$\sum_n c_n|u_n\rangle\,|v_n\rangle \tag{23–4}$$

Let $|b_i\rangle$ describe the possible states of instrument B, and let $|a_k,{}_l\rangle$ describe those of instrument A. In $|b_i\rangle$, the value $i = m$ refers to the state into which B would evolve as a consequence of its interaction with V if all the c_n were zero except just one, labeled m. In $|a_{k,l}\rangle$, the index k has the same significance (only, now, with respect to A and to U) and the index l refers in a similar way to the measurement made by A on B: it takes the value $l = p$ if B is observed by A to be in the state $|b_p\rangle$. Then, if the initial state is

$$|a_{00}\rangle\,|b_0\rangle\sum_n c_n|u_n\rangle\,|v_n\rangle \tag{23–5}$$

the states just after the first, second, and third measurement processes are, respectively, described by

$$\sum_n c_n |a_{n,0}\rangle |b_0\rangle |u_n\rangle |v_n\rangle \qquad (23\text{–}6a)$$

$$\sum_n c_n |a_{n,0}\rangle |b_n\rangle |u_n\rangle |v_n\rangle \qquad (23\text{–}6b)$$

$$\sum_n c_n |a_{n,n}\rangle |b_n\rangle |u_n\rangle |v_n\rangle \qquad (23\text{–}6c)$$

this being a mere consequence of the Schrödinger-like time evolution of the complete state vector.

The fact that the *same* index n appears twice in Eq. (23–6c) shows that indeed the two results obtained by A do agree, as they should.

In the conventional formalism, an external observer is considered, and the coefficients a_n or c_n that appear in the formulas above are interpreted as probability amplitudes for observations carried out by that observer. Here, however, no external observer is considered. All the possible observers are systems such as A and B (i.e., are endowed with relative state vectors), and consequently, as already mentioned, we have no a priori statistical interpretation for the coefficients a_n or c_n. It has been stated [3, 4] that the statistical interpretation of coefficients such as these *emerges from the formalism itself*, when an interaction of an instrument with a large number of identical system is considered. Let us show that this is indeed the case, provided that the very natural assumption formulated below is made.

To show this,* let us consider a large number M of systems $S_l . . . S_k . . . S_M$, which, initially, are all in the same state:

$$|\psi\rangle = \sum_n{}' c_n |s_n\rangle \qquad (23\text{–}7)$$

with

$$\sum_n |c_n|^2 = 1$$

Let us, moreover, consider an instrument A, described by

$$|a_{\alpha_1,\alpha_2 \cdots, \alpha_k, \cdots, \alpha_M}\rangle \qquad (23\text{–}8)$$

which interacts exactly once with each system S. Initially the value of every α_k is zero, and it evolves to $\alpha_k = n_k$ when A interacts with a system S_k, which is in the state n_k. Then, after all the interactions have taken place, the complete state vector is

$$|\Psi\rangle = \sum_{n_1,n_2,\ldots n_k\ldots} c_{n_1} c_{n_2} \cdots c_{n_k} \cdots |s_{n_1}\rangle$$
$$|s_{n_2}\rangle \cdots |s_{n_k}\rangle \cdots |a_{n_1,n_2,\cdots n_k,\cdots}\rangle \qquad (23\text{–}9)$$

*According to De Witt [4], the following derivation is due to R. N. Graham.

This shows that the instrument A does not generally record a sequence of identical values in the course of its observations, even within a single element of the superposition. A certain distribution of values of the system observable —which is in a one-to-one correspondence with the index n and can be replaced by it—thus emerges. This distribution is characterized by the *relative frequency function*

$$f(n; n_1, n_2, \ldots, n_k, \ldots n_M) = M^{-1} \sum_{k=1}^{M} \delta_{n,n_k} \qquad (23\text{--}10)$$

which obviously measures the relative number f_n of indices n_k that are equal to a given value n. When the number M of systems goes to infinity, we expect this relative number f_n to become equal to $|c_n|^2$ if the theory is to reproduce correctly the usual statistical predictions of quantum mechanics. To show that this is indeed the case, let us introduce the function

$$\delta(n_1, \ldots n_M) = \sum_n [f(n; n_1, \ldots n_M) - |c_n|^2]^2 \qquad (23\text{--}11)$$

and let ε be an arbitrarily small positive number. Then let us remove from superposition (23–9) all the sequences $n_1 \ldots n_k \ldots n_M$ for which $\delta \geq \varepsilon$. Let us denote the result by $|\Psi(\varepsilon)\rangle$, and let us consider

$$|\chi^{(\varepsilon)}\rangle \equiv |\Psi\rangle - |\Psi^{(\varepsilon)}\rangle \qquad (23\text{--}12)$$

$$= \sum_{\substack{n_1,\ldots n_k,\ldots \\ (\delta(n_1,\ldots n_k,\ldots)\geq\varepsilon)}} c_{n_1} \ldots c_{n_k} \ldots |s_{n_1}\rangle \ldots |s_{n_k}\rangle \ldots |a_{n_1,\ldots n_k,\ldots}\rangle \qquad (23\text{--}13)$$

As is apparent from (23–3), all the relative states which have different indices n (indices that label the eigenstates of the measured observable on the system) are orthogonal to one another. Thus the states $|a_{n_1}, \ldots n_k, \ldots\rangle$ which differ by the values of one or several n_k, are mutually orthogonal and therefore

$$\langle \chi^{(\varepsilon)} | \Psi^{(\varepsilon)} \rangle = 0 \qquad (23\text{--}14)$$

and

$$\langle \chi^{(\varepsilon)} | \chi^{(\varepsilon)} \rangle = \sum_{\substack{n_1,\ldots n_k,\ldots \\ \delta\geq\varepsilon}} |c_{n_1}|^2 \ldots |c_{n_k}|^2 \ldots$$

$$\leq \varepsilon^{-1} \sum_{n_1,\ldots n_k,\ldots} \delta(n_1, \ldots n_k, \ldots)|c_{n_1}|^2 \ldots |c_{n_k}|^2 \ldots \qquad (23\text{--}15)$$

Now, a general property of the function $f(n; n_1, \ldots, n_M)$ defined by (23–10) is that, for any family of numbers $w_1 \ldots w_n \ldots$ that add up to unity,

$$\sum_{n_1,\ldots n_k,\ldots} \delta(n_1, \ldots n_k, \ldots n_M)\, w_{n_1} \cdots w_{n_k} \cdots w_{n_M}$$

$$= M^{-1}\left(1 - \sum_n w_n^2\right) + \sum_n \left(w_n - |c_n|^2\right)^2 \qquad (23\text{–}16)$$

This formula is easily proved. For example, we may consider at first N independent real variables $w_1 \ldots w_N$, and we may define

$$F(w_1, \ldots w_n, \ldots) \equiv \sum_{n_1,\ldots n_k,\ldots n_M} w_{n_1} \cdots w_{n_k} \cdots w_{n_M}$$

$$= (w_1 + \ldots + w_n + \ldots + w_N)^M$$

The left-hand side of (23–16) is then easily shown to be equal to

$$M^{-2} \sum_n \left[w_n \frac{\partial}{\partial w_n}\left(w_n \frac{\partial F}{\partial w_n}\right) - 2M|c_n|^2\, w_n \frac{\partial F}{\partial w_n} + M^2|c_n|^4 F \right] \qquad (23\text{–}17)$$

When the derivations are performed and the supplementary condition $\sum_m w_m = 1$ is taken into account, (23–17) reduces to the right-hand side of (23–16).   Q.E.D.

Since $\sum_n w_n(1 - w_n) = 1 - \sum_n w_n^2 < 1$,
(23–16) gives

$$\langle \chi^{(\varepsilon)} | \chi^{(\varepsilon)} \rangle < (M\varepsilon)^{-1} \qquad (23\text{–}18)$$

In other words, no matter how small we choose ε, we can always find M large enough so that the norm of $|\chi^{(\varepsilon)}\rangle$ becomes smaller than any positive number. This means that the distance between the two state vectors $|\Psi\rangle$ and $|\Psi^{(\varepsilon)}\rangle$ goes to zero when M approaches infinity. If we further assume that state vectors with zero norm correspond to nonexisting branches, then, when M goes to infinity, all the branches which would correspond to $\delta \geqslant \varepsilon$ vanish. The remaining branches correspond therefore to $\delta < \varepsilon$, that is, to $f(n; n_1, \ldots, n_M) - |c_n|^2 < \varepsilon$.   Q.E.D.

If we call "measurement" each separate interaction of A with any one of the systems $S_1 \ldots S_M$, the statement just made means that under the conditions studied here the results of a relative number $|c_n|^2$ of measurements agree with one another .This is precisely the essence of the conventional probability interpretation of quantum mechanics. However, the statement here emerges as a *result* of the formalism plus *one*, very general assumption, instead of having to be postulated as a separate axiom.

23.2 DISCUSSION

The relativity of state theory obviously has extremely attractive aspects, along with some other features that are very hard to accept. One of its weaknesses is, in a way, a consequence of its success: as shown above, it is a theory that cannot be disproved by any specific experiment, since we cannot observe the split. For this reason it can rightly be called a "metaphysical theory," provided that the prefix "meta-" is given roughly the same sense as in *meta-mathematics*. It provides us with a consistent, logical framework of rules and concepts and shows that real physics can be inserted completely into this framework. Seen from a somewhat different standpoint, this theory also has the very unpleasant feature that, in a way, it runs counter to the principle of economy of assumptions (Occam's razor), which is otherwise known to be so important in science. Indeed, it does so without any restraint, since it goes as far as to postulate infinities of completely unobservable worlds or branches thereof. An answer to this objection is that if the theory thus lavishly multiplies universes (or, at least, *branches* of the Universe), it economizes on the fundamental principles, as we have shown. Whether we should economize preferably on universes or on principles is likely to remain a matter in which each of us can follow his own preferences.

Although the above-mentioned advantages and drawbacks are more or less obvious and (though puzzling) do not easily lend themselves to extensive analysis, other aspects of the theory also require our attention.

The basic idea of the scheme is, as already mentioned, the assumption that a one-to-one correspondence exists between reality and the state vector that represents it. Hence, if we imagine that two independent worlds exist, each of them being composed of a system $S^{(k)}$ and of an instrument $A^{(k)}$ ($k = 1, 2$) similarly constituted and obeying the same laws of force, and if we further postulate that their initial state vectors are the same, then their final states, described by

$$\sum_{n=1}^{N} c_n |a_n^{(1)}\rangle |s_n^{(1)}\rangle \qquad (23\text{--}19)$$

and

$$\sum_{n=1}^{N} c_n |a_n^{(2)}\rangle |s_n^{(2)}\rangle \qquad (23\text{--}20)$$

respectively, are necessarily identical in regard to all their physical attributes. The same holds, of course, if we consider the systems $S^{(1)} + A^{(1)}$ and $S^{(2)} + A^{(2)}$ as composing, not two distinct worlds, but one and the same world (at any rate, as long as the notion of the final states in which these two composite systems evolve preserves a meaning), since this is but another way of formulating the same physical assumptions. Then, however, we may ask, "How is this identity of the two physical entities described by expressions (23–19) and (23–20) to be reconciled with the fact that $S^{(1)} + A^{(1)}$ can be, for example, in the final state $n = 1$, while $S^{(2)} + A^{(2)}$ can be in the final state $n = 2$, which is different from the former?"

The answer obviously is that in (23–19), as well as in (23–20) or in similar expressions, it is not the case that one branch alone exists: all of them must be considered as separately existing. The latter statement can, however, be understood in two different ways. First, we can understand it in the very material sense that neither the number of instruments nor the number of systems is constant during the process: in each measurement-like process, each one is multiplied by N, the number of branches in the process. This solution [discussed below as "solution (a)"] is the one that seems to be favored by the proponents of the theory. Another possibility is to assert that the numbers of systems and instruments are always conserved. The identity of expressions (23–19) and (23–20) then compels us to assert that the two *systems* $S^{(1)} + A^{(1)}$ and $S^{(2)} + A^{(2)}$ are themselves identical to one another.

The latter view, discussed below as "solution (b)," is one that many of us would accept without qualms as long as S and A are two quantum systems: if a one-to-one correspondence is postulated between state vector and reality, then two $S + A$ systems that have the same state vector must clearly be identical, and it seems that not much more remains to be said. Considering that instruments are but complex quantum systems, some physicists would even go as far as to extend their approval of such a view also to the case in which A is an instrument.

But when it comes to the question of whether this view should finally be extended to systems that incorporate a conscious observer, a difficulty appears, which is connected with a fact that was stressed in the preceding chapter: the existence of our impressions is something of which we are absolutely sure. Even if it is true that some of our impressions are illusions, their very existence and specificity is an unquestionable fact. On the other hand, if it is assumed that the state vectors are in a one-to-one correspondence with the whole of reality, including the impressions of the observers, then in our example these impressions cannot differ from one observer to the next, for, if they did, the systems $S^{(k)} + A^{(k)}$ would not be identical to one another in every respect. Then, if one of the observers, $A^{(1)}$, for instance, had a definite impression (that of seeing a definite signal, corresponding to some particular value n' of n), the other observer would necessarily have the same impression. However, this is impossible because of the symmetrical way in which all the values of n appear in the formalism: nothing singles out the particular value n' (as a matter of fact, it would necessitate large statistical fluctuations occurring even in arbitrarily large ensembles, whereas we do not even have indeterminism at our disposal). Thus we are led to the conclusion that, if the impressions of the observers are just physical properties of these systems—and, as such, are described by the state vectors—then these impressions *must* be blurred in the kind of experiments studied here. This conclusion is, however, in obvious contradiction with our immediate experience, since under the conditions of these experiments every one of us knows with certainty that he always has the definite impression of seeing, or not seeing, a signal.

Keeping these considerations in mind, let us discuss in detail solution (a) and solution (b) mentioned above.

Solution (a)

This runs as follows. In a measurement-like process that incorporates an observer, we can assume that *all* the branches have a reality of the type familiar to us through our personal experience. That is, we can assume that each of these branches contains an observer whose consciousness is in a well-defined state (of "having seen" such or such signal). This implies that if, before the process took place, our simplified world incorporated one observer only, after the measurement-like process it incorporates altogether several observers, namely, one in each branch.

It can easily be verified that this interpretation of the formalism does indeed offer a solution to the above-mentioned difficulty (essentially this comes about because the "observed" diversity is then only apparent). However, it is clear that this process of physical multiplication should then also take place in phenomena in which we have instruments instead of observers. Then, since, by assumption, the interactions of systems with instruments are not governed by special laws, but rather by the general laws of physics, the multiplication alluded to above should take place also in any process in which the evolution of the wave function follows the general pattern of a measurement-like process. For instance, it should presumably also occur in the case considered in Chapter 15, in which a particle whose spin points along O_x is shot inside a Stern-Gerlach device whose magnetic field gradient is directed along O_z. In that case a particle with spin equal to S should be transformed into $2S + 1$ particles, one in each "branch" of the universe, each of these particles pertaining to a given beam and having a definite spin component S_z along O_z.

One of the difficulties raised by this interpretation is that it is hard to understand how these multiplication processes could be restricted to the measurement-like processes, which (as we saw in Chapter 15, 16, 17, and 18) do not constitute a fully specific and logically well-defined category. Indeed, they should presumably be extended to any *interaction* process whatsoever. However, this in turn seems to require that a sharp distinction be made between what we call one (composite) system (no *interaction* proper, no multiplication, possibility of interferences) and what we call two or more systems in *interaction*. Unfortunately such distinctions seem artificial to a large degree.

Another difficulty raised by this same interpretation occurs if we consider, for instance, the decay of a hypothetical spin-zero dineutron system into two neutrons, U and V, assuming that this decay occurs by means of a parity-conserving interaction. If, for the sake of definiteness, we assume, moreover, that the parity of the dineutron system is positive, the emerging composite system is in a state of total spin-zero. It is then described by

$$|\psi\rangle = |\psi_1\rangle \equiv |u_+\rangle\,|v_-\rangle - |u_-\rangle\,|v_+\rangle$$

but that state vector can be written in many other ways, such as

$$|\psi\rangle = |\psi_2\rangle \equiv |u_+'\rangle|v_-'\rangle - |u_-'\rangle|v_+'\rangle$$

(the unprimed and primed symbols stand for eigenvectors of S_z and of S_x, respectively). If, in the interpretation under discussion a duplication of the original systems occurs, then after the decay has taken place we have altogether *four* neutrons, two in each branch. However, if we look at $|\psi_1\rangle$, we must assert that this duplication results in the existence of one particle U with $S_z = +1$ and one particle U with $S_z = -1$ (*plus* the corresponding particles V), whereas, if we look at $|\psi_2\rangle$, we must assert that the duplication results in the existence of one particle U with $S_x = +1$ and one particle $S_x = -1$ (*plus* again the corresponding particles V). Apparently, of these two (in fact, of an infinity of) possibilities, only one should be considered as being true. However, this amounts to giving a privileged role to one specified axis in ordinary space, and this, obviously, is absurd in the present case, since there is no privileged direction in the problem. On the other hand, if we take the position that there is *no* duplication of the neutrons in the present case, it becomes hard to understand how duplication could occur in the more general case of a mutual scattering of two objects, and consequently also in a measurement-like process, since the separation between the cases in which multiplication occurs and those in which it does not would seem to be necessarily qualitative and sharp.

We are thus led to investigate the other solution, solution (*b*), to our original difficulty.

Solution (b)

This consists in assuming that observers, instruments, and so forth, instead of being *multiplied*, as was the case in solution (*a*), are simply *split* between the various branches. In other words, we now assume that, when *one* atom is shot through a Young two-slit device or through a Stern-Gerlach apparatus, there is only *one* atom at any given time. However, we assume that the reality of this atom (whatever this may mean) is somehow *split* (e.g., wavelike) among the two slits or among the beams. Moreover, we extend this assumption to macroscopic systems, instruments, and observers also. In that case, the difficulty analyzed at the beginning of this section [before and after Eqs. (23–19) and (23–20)] is indeed a real one, as long as it is assumed that the state vector is in a one-to-one correspondence with the totality of reality, including the impressions of the observers. The same assumption also implies (see, e.g., Ref. [6]) that, if an outside observer could ascertain the state vector of the universe, he would find that my macroscopic observables have no definite value; in fact, he would find that I am in a linear combination of a large number of states, each of which is correlated with the states of immensely many objects. As previously noted, we are not conscious of such a split. Hence the theory can be retained only if it is assumed that consciousness is a property of physical systems which is, at any rate, very different from all the other properties in that it is *not* described by the state vector, and which, consequently, is a *supplementary variable* in the sense in which that concept is defined in Chapter 7. This new standpoint should not be frowned upon by even the most traditionalist of scientists, since—as we already observed in Chapter 22—it bears an unquestionable resemblance to the old idea of describing consciousness as an "epiphenomenon," that is, as a phenomenon distinct from every other one in that it does not imply reactions. Yet we must grant that we are led here, once again, to the idea of an object-subject separation, whereas we had somehow hoped to avoid it in the theory under discussion.

Thus, a last but important question is the following one: "If we accept the view that consciousness is a supplementary variable of some kind, does this really solve the difficulties of the theory under discussion?" As an answer, let us show that such an assumption is not yet entirely sufficient in itself. More precisely, let us prove that, if the assumption considered is not accompanied by other ones, it leads to a strange, though not disprovable, result.

For this purpose, let us describe the evolution of a composite system $S + A$, where A is endowed with a consciousness C, by the system of relations

$$\sum_n c_n |\psi_n\rangle |a_0\rangle \to \sum_n c_n |\psi_n\rangle |a_n\rangle \tag{23-21a}$$

$$C_0 \to C_m \tag{23-21b}$$

We have thus constructed a theory that really parallels quite closely the metatheory of classical epiphenomenism alluded to above: consciousness C is affected by the evolution of the "material world" represented by the state vector, but is unable to affect it.

If a somewhat complex case is considered, such as that of one observer making measurements on several identical systems, as was discussed previously [Eqs. (23–8) to (23–18)], the theory described by (23–21) gives the following results. The state of consciousness of the observer is, after the interaction, in correspondence with *one* branch only of the superposition (23–21). In other words, if we write (23–21a) as

$$(\sum_n c_n |s_n^{(1)}\rangle) \ldots (\sum_n c_n |s_n^{(M)}\rangle)|a_0 \ldots {}_0\rangle$$

$$\to \sum_{n_1,\ldots n_k,\ldots n_M} c_{n_1} \ldots c_{n_k} \ldots c_{n_M} |s_{n_1}\rangle$$

$$\ldots |s_{n_k}\rangle \ldots |s_{n_M}\rangle |a_{n_1}, \ldots n_k, \ldots n_M\rangle$$

The state of consciousness is in correspondence with only one of the $|a_{n_1}, \cdots n_k \cdots, {}_{n_M}\rangle$ Since we have shown that the relative number of indices n_k which have a given value m is $|c_m|^2$, this result implies that the observer's consciousness (memory) contains $M|c_m|^2$ "marks" indicating that it has had the impression of interacting with a system S in state $|s_m\rangle$ (for any m). Hence this model exactly reproduces the predictions of the conventional theory. Up to this point, the metatheory described by (23–21) is therefore a success. And it even appears as the conceptual framework that should tacitly underlie the relativity-of-states theory.

However, let us now consider a phenomenon of correlation at a distance, in which two distant observers, A and B, participate, as in the example described above by means of Eqs. (23–5) and (23–6). We must then specify our model somewhat more if we want to obtain unambiguous predictions. Now, if we remember that the wave packet is never reduced, and if we want to preserve the locality of the acts of observation, we are led quite naturally to make such a specification as follows: the state of consciousness of A is in correspondence with one of the state vectors in the superposition, and the state of consciousness of B is in correspondence with *any* other one. It could hardly be otherwise, since, at the time when B observes, all these state vectors are still present, and since—at least when the interval between the two measurements is spacelike—B has no way to become informed of the result obtained by A. The same holds in the case of repeated measurements, such as the one considered above. We then immediately notice that A and B can now develop states of consciousness that are *not* matched to one another. They correspond to *different* values of n. This occurs because of the fact that expression (23–6a) is now replaced by the system

$$\sum_n c_n|a_{n0}\rangle\,|b_0\rangle\,u_n\rangle\,|v_n\rangle, \qquad C_m^{(A)}, \qquad C_0^{(B)}$$

where $C_m^{(A)}$ means that the state of consciousness of A is the one that corresponds to the index m. Equation (23–6b) is then replaced by

$$\sum_n c_n|a_{n0}\rangle\,|b_n\rangle\,|u_n\rangle\,|v_n\rangle, \qquad C_m^{(A)}, \qquad C_{m'}^{(B)}$$

where m' has no reason to be equal to m. Nevertheless, as (23–6c) shows, A can never *know* that B does not have the same impression as he has. This comes from the fact that any transfer of information from B to A—for example, any answer made by B to a question asked by A—unavoidably proceeds through physical means. Therefore it necessarily takes the form of a measurement made by A on B. And we know that under these conditions A necessarily gets a response (answer) that agrees with his own perception. But in the present conceptual framework, there is no reason that this apparent agreement should be real.

The problem of the intersubjective agreement is an old one. Many philosophers of previous ages have tried to cope with it, and it is interesting to observe that it appears again, and with considerable force, at a sensitive point of contemporary (meta) science. Of course, poets also have repeatedly emphasized the old, distressing truth that we are all like small isolated islands. One of their main messages is that, in regard to what is most essential to us, we all feel in a unique way, differently from each other. That we should extend these wise but apparently lofty views to events of everyday life is, however, difficult to believe. As shown above, the idea that we should be able to take tea together while localizing differently the same teapot is a logical possibility. It is, however, one that is somewhat hard to accept.

REFERENCES

[1] E. Schrödinger, *Naturwissenschaften* **23**, 807 (1935).
[2] H. Everett III, *Rev. Mod. Phys.* **29**, 454 (1957).
[3] J. A. Wheeler, *Rev. Mod. Phys.* **29**, 463 (1957).
[4] B. De Witt in *Battelle Rencontres, I, 1967: Lectures in Mathematics and Physics*, M. De Witt and J. A. Wheeler, Ed., W. A. Benjamin, New York, 1968.
[5] L. N. Cooper and D. van Vechten, *Am. J. Phys.* **37**, 1212 (1969).
[6] E. P. Wigner, Observations at the End of the Conference on the Epistemology of Quantum Mechanics held in London, Ontario, March 1971 (preprint).

Further Reading

The Many-Worlds Interpretation of Quantum Mechanics, B. De Witt and N. Graham, Eds., Princeton Series in Physics, 1973.

PART SIX

SUMMARY AND OUTLOOK

It sometimes happens that a scientific study has nonscientific or, at any rate, parascientific implications. When this is the case, a suitable procedure may be to separate as well as is possible the scientific conclusions, if there are any, from the parascientific outlook and perspectives, that is, from all the considerations which (albeit they are justified by the premises) are of the type that do not admit proofs. This is what is attempted in this last, concluding part. The first of its two chapters is just a summary of a few salient points that seem to emerge in an unquestionable way from the discussions reviewed or completed in the present essay. The second one is both more general and less precise. It presents views that are quite definitely not binding but that, in the light of all the foregoing considerations, appear to contain a certain rationality.

CHAPTER 24

SUMMARY

As is apparent from all the investigations described in the foregoing chapters, quantum measurement theories are interesting mainly because of the fundamental questions they oblige us to consider.

Since quantum mechanics is primarily a theory of small, simple systems, it could a priori be expected that a full description of this formalism would be possible in a language that would refer only to the objective properties of such small systems. If this were the case, we could look forward to a scientific picture of the world in which the large-scale effects would essentially result from the conjunction of very small and numerous independent objective causes, in full conformity with the statistical-mechanical standpoint that was long considered as the most reasonable one.

As we saw, this is *not* the case. In other words, the limitations in separability and/or reality of quantum systems introduced by the Copenhagen physicists (or the introduction of an external observer by von Neumann) cannot be considered as stemming from a priori philosophical options. Unless quantum physics is objectively false in some of its predictions (which are observable in principle if not in practice), limitations of such a kind are unavoidable, as we verified in detail. Indeed, we observed that such limitations extend to systems of arbitrary finite size and of arbitrary complexity.

As a matter of fact, this is the first and perhaps the most significant conclusion in our investigation of the conceptual foundations of quantum physics. The world cannot consistently be described as essentially a collection of physical objects of finite size (or of finite complexity) which would each possess its own specific attributes, even if these attributes—size, location, velocity, and so forth—are only approximately determined, and even if the systems are allowed to interact through arbitrarily complex, long-range forces that decrease when distance increases.

What can we assert about the systems that are truly infinite ones? This is a much more difficult question, and it is fair to acknowledge that no definite conclusion concerning it has been reached as yet. The reason for this is simply that the theory of truly infinite systems is still under construction: as has been pointedly observed [1], science differs from the art of house building in that the foundations of a discipline can be laid only *after* the discipline in question has acquired its full existence. Now, in regard to truly infinite systems, their theory can be viewed from several different standpoints. One of them consists in simply extrapolating to the infinite case the results obtained for finite systems. The formalism, then, is well known, but, as we saw, serious difficulties exist.

An equally tenable standpoint is to assert that truly infinite systems have special features of their own that make them fall outside the realm of elementary quantum mechanics. Recent advances show that this is indeed a possibility to be considered seriously. A new and very interesting field of study is thereby opened; and when the theory of these systems is better known, it will become extremely important to investigate its conceptual foundations. In particular, it will be suitable to try to construct, with the help of this formalism, a good, reasonably consistent theory of the measurement-like processes. For the time being, however, we can only be guided by the observation of Chapter 17 that, if the description of ensembles by means of state vectors and density matrices is retained, the difficulties do not disappear. Apart from nonlocality we cannot say what results will emerge in future theories that would operate with more general algorithms. Perhaps even the objection that presently seems the most worrisome—namely, the noninvariance of the reduction process—could somehow be circumvented. Indeed, the *hope* that many present problems will then be solved is certainly a healthy one, since it fosters further research. At the same time, let it be stressed that this research will deal with a discipline that in a way would be distinctly different from the conventional, present-day quantum mechanics.

For the time being, since such a theory is far from having reached as yet any definitive status of existence, we must fall back on what we know. But what we know is already quite substantial. It consists in the usual formalism of state vectors and density matrices (which is adapted to quite complex situations) on the one hand and in nonseparability on the other hand. The formalism strongly suggests, and nonseparability demands, a rejection of the natural philosophy we called "macro-objectivism." For the best-informed theorists, this, of course, is not a surprise, since the views embodied in the so-called orthodox standpoint—or Copenhagen description—necessarily depart from the macroscopic objectivism. In Part Five we traced this departure to the fact that in Bohr's approach the role of the observer, though somewhat implicit, is nevertheless primeval, since the observer serves in defining the instruments.

At the same time, it is mainly because of this aspect of the Copenhagen description, and because of the vagueness or nonexistence of the corresponding ontology, that Bohr's standpoint never gave rise to a universal consensus of the experts. Even Heisenberg may in a sense be said to have gone beyond it when he introduced his philosophy of the potentia, as we have already observed.

The alternative offered by the hidden variables theories is interesting, but has its own, rather well-known, difficulties (see Chapter 11). Another one is presented by the school of thoughts of von Neumann, London and Bauer, and Wigner. For these authors, the role of the observer is both more explicit than, and different from, what it is for Bohr and Heisenberg. Still another possibility is represented by the work of Everett and Wheeler. Both these standpoints differ from that of Bohr in that they are more definite about what they assume concerning reality. However both, as a result, are so unfamiliar and conceptually strange that neither of them has up to now been universally accepted, even as a basis for further investigations.

As a consequence of the situation just summarized, many theorists have yielded to the temptation offered by logical positivism. This philosophy is sometimes also called scientific empiricism or phenomenism. In a way, it is an extrapolation issued from the very fruitful positive *method*. According to it, science is *exclusively* a discourse that bears on our own, intersubjective impressions.

In order—presumably—to preserve the fundamental character of science, most positivists associate the conception described above with the opinion that the scientific modes of thought come nearer to any possible truth than any other conceivable ones. Obviously, this association is not a logical necessity. For the sake of definiteness, we may call it the *purely linguistic standpoint*. The reason for this name is, of course, that—implicitly or not—the standpoint under discussion reduces reality to a universe of discourses; discourses made by men, ultimately about themselves and other men. The discussion of this conception is deferred to the next section.

We must conclude this summary of what can be regarded as established in a noncommittal way. If we exclusively consider the requirement of strict consistency with the quantum rules (regarded as universally valid), then we are not forced to adhere to one conception and to reject the other ones. On the contrary, the purely linguistic standpoint, the hidden variables concept, Bohr's approach, and the general ideas of the relativity of state theory all constitute systems that admittedly are not always definite, but that nevertheless may all be considered, on the whole, as legitimate.

Outwardly, these conceptions are extremely different. Some of them are based merely on weak objectivity, whereas others satisfy strong objectivity also. On further reflection, however, they are perhaps not as divergent as it may seem. Because of the unavoidably contextualistic and nonseparable structure of the hidden variables theories, these theories are to some extent meta-theories, just like the relative state theory; and it has indeed been asserted (see

Bell [2] and compare Section 23.2) that the latter theory is in a way a disguised version of the former ones, at least when the idea of a real multiplication of universes is abandoned. On the other hand, the difficulties of the measurement theories in "orthodox" quantum mechanics (i.e., the difficulties that we have found to be present when Assumption Q of Section 4.2 is made) are all associated with the idea that instruments are in a way describable as quantum systems, an idea which implies, among other things, that the approximation of considering such systems as isolated from the rest of the world is acceptable, even at a fundamental level. Such a view is questionable, however, since the enormous density of the energy levels of a macroscopic system makes it practically impossible to isolate a macroscopic object from its surroundings. It has been pointed out that transitions between the quantum-mechanical states of a macroscopic system can be caused by a single particle located miles away from that system [3]. It is conceivable that such a remark would offer a way around the difficulties of the measurement theories just alluded to. On the other hand, the particles located within "any finite region of space containing the instrument which we are using" can always be considered as being part of that instrument. Hence a solution of the fundamental problem of quantum mechanics developed along these lines could be acceptable only—at best—if it were based on due recognition of the fact that no part of the universe can be thought of in isolation, so to speak, that is, as separated from the other ones (including ourselves). Such a conception is akin to nonseparability. Hence it gives further support to the view (see Chapter 12) that indeed nonseparability governs anything that we may call (primitive) "Reality." Implicitly or explicitly, nonseparability is indeed present in all the various interpretations of the quantum rules which have been set forth, and, as shown in Chapter 12, it is even more general than the set of these rules.

REFERENCES

[1] H. J. Groenevold, *Foundations of Quantum Theory, Proceedings of the Salzburg Colloquium on Philosophy of Science, 1968*, International Union of History and Philosophy of Science. This work contains very useful developments on measurement theory. From the same author see also: Skeptical Quantum Theory, *Phys. Repl.* **11C**, 329 (1974).

[2] J. S. Bell, The Measurement Theory of Everett and de Broglie's Pilot Wave, CERN Th 1599 (1972).

[3] H. D. Zeh, *Found. Phys.* **1**, 69 (1970)

CHAPTER 25

OUTLOOK

With this final chapter we boldly cross the borderline between what can be proved and what appears to be plausible. Scientists should presumably not wander too far into the wild country we are now entering, and hence we shall not linger within its boundaries. Nevertheless it is, let us hope, not too daring to incorporate in our general conclusions a few arguments gleaned there.

A discussion of the *purely linguistic standpoint* defined in Chapter 24 can serve as an introduction to the quest on which we want to embark. Admittedly, this standpoint meets, as does every other one, with a few technical problems. These are, for instance (as discussed in Chapter 20), the justification of the inductive inference or the meaning of past events. Our present interest lies, however, with two difficulties of a more general nature. Although they are not specific to our present field of study, it seems appropriate that attention be drawn to them.

The first of these difficulties is internal to the development of science itself. It is most clearly perceived when the purely linguistic standpoint is contrasted with the other ones. When, as in all *other* views, we imagine the world as having its own existence and therefore its own set of laws, both independent of ourselves, we automatically make a fruitful separation in our concepts between the notion of physical laws and principles—which, we hope, ought to be quite few in number and which should appear as simple to the human intellect—and the numerous approximations that we must usually employ. The latter are necessary either because we cannot calculate otherwise or because we are still ignorant of the true, fundamental laws; and, admittedly, it may be that our limited abilities will in fact never allow us to reach these fundamental laws. Even this remark, however, does not destroy the fruitfulness of the conceptual separation that we usually instinctively make between what *is* and our own images, constructs, and efforts. Indeed, most of the men who effectively built up science intuitively strived at reaching a greater simplicity, a greater generality, and a greater independence—in their formulation of the laws—from the practical conditions of the actual calculations. They were convinced that, if they strayed from these guide lines, they would also stray from the direction in which the *true* explanation is eventually to be found.

Now, in the purely linguistic standpoint, the source of all the foregoing convictions disappears. We no longer have any reason to distinguish the laws or principles from the (usually approximate) rules through which we work out our empirical predictions. Indeed, even the qualificative "approximate" loses much of its usual sense. We can thus easily be led to a proliferation of models (sometimes even called "techniques") which are mutually incompatible, each of which leads to a few successes and also to a few failures, and out of which no general idea emerges. The many brilliant advances of fundamental physical theory in the past two decades cannot completely blind us to the existence of a real danger of this kind.

Independently of this objection, another one can be formulated against the purely linguistic standpoint. It is that, if a day comes when the latter is really taken seriously, a danger exists that it will deprive the scientists of the motivations they can have in their work. As already stressed, when a geologist tries to learn more about the ancient mountains of the earth, the reason why he takes interest in his subject is that he believes these mountains really did exist. And here the word "really" unquestionably means "independently of himself." If this geologist happened to become convinced that the *only* legitimate motivation for his activity was to write some papers that had a good chance of agreeing with other papers—or eventually to find oil—he would almost certainly quit. As for the high-energy physicists, their situation is even worse, for, whereas their activity appears as essential as long as we believe in the independent existence of fundamental laws that we can still hope to know better, it loses practically its whole motivation as soon as we believe that the sole objective of these scientists is to make their impressions mutually consistent. These impressions are not of the kind that occur in our daily life. They are extremely special, they are produced at great costs, and it is doubtful that the mere pleasure their harmony gives to a selected happy few is worth large public expenditures.

Let us now assume that the foregoing arguments have somehow carried conviction, and that the necessity of evading the ethereal vacuity of the purely linguistic standpoint is acknowledged. The problem is to know how this can possibly be done.

In that respect, the lesson to be drawn from all the attempts at edifying measurement theories is presumably the following one: we must adhere to the idea that, whatever reality is, science is presumably not in itself a sufficient tool for gaining full access to it. If we remember that our intellect is a mere product of evolution, and that it is therefore essentially a tool for action, we are of course quite well prepared to accept such a conclusion. This is especially true if we also remember (*a*) that man's evolution is perhaps not finished, and (*b*) that, in any case, the number of his cortical interconnections is limited. A question which is, of course, much more difficult to study is whether or not there exists any other means of investigation that could support science in its quest for reality. Intuition, feeling, meditation, and instinct have been proposed. There are justifications for each one, and we would certainly contradict our own principles of scientific objectivity if we were to reject all these means without producing good reasons. Nevertheless, we can legitimately observe here that their investigation would soon lead us quite far apart from our subject.

Once we have decided that a Reality somehow exists, and once we have acknowledged the essentially limited scope of present-day science—and, more generally, of our present-day intellect—in providing knowledge of it, we can consider in a somewhat new light the considerations developed in this book.

In the first place, since our abilities to form a positive knowledge of what *really is* are thus severely limited, we must consider as extremely enlightening any piece of information that shows with certainty some idea previously accepted to be erroneous or too naive. In this perspective, the rejection by modern physics of what is still often referred to as "scientific materialism" acquires special importance. As a matter of fact, such an association of words could legitimately be made during the second part of the last century, for then physicists thought they were able to define matter (as the collection of all the atoms *plus* the fields) and believed they could formulate their science without any reference—even an implicit one—to states of consciousness of observers. Consequently, thinkers at that time legitimately believed that "matter" thus defined was indeed the sole and primeval reality. Nowadays, however, the situation in this respect is obviously completely different. Although biologists are—in a limited sense [1]—successfully explaining life by means of physics, at the same time the principles of physics itself have undergone such an evolution that they cannot even be formulated without referring (though in some versions only implicitly) to the impressions—and thus to the minds—of the observers.* As a consequence, materialism is bound to change.

And, indeed, the modern version of materialism, namely, dialectical materialism, accepts more and more the idea that our consciousness has some remarkable "specificity" of its own [2]. However, the presence of the word "materialism" in the very name of this philosophy correspondingly becomes less and less adequate, and the danger progressively increases that the name will create misunderstandings, particularly among nonexperts. As a matter of fact, when used by anybody except perhaps experienced dialecticians, the word "materialism" suggests a definite ontology that is a special case of micro- or macro-objectivism. Now, we know from all the works summarized above that

---

*A successful (and hence nonseparable) hidden variables theory should precisely avoid this. Nevertheless, even in such a theory the relationship of the nonseparable reality to the (separable) phenomena might necessitate reference of this kind.

the quantum-mechanical limitations that bear on the concept of physical objects precisely amount to nothing less than a rejection of both objectivisms as fundamental ontologies. For all these reasons, the use of the expression "scientific materialism" should nowadays be tolerated only with reference to a set of methods or to an attitude of mind. With reference to a general conception of the world, it has become a meaningless association of words.

The foregoing remarks are elementary. However, they introduce us naturally to a problem that is much less so and that indeed is not definitively solved as yet. We can formulate it as follows. It being granted that the concept of a classical physical object is not an adequate basis for an *ontology* that would be consistent with quantum mechanics, should we nevertheless consider it as an a priori concept of our *epistemology*?

As already stressed in Chapter 21, the whole Copenhagen interpretation of quantum physics is based on a positive answer to this question. On the other hand, two facts seem to favor a negative answer. One of them—acknowledged, as we saw, both by Bohr and by Heisenberg—is simply that the concept of a classical object has only a limited, ill-defined domain of application. The other one must be borrowed from an altogether different science, namely, the experimental psychology of children. As is well known, Piaget's experiments [3], for instance, show quite convincingly that the concept of a physical object is not in the least an a priori of our minds. Instead, it is progressively construed by the young child in the first few months of his life. In a sense, this concept should therefore be considered just as the central element of a physical *theory*. This is a theory that every infant builds up for his own use. Since it proves to be extremely efficient in accounting for the events of daily life, most of us never feel a need to alter its conceptual frame. However, the situation here does not differ radically from the one that prevails in regard to such notions as the celestial sphere or the atomic electron orbits. The mere usefulness of the concept of a macroscopic body is no more a proof that the concept is an a priori element of men's sensibility (to make use of Kantian language) than it is a proof of the absolute, primeval existence of the reality it is supposed to designate.

Although they are qualitative, these considerations should not be taken too lightly. To a certain extent, they presumably justify a feeling that seems to prevail in some circles of physicists: that somehow the situation is now ripe for a slight transgression of the strict Copenhagen orthodoxy. This could perhaps be done by considering two (or more) levels of knowledge.

One of these levels would be the one of proper science, in the strict and restrictive sense that is nowadays commonly given to this word, at least among theoretical physicists. At that level, all statements ultimately refer to observable *phenomena*, and correspondingly must all be such that they could possibly be shown to be correct or to be wrong. However, at that level some reference must be made to the states of mind and abilities of the observers. This can be done directly, as by von Neumann et al. It can also be done indirectly, as in Bohr's work, by considering the classical concepts as a priori's of our sensibility (though, in fact, we know they are not). If we choose the second standpoint, we discover a surprising coincidence between the scientific knowledge at that level and the ordinary, commonsense knowledge. This is that both, as Heisenberg [4] observes, must attribute a considerable degree of meaning to some usual, familiar notions, even though they are ill defined. This fact is of paramount importance. It shows that no scientist should scoff at people who believe in the realities described by the good old concepts that have proved useful so far, even if they are ill defined, as is usually the case, and even if we cannot produce any better definitions of them. Matter and cause, but also sensibility and finality, should therefore be believed in by any up-to-date scientist.

However, another level should simultaneously be considered, precisely because of the lack of coherence of the first level—because of the fact that, in particular, its basic concepts are neither truly a priori nor well defined, and that consequently it does not allow any self-consistent description of reality. Indeed, as we already noted, some support in favor of the necessity of this second level can be found in the works of Heisenberg himself. Heisenberg acknowledges the fact that the classical concepts he regards as the a priori elements of our scientific descriptions are relative in two aspects. First, they have—as repeatedly stressed above—a limited field of application; and, second, they are still—though indirectly—related to our experience in that they progressively emerged in man's mind as a consequence of evolution. Under these circumstances, it is hard to imagine imperative reasons that would forever force men to remain bound to a privileged use of these concepts. If this analysis is cogent, we may legitimately regain some of our instinctive hopes in the future existence of a scientific description that would be centered on reality instead of on our impressions. Indeed, the relativity of state theory described in Chapter 23 is, in a sense, an attempt in this direction. So is, in a less precise way, the theory of the Aristotelian "potentia" that can be found in Heisenberg's books. Although Heisenberg does not classify his potentia among the "physical realities" proper, still, since they are effective, they must necessarily have an existence of some sort: they must therefore constitute an aspect of the reality that we are now considering.

The great difficulty with all theories of this kind is, of course, that they are in fact *metatheories*. And, as we saw in the case of the relativity of state model, they cannot be directly tested. Admittedly, this represents a considerable weakness. It is not, however, one that is necessarily lethal; it could well happen, for instance, that one of these theories would exhibit a quite remarkable internal consistency not present in the other ones. If, moreover, this particular scheme allowed for successful treatments of situations that more or less defy our usual descriptions—quantum general relativity, for instance, when applied to cases in which classical concepts fail—it would thereby acquite a large degree of plausibility.

At the present time, we are still quite far from any such achievement. In other words, although we can produce such metatheories, we can select none of them except on extrarational agruments. To illustrate, we could, for instance, easily cure the disease of the relativity of state theory disclosed in the discussion presented in Section 23.2. This difficulty, let it be recalled, is that the intersubjective argeement has no reason to hold *really*, although no individual can verify that it does not (even by receiving answers to questions he asks, since these answers do not reflect the true impressions of the speaker). Now should we consider this difficulty as final, in the sense that it would show some internal contradiction in the theory and that no solution to it would therefore be conceivable? This is obviously not the case. For instance, we can imagine that the individual consciousnesses have extraphysical interactions. Alternatively, we can conceive that some superconsciousness acts as a monitor for them all. Still other avenues of escape, borrowed from different philosophies, can be thought of. Admittedly, these solutions all look so easy and so arbitrary that nobody would seriously claim they constitute any scientific advance. The point, however, is that they do exist and do fill a gap, be it in their own, seemingly uncouth way: we cannot reject them simply on the grounds of uselessness. And, as scientists, we can, of course, no more declare a priori that these solutions are false or stupid than we can make a choice and select a particular one out of them all. Both standpoints would involve an equal share of dogmatism.

Such a situation does not appear as special to the relativity of state theory, which is used here merely as a clear-cut example. Indeed, the considerations developed in the foregoing chapters seem to make it clear that, quite generally, the whole of the world, including our own impressions, cannot be described consistently by modern physics alone. Consistency apparently requires the help of some assumptions that in one way or other must be borrowed outside this science.*

This state of affairs has a number of consequences. Two of them are worth mentioning.

The first one is of a general nature. It consists in the legitimacy of the hope that—in spite of all the semantic difficulties that are now accumulating— some kind of intellectual exchange will be actively pursued between science and philosophy, or at any rate between the parts of these disciplines that are not predominantly engaged in a search for their own language. Of course, such an exchange of ideas could hardly contribute to the advancement of most

---

*This may remind us of Laplace's well-known assertion: "I do not need this hypothesis," which was directed precisely against a special assumption of this kind. If the foregoing considerations hold, they show that in our times Laplace's proud statement would seem preposterous.

parts of our present science. However, it could open some perspectives, as it has already done a few times in the past. Moreover, it probably represents the only method by which a man can hope to really apprehend—and bring to some semblance of unity—the multiple aspects of our age.

The other consequence of the observation stated above is of more immediate use. It is that in our present general conclusions we must necessarily remain somewhat vague; otherwise we will fall into mere arbitrariness. How can we therefore summarize, without unmotivated specifications, the standpoint that we have come to acknowledge as reasonable? For example, through the following propositions.

Proposition (i). The verb "to exist" has a meaning, irrespective of the difficulty we usually encounter in ascertaining whether or not a given concept qualifies as one of its possible subjects.

Proposition (ii). Reality—defined as the totality of what exists—is essentially independent of us in its behavior. In other words, though we are parts of it, we are definitely not its regulators, in any sense.

Proposition (iii). What we presently know is not in contradiction with the hypothesis that this reality has fundamental laws [they are, of course, independent of us as a consequence of proposition (ii)].

Proposition (iv). What we know is *also* not in contradiction with the assumption that we can reach an increasingly good knowledge of the *general structure* of these laws. Indeed, this assumption is free from contradictions stemming from the measurement theory, since the difficulties of this theory bear on *individual*, not on *general*, attributes. Moreover, it accounts in a rather natural way for the regularities we observe. There is thus no ground for discarding it.

Proposition (v). What we presently know *does* contradict, however, the assumption that reality as defined above coincides in every respect with *empirical reality*. By "empirical reality" we mean here a *description* which is made in terms of particles, fields, classical physical objects, and so on, and in which individual attributes appear. In the last resort the circumstance that no such description coincides with reality is due, as we saw, to the fact that in quantum physics this description unavoidably disregards the specificity and locality of the states of consciousness. Thus it seems that the least unsatisfactory image we can form of reality is one in which physical (i.e., empirical) reality and consciousness are not, as yet, differentiated. This is, therefore, what we suggest.

Proposition (vi). According to this standpoint, consciousness and physical (or empirical) reality should be considered as two complementary aspects of reality. Their complementarity consists in the fact that each of them contributes decisively to a greater characterization and to a greater specification of the other one.

Considered as a model, the relativity of state theory can help to visualize this to some extent. In it, the relative state plays the role of empirical reality; and this relative state is determined by the state of the corresponding observer, just as it also influences observers. Although at the present time there would be too great an amount of arbitrariness in accepting the relativity of state theory at its face value for a description of reality, the views it suggests of the relationships between empirical reality and observer can be thought of without contradiction as being substantially correct.

Proposition (vii). Neither space nor time nor even space-time has a primitive existence. They are not parts of reality as the latter concept is defined in proposition (ii). They belong to *empirical reality*, that is, they are modes of our sensibility.

Proposition (viii). In the present description, a fact that has been considered by many thinkers as an extremely puzzling one appears as rather natural. This is simply the circumstance that *empirical reality—the world—can be understood by our minds or, in other words, is amenable to a mathematical description.* To the extent that our minds and empirical reality are complementary sides of one and the same reality, it does not appear as highly surprising that the general structures of this reality should, on the one hand, be reflected in the mathematics we build up and, on the other hand, manifest themselves in empirical reality.*

Of course, the view that the physical descriptions we make bear directly, not on reality, but on empirical reality does not invalidate proposition (iv) that our increasing knowledge of the physical laws truly bears on *real structures*. As already stressed, this is due to the fact that this knowledge is concerned only with general, not specific, attributes. Thus reality and knowledge of it bear to one another more or less the same relationship that Russell exemplified by means of the relationship between a concert and its registered record. The abstract structures are the same, although their supports are altogether different. Indeed, the plausibility of the assumption that our scientific laws are steadily approaching those of reality itself is, of course, increased by the observation made above [proposition (viii)] that we can understand the world.

---

*The present suggestion differs from Bohr's generalized complementarity. The latter author never considered a fundamental reality possessing knowable structures.

We think that the great majority of scientists have an immediate intuition of the fact that the reality in which they are interested, although it is *known* through experience, is not *created* by experience. Hence, when these scientists are led to consider a necessity that they most willingly forget about, namely, that of making a choice between the two categories of *acts of faith* described in chapter 20, they instinctively prefer that of the realists to that of the positivists. Under these conditions, we shall not insist on the arguments which speak in favor of the set of propositions stated above, since these arguments would be obvious to many. Barring theories such as micro-objectivism or macro-objectivism, which are contradicted by modern physics, these propositions constitute what seems to us to be the only—or at least the most satisfactory—formulation of a fundamentally realistic outlook. That this outlook, in a sense is Platonic, or rather Pythagorean, is not due to any arbitrary decision on our part. It proceeds from the plain necessity of not contradicting one of the most successful and most far-reaching of physical theories.

In our most optimistic moods we may even conceive that the views expressed above could have some bearing, not only in the field of general ideas, but also in regard to the evolution of our interest in a particular line of research. This is an opinion that should not be presented without caution, for speculations about the future interests of science are most hazardous in all cases. It may be conjectured, however, that a part of the difficulties modern theory encounters is due to the existence of a somewhat excessive arbitrariness in the choice of possible models. This arbitrariness would then presumably reflect a lack of sufficiently stringent guiding principles in our search. Then the mere assumption that a reality exists and is subject to simple laws may help, because by decreasing the interest of fragmentary techniques and models it does indeed yield guiding principles of this kind.

Another task that the foregoing view; seem to offer to the theorists is to parallel the investigation of the *general* structures of reality by research of a somewhat different type. Specifically, what we have in mind is—quite ambitiously—a search for the entities of theoretical physics that might possibly be in some one-to-one correspondence with reality itself, also in more individual aspects.

This, as we know, is an extremely difficult program. All the developments made above seem to indicate that for such a purpose we should rely on indirect arguments, on views based on internal consistency only, because of the very likely absence of direct tests of any kind whatsoever. Moreover, if these developments are correct, we must find a working assumption about reality that differs from conventional mechanism, since the latter appears more and more as a blind alley; and this is not easy, as shown by the fate of the separable or noncontextualistic hidden variables theories. However, in spite of all these stumbling blocks, it is conceivable that the time has now arrived when such a transgression of positivistic taboos would be both manageable and fruitful.

To keep such purposes in sight, one of Einstein's suggestions should now perhaps be remembered. Its essence can be expressed in one sentence: It is a goal that is neither out of reach nor unworthy of the efforts of physicists to try to discover *elements of reality*.

REFERENCES

[1] W. Elsasser, *Atoms and Organisms*, Princeton University Press, Princeton, N.J., 1966.
[2] R. Garaudy, *Marxisme du XXème Siècle*, La Palatine, Lausanne and Geneva, 1967.
[3] J. Piaget, *La Construction du Réel chez l'Enfant*, Delachaux et Nietle, Lausanne and Geneva, 1937.
[4] W. Heisenberg, *Physics and Philosophy*, Harper, New York, 1958.

AUTHOR INDEX

Numbers in parentheses indicate numbers of references cited in the text without author's names. Numbers set in *italics* designate page numbers on which complete literature citations are given.

Aharonov, Y., 80(5), 86, *91*, 154(2), *156*
Araki, H., 209, 210(2), *226*, 236(1), *240*
Armenteros, R., 85(4), *91*

Bauer, E., 263, *265*
Belinfante, F., 106, *120*, 203(3), *206*
Bell, J. S., 104, 104(4), 105(3), 109, 110(7), 111(4), 112, 113, 115(7), 118(4), *120*, 132(7), (8), 135(7), 141, 145(17), *147, 148,* 168(3), 172(3), *172*, 195(12), *196*, 228, 229, 282(2), *283*
Bergmann, P. G., 154(2), 156
Birkhoff, G., 47, *55*, 124(1), *147*
Bohm, D., 26, *29*, 80(5), 86, *91*, 103, 113, *120*, 145(17), *148*, 181, 181(3), 182, *185*
Bohr, N., 85, *91*, 94, *95, 96,* 250, 253(3), 254, 256(3), *259*
Born, M., *172*
Broglie (de), *see* de Broglie
Bubb, J., *120*

Carnap, R., 244, *249*
Chew, G. F., *196*
Clauser, J. F., 104(5), 111, 115(8), *120*, 132(7), 137, 141(11), *147*
Cooper, L. N., 266, *278*
Costa de Beauregard, O., 144(15), *148*

Daneri, A., 192(6), *196*
de Broglie, L., *120*, 145(16), *148, 172*
d'Espagnat, B., 52(11), 54(11), *55*, 61(4), 65(7), *71*, 85(3), *91*, 128(5), *147*, 149(1), 154(1), *156*, 171(6), *172*, 201(2), *206*, 213(7), 224(14), 225(14), *226*, 229(14), 239(2), *240*
De Witt, B., 266, 269(4), *278*
Doncel, M. G., 52(10), *55*

Earman, J., 213(8), *226,* 257, *259*
Einstein, A., 75(1), 81, *91,* 112(9), *120,* 132, *147*
Elsasser, W., 286(1), *293*
Espagnat (d'), *see* d'Espagnat
Everett, H., 225, *226, 266, 278*
Ewing, A. C., *249*

Faraci, G., 115(13), *120,* 141(3), *147*
Fehrs, M. H., 215, *226*
Ferretti, B., *96*
Feyerabend, P. K., *96,* 174(2), *185*
Feynman, R. P., 245, *249*
Fine, A., 215, *226*
Finkelstein, D., 98, *102*
Fok, V. A., *185*
Freedman, S. J., 115(12), *120,* 141(11), *147*
Furry, W. H., *91,* 157, 169(5), *172,* 217, 218, *226*

Garaudy, R., 286(2), *293*
George, C., 193(10), *196*
Gleason, A. M., 47(9), *55*
Gottfried, K., 13(2), *13*
Graham, N., 269, *278*
Green, H. S., 217, 218(12), *226*
Groenevold, H. J., 183, *185,* 281(1), *283*
Guenin, M., 64(6), *71*
Gutkowski, D., 115(13), *120,* 136, 141(13), *147*

Haag, R., 47, *55*
Hanson, N. R., *96*
Hartle, J. B., 98, 99(2), 100(2), *102*
Heelan, P., *259*
Heisenberg, W., *249,* 257, *259,* 288, *293*
Hempel, C. G., 244, *249*

SUBJECT INDEX

Abilities (human), 185, 194, 236
Action, 285
  and reality, 119
Antiprotons, 80, 85
Antisymmetrization, 27, 28
Approximate statements, 195
  see also Property (properties)
A-priori, 256
Assumption Q, 22-24, 25
  and Condition R, 67
  and improper mixtures, 60-62
  and measurement theories, 160
  and nonseparability, 79, 92, 108
  and wave packet reduction, 90
  versus supplementary variables assumption, 26
Assumption Q', 47, 67
Assumption Z, 66, 154
Atomism, xiii, 234
Axiomatic formulations, xii, 1, 47
  see also Calculus of propositions

Being, xxiv - xxviii
  unchanging, xxx
Bell inequalities, 110, 111
  and experiment, 115
  and quantum propositions, 133-147
  and separability assumptions, 116-119
  generalized, 110, 135, 138
Bohrs' epistemology, 94, 208, 250-256
  see also Copenhagen interpretation
Bosons, 28
Branches of Universe, 267

Calculus of propositions
  Boolean, 121-124
  classical, 121-124
  non Boolean, 124-127, 129, 143
  quantum, 124-127, 133, 143-144
Cat (Schrödinger's), 205
Causality Principle, 93, 94, 95, 238

  see also Instructions (propagation of),
    Separability (questioned concept)
Classical instruments, *see* Instruments
  (classical)
Classical measurements, *see* Measurements
  (classical)
Classical terms, 251
Collapse (of wave function) 19, 90, 92
  see also Reduction of wave packet
Communicability, 242, 250
Complementarity, 252
  generalized, 254, 291
Complement (orthogonal), *see*
  Orthogonal complement
Condition R, 65, 67, 154, 199
Consciousness, 262-265, 286, 289, 291
  in relative state theory, 276-278
Contextualistic theories, 106, 118
Copenhagen interpretation, 23, 129, 224,
  250
  see also Bohr's epistemology
Correlations, 50-54, 64, 99
  long range 237-238
  see also Einstein, Podolsky, Rosen problem
Correspondence Principle, 11
  extension of, 82
Crow argument, 247

Degeneracy, 17
Degrees of freedom (infinite number of),
  194, 236
Democritus, 234
Density matrices, 39-45, 48-50
  measurable?, 57, 202, 213
  nonobjective (Bohr), 253
  of systems *plus* instruments, 199-204, 213,
    216, 217-224
  time dependence of, 45
Density repartition, 7
Descartes, 261

297

9 780738 201047